Programming and Customizing the BASIC Stamp Computer

Programming and Customizing the BASIC Stamp Computer

Second Edition

Scott Edwards

McGraw-Hill

New York   Chicago   San Francisco   Lisbon   London
Madrid   Mexico City   Milan   New Delhi   San Juan
Seoul   Singapore   Sydney   Toronto

Library of Congress Cataloging-in-Publication Data

Edwards, Scott (J. Scott)
    Programming and customizing the BASIC Stamp computer / Scott Edwards.—2nd ed.
       p.   cm.
    ISBN 0-07-137192-3
    1. BASIC Stamp computers—Programming. I. Title.

QA76.8B33E39   2001
005.265—dc21   2001030029

McGraw-Hill

A Division of The McGraw·Hill Companies

1 2 3 4 5 6 7 8 9 0 DOC/DOC 0 7 6 5 4 3 2 1

P/N 137193-1
Part of
ISBN 0-07-137192-3

The sponsoring editor of this book was Scott Grillo. The editing supervisor was Sally Glover, and the production supervisor was Sherri Souffrance. It was set in Times New Roman PS following the TAB4 Design by Joanne Morbit of McGraw-Hill's Professional Book Group composition unit, Hightstown, New Jersey.

Printed and bound by R. R. Donnelley & Sons Company.

This book is printed on recycled, acid-free paper containing a minimum of 50% recycled, de-inked fiber.

CONTENTS

APPENDICES

INTRODUCTION

Computers are old news. As their power has soared, their fun factor has plunged. If you weren't around for the days when Apple operated out of a garage and Microsoft was truly micro, you missed all the fun.

Wrong!

Just in the nick of time, some California mavericks have appeared on the scene with a new kind of computer—the BASIC Stamp. This time, the fun won't fade. The Stamp is a microcontroller—an inexpensive speck of computer intelligence that you can program and embed in almost anything: homemade robots, toys, new-wave artwork, wacky inventions, custom cars, anything!

This book will show you how to get started building fantastic projects with the Stamps. First, you'll pick up essential programming fundamentals and electronic assembly techniques.

Then you'll dive right in and build working projects, including:

- Magic message display
- Mobile robot with sensor whiskers
- Sonar rangefinder
- Wireless household remote controller
- Desktop time/temperature display
- Data logger
- Network communication terminal

You'll also discover a worldwide community of Stamp enthusiasts whose restless hands and relentless imaginations have created hundreds of fascinating Stamp applications that you can build and customize.

Requirements

You can use this book as an introduction to microcontrollers even if you never build or program one yourself. To get the greatest benefit, you'll need

- A PC running Windows 95 or better
- A BASIC Stamp I or II (or equivalent)
- Stamp programming software and cable
- Access to assorted common electronic parts and simple tools

If you don't have these items and are unsure about specifically what to buy, read on. This book is crammed with advice and information to help put you on the right track. But first, let's answer a fundamental question: what is a microcontroller?

Microcontrollers

Microcontrollers are often inaccurately called computers or microcomputers. But there's a difference. The main purpose of a microcomputer like a PC is to communicate with a human being. The PC accepts instructions and information through a keyboard, mouse, and in some cases a microphone. It responds through a video screen, printer, and speaker.

Microcontrollers are different. Their main purpose is to communicate with electronic devices, not people. Microcontrollers are the hidden, hardworking computers that inhabit our homes and cars in devices like VCRs, microwave ovens, alarm systems, fuel injectors, exercise equipment, even toasters. In fact, as an example, let's look at how a microcontroller might improve a toaster.

An old-fashioned toaster is controlled by a strip of metal that warps when it gets hot. The longer the toast is down, the hotter this thermostat strip gets, and the more it warps. When it warps enough, it releases a catch and lets the toast pop up.

While this is ingenious, the warping of a thermostat strip doesn't directly indicate how dark the toast is. Toast may be OK, but bagels burn and pastries stay soggy. The thermostat responds to heat, not doneness, and the two don't always match.

Suppose we wanted to create a smart toaster using a microcontroller. By smart, we mean a toaster that makes toast by more or less the same methods that a person would. A logical first step might be to observe a person toasting bread and make two lists of facts concerning the toast-making process: what the person must know, and what he or she must be able to do. Here are my lists:

In order to make toast:

I HAVE TO KNOW THESE THINGS (INPUTS)	I HAVE TO BE ABLE TO DO THESE THINGS (OUTPUTS)
Whether there is toast in the toaster.	Control the heating element (at least on/off).
How dark/light the toast should be when done.	Pop the toast up when done.
How done the toast is at a given time.	

From the standpoint of a microcontroller, the know list translates to inputs and the do list to outputs. We can sketch a block diagram of our microcontroller toaster (Figure 1-1) from just that information.

The sketch shows the hardware portion of the toaster design. By hardware, we mean the physical components of the toaster. The other part of a microcontroller design is *software*—the program that interprets the inputs and controls the outputs. Software is called

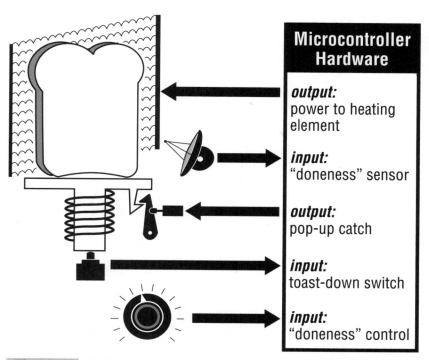

Figure 1-1 Hardware for the hypothetical microcontrolled toaster.

soft because it is just a pattern of electrical energy stored in the microcontroller. It has no more substance than a shadow on the wall, but it dictates everything that the microcontroller does. In microcontroller jargon, software is often called *firmware* because it is permanently stored inside a memory device.

Programming Fundamentals

No matter whether you call it software or firmware, a program is list of precise instructions for the microcontroller to follow. A typical program looks at the inputs, processes them, and uses the information to decide what the states of the outputs should be.

Several factors distinguish a program from an ordinary list of instructions that we might write for ourselves (a to-do list) or others (instructions for evacuating a building). Programs:

- Are written in a language that the microcontroller understands, or one that can be converted into a microcontroller language.
- Are very precise, leaving nothing to chance.
- Must cover every possible combination of inputs.
- Must never produce outputs that might be dangerous.
- Can employ fast, accurate calculations that would not be practical for human beings to perform.
- Can include instructions that alter the order in which other instructions are carried out, allowing the program to mimic human decision making, or to repeat a set of procedures until a desired effect is achieved.

Although a program must eventually be translated into a language the microcontroller understands, most programs start out as human-readable outlines. For example, our toaster application might go something like this:

(1) Is toast-down switch on (indicating that bread has been put into toaster)? If not, go to step (1); otherwise continue with step (2).

(2) Engage catch to hold bread down.

(3) Turn on power to heating element.

(4) Read the doneness control.

(5) Read the doneness sensor.

(6) If doneness sensor reading is less than the doneness control setting, go to step 4; otherwise continue with step (7).

(7) Turn off power to heating element.

(8) Release catch to let toast pop up.

(9) Go to step 1.

That simple procedure has most of the elements of a typical microcontroller program. Let's pick up some microcontroller jargon by analyzing the toast program in detail.

First of all, a general rule: Unless instructed otherwise, a microcontroller executes a program in order from beginning to end. In terms of human-readable program listings, this means from top to bottom and left to right. Unlike a person working through a to-do list, the micro cannot elect to do step 4 before step 3 because it's convenient. It must start with the first step and go step-by-step.

In the toaster program, step (1) checks the switch that indicates whether bread has been put into the toaster. If it hasn't, the program repeats step (1). Repeating an instruction or section of a program is called *looping*. Repeating step (1) until the switch is set makes the toaster wait for a piece of bread before carrying out the rest of the program.

Once bread is detected, the program continues with steps (2) and (3), locking down the bread and turning on the heating element.

The next two instructions read a control and a sensor. "Read" is an oversimplification, since the controller actually gets the data and stores it in memory. A memory location used to hold such values is called a *variable*. You may recall the term from math, where variables are symbols that stand for the numbers in a formula.

For instance, you compute the area of a rectangle with the formula height × width = area. Substituting actual numbers for the variables height and width allows you to compute the area. In programming, named variables represent actual locations in memory in which data is stored.

Just like a formula, a program can specify math or other operations to perform on variables, and the values stored in the corresponding memory locations will be used in the computation. This is another source of great power in programming. As in our example, the controller can read controls and sensors, perform computations with the readings, and make decisions based on the result.

That's what the toaster does in step (6). It compares the sensor reading to the control setting and, if the sensor reading is less (meaning the toast isn't done yet), it loops back and reads the sensor and control again. If the sensor reading is equal to or greater than the control setting, the program continues with the next step.

Notice the fail-safe logic of step (6). It says that the bread should keep toasting if it's lighter than the control setting calls for. Why not have it toast until the sensor reading equals the control setting? Because that might get you into trouble! Suppose somebody wanted to reheat a piece of toast that was already more toasted than the control setting. Its doneness would never equal the control setting, because it would already be too dark and getting darker. This small mistake in logic could start a fire, or at least make some very black toast.

The way the program is written, the toast would just pop back up after the first sensor/control reading. The person using the toaster might notice that the control was set lighter than the toast, change the setting, and try again.

Once the toast is done, the program turns off the heat and pops up the toast. Its final step is to go back to step (1), in which it waits for another piece of bread. This is another common characteristic of most control programs; the entire program is a big loop. There's no quit or exit command as there would be in a computer program. As long as the controller has power, it's either working or waiting to go to work. After all, what else does it have to do?

Programming Languages

The major flaw in our toaster program is that no microcontroller could run it. It's written in ordinary English, which computers and controllers don't understand. Before we can get a controller to run our program, we must translate it into a programming language.

A programming language is a set of words and rules for combining those words, like the vocabulary and grammar of a human language. What makes it different from human languages is that a programming language is either understandable to a microcontroller, or readily converted to a form that the micro understands.

Just what does the microcontroller understand? A microcontroller is an electronic component made up of thousands or millions of electronic switches. These switches have only two states, ON and OFF, represented by 1 and 0 respectively, and stored as two different electrical charges. A program is just a sequence of these 1s and 0s.

The actual microcontroller hardware is an intricate mechanism that retrieves the patterns of 1s and 0s that make up a program, and carries out a small set of very simple instructions. For example, the pattern 011010101 might mean "increment (add 1 to) the value stored at memory location 10." Another pattern would move data around in the controller's memory, and another would perform simple arithmetic.

The language the microcontroller understands is called *machine language*. It's possible to write programs in machine language with the help of a program called an *assembler*. An assembler is an electronic dictionary that looks up more-or-less readable symbols like MOV and LD (move and load) and outputs the appropriate 1s and 0s. Even with the aid of an assembler, writing programs in machine language is tedious. The instructions of machine language perform very small tasks, and it can take hundreds of them to perform even the simplest job, like our toaster, for instance.

Since an assembler is a program that helps you write programs, isn't it possible to write a more complicated program that would make it much easier to write programs? Yes! Programming languages that require this kind of processing are called *higher-level languages*. The scale from low to high is easy to understand; at the lowest end is machine language, and at the high end is human language. Programming languages are classified by their relative position between hardware and human beings on this scale.

There are two kinds of high-level languages: *compilers* and *interpreters*. Let's borrow an example from human language. Suppose you are working at an embassy in a foreign country whose language you don't understand. You need to keep up on current events. Each night, your staff of bilingual clerks reads the local newspapers and compiles a summary in English for you to read in the morning. Very efficient.

Another possibility: The embassy assigns you a bilingual assistant to act as an interpreter. You receive information in the local language, and the interpreter translates it into English on the spot. Somewhat less efficient, because you have to wait for the interpreter to translate each item. But very flexible, since you can interact with the interpreter and get items of interest translated without waiting for the overnight compilation.

That's how it is with compiled and interpreted programming languages. A compiler takes high-level instructions and converts them to machine language that the microcontroller can use directly.

An interpreter is a program that runs on the microcontroller and translates each high-level instruction into machine language on the spot. This requires more processing on the part of the microcontroller, and tends to run programs more slowly than those written in or compiled to machine language. But it has the advantage of immediacy; the programmer can change a program and see the effects of those changes without a compilation step.

Now that we know a little about how higher-level languages are created, let's look at a specific example, a language aptly named BASIC.

The BASIC Programming Language

One of the most popular high-level programming languages is called BASIC, which stands for Beginner's All-purpose Symbolic Instruction Code. It seems likely its creators started with the name BASIC, and filled in the words later! BASIC was created in the 1960s as a gentle introduction to programming. The idea was that once you had cut your teeth on BASIC, you could go on to tackle more serious programming languages.

But the simplicity that made BASIC easy to learn also made it hard to leave behind. In 1975, Microsoft founder Bill Gates got started in the computer business by helping to create a version of BASIC that is often cited as the first software package ever sold for a personal computer, the MITS Altair. Before BASIC, Altair owners programmed in machine language by flipping toggle switches.

Now, more than two decades later, BASIC lives on in Microsoft Windows products as a way of automating repetitious tasks. Other versions of BASIC enjoy great popularity with a wide range of users from hobbyists to engineers and scientists. During the same period of time, hundreds of high-level programming languages were created, enjoyed a flush of popularity, and then were abandoned and forgotten. BASIC keeps right on going.

BASIC's enduring appeal is no accident. Its creators were gifted teachers who set out to create a language that would seem familiar and comfortable to students who were learning about programming at a time when computers were mysterious and scary. To understand how well they succeeded, let's look at the toaster program as it might be written in BASIC. See Figure 1-2.

One of the first things you should notice about the program listing is the inclusion of comments. Anywhere you see the tick mark ('), the text that follows it (to the end of the line) is a comment. The computer ignores comments completely; they're solely for people to read. So why include them? Even an easy language like BASIC and a simple program like Toaster deserve some explanation. Comments provide it. Throughout this book, and throughout any well-written programs you encounter, you will see lots of comments. When you begin writing your own programs, you should also include plenty of comments, even if you never plan to show your programs to anyone else. Writing comments helps clarify your thinking. And since most programs are not written in a single sitting, comments help you pick up where you left off.

The first line in the program that's not a comment is the label WaitForBread:. Labels serve two purposes in a BASIC program:

- Like comments, they can provide a hint to human readers about the purpose of a section of code.
- Like the step numbers in our first draft of the program, they serve as markers for instructions that change the order in which the program is carried out. (In fact, the early

```
'TOASTER.BAS (Program to control a toaster)
'Wait for bread to be put into toaster. ToastDown is a switch
'that outputs 0 for no bread and 1 when bread is inserted.
WaitForBread:
 IF toastdown = 0 THEN WaitForBread
'Once bread is detected, turn on the output to popUpCatch in
'order to hold the bread down and turn on the heater. In both
'cases, 1 = ON.
LET popUpCatch = 1      ' Hold bread down.
LET heater = 1 ' Begin toasting.
'The instructions within ToastingLoop compare the darkness of
'the toast to the setting of the control. If the toast is lighter
'than the control setting, the loop repeats. All the while, the
'heating element and pop-up catch remain on. The words "setting"
'and "toast" are storage elements called variables which hold
'readings from the control and sensor so that they can be
'compared.
ToastingLoop:
 setting = controlInput
 toast = sensorInput
IF toast < setting THEN ToastingLoop
'The toast is done. Turn off the heater and release the
'catch to let the toast pop up.
heater = 0        ' Turn off the heat.
popUpCatch = 0 ' Pop the toast up.
'Wait for another slice to toast.
GOTO WaitForBread
```

Figure1-2 **BASIC program to control the toaster hardware.**

versions of BASIC used line numbers, but programmers got sick of renumbering their programs whenever they made a change.)

The next line shows why the label is necessary; it serves as a destination for the IF... instruction. Since we'll get into the details of the various instructions in detail later, let's just skim the high points of the remainder of the program.

■ Expressions like LET popUpCatch = 1 assign a value (1) to a variable—a storage location in memory. Some variables are more than storage, though. They are both storage within the controller and an electrical input or output to other circuits. That's the case with popUpCatch. Putting a 1 into this variable not only stores it in memory, it outputs an electrical signal that controls a physical device, the latch that holds the bread down. Don't worry about how inputs and outputs work; we'll discuss it in detail later and see it in every project we build!

■ Once a value is put into a variable (or a variable that also serves as an output, like popUpCatch) it stays there until changed. Look at the instructions between ToastingLoop: and IF toast < setting... They don't mention the pop-up catch or the heating element at all. Yet the bread is held down and the heater is on, because the program put ones in these variables and hasn't changed them.

■ BASIC uses common math symbols like < to perform comparisons. The instruction IF toast < setting... means IF toast is less than setting...

■ The instruction IF toast < setting... illustrates the guiding rule that programs execute in order unless an instruction says otherwise. As long as the condition (toast < setting) is true, the program keeps looping back to ToastingLoop:. When it's no longer true (because the toast is done), the program continues with the next instruction.

■ The last program line is the motto of a control program; `when your work is done, go back to the beginning and get ready to do some more work (GOTO WaitForBread).'

That's the end of our whirlwind tour of programming and BASIC. Now that you know a bit about microcontrollers and the BASIC programming language, it's time to introduce the BASIC Stamps; inexpensive, friendly microcontrollers that you can program in BASIC.

INTRODUCING THE BASIC STAMPS

CONTENTS AT A GLANCE

From the Introduction, we know what a microcontroller is—a hardworking kind of computer that labors away inside our high-tech gadgets. And we know what BASIC is—a programming language designed from the ground up to be easy to use. Now we're ready to answer the big question:

What is a BASIC Stamp?

A BASIC Stamp is an inexpensive (less than $50) microcontroller with a built-in BASIC interpreter. If you're new to microcontrollers, you should know that the majority of controllers are industrial-strength devices with hefty price tags ($100s) and steep learning curves designed for use by professional engineers. The original BASIC Stamp, introduced in 1993 by California-based Parallax, Inc., was the first to break this mold. The Stamp was meant from the beginning to appeal to hobbyists, students, basement inventors, and, of course, engineers.

At the time of this writing (summer 2000) there are two Stamp families known as BASIC Stamp I (BS1) and BASIC Stamp II (BS2). BS2s are further divided into standard and high-performance (-SX and -E) types.

Let's take a look at some of the features and capabilities of the current Stamps. Don't worry if some of the jargon is new to you; later on our projects will show you the how and why of the Stamps' various features.

One more thing. You may be wondering how the Stamp got its name. The components that went into the original BASIC Stamp fit into an area about the same shape and size as a commemorative postage stamp. Since the creators were looking for a friendly, low-tech name to reflect the product's ease of use, BASIC Stamp just jumped out at them.

Stamp Hardware: Overview

Figure 2-1 is a family portrait of the currently available Stamps, divided into the BS1 and BS2 sides of the clan. All Stamps share these family traits:

- Small size.
- BASIC interpreter firmware built in.
- Powered by a 9V battery for periods ranging from days to months, depending on the application.
- BASIC program storage in permanent (but erasable) memory. Whenever the battery is connected, Stamps run the BASIC program in memory. Stamps can be reprogrammed at any time by temporarily connecting them to a PC running a simple host program. Type in the new program, press a key, and the program is loaded into the Stamp.
- Input/output (I/O) pins that can communicate with other digital devices, sense switches, and even directly drive small loads like LED lights.
- Based on an inexpensive but powerful PIC processor made by Microchip Technology Inc., allowing operation at up to 5 million machine-language instructions per second.
- Available as either assembled modules (as shown in Figure 2-1) or as individual components, "chipsets," that can be built into custom products.

Figure 2-2 shows the major components and hardware functions of the BS1 and BS2. Figures 2-3 and 2-4 are schematic diagrams of the Stamps.

BS1-1C

BS2-1C

OEM BS1

BS1 FAMILY

OEM BS2

BS2 FAMILY

Figure 2-1   BASIC Stamps I and II.

Stamp Hardware: Detailed Tour

Electronics neophytes may run into some unfamiliar terms and concepts in this section. Just smile and nod. The great thing about Stamps is that pretty soon you will know what I'm talking about, and can return to this section with a new appreciation of the Stamps' clever design.

This description is keyed to the part numbers from the schematic diagrams.

PBASIC INTERPRETER CHIP (U1)

The brains of the Stamps are PIC- or SX-series microcontroller chips (U1), custom programmed by Parallax, Inc. Other companies[1] make the physical components, but it is Parallax's firmware that turns them into PBASIC interpreter chips. Because they are interpreters, the Stamp controllers have the entire PBASIC language permanently programmed into their internal program memory (one-time-programmable read-only memory, abbreviated OTP ROM, or OTP EPROM).

1. Microchip Technology Inc. makes the PICs; Scenix, Inc. makes SX microcontrollers.

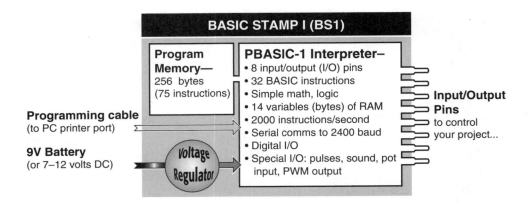

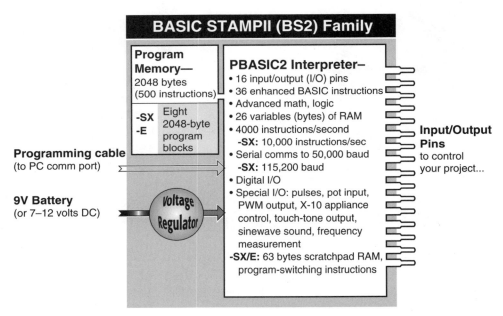

Figure 2-2 Block diagrams of the BS1 and BS2.

In the BS1, U1 (PIC16C56) runs at a clock speed of 4 MHz (4 million cycles per second; MIPS). A PIC executes one machine-language instruction every fourth cycle, so that amounts to 1 million machine-language instructions per second. Since one BASIC instruction can stand for many machine language instructions, the BS1 executes about 2000 BASIC instructions per second.

In the standard BS2, U1 (PIC16C57) runs at 20 MHz for 5 MIPS in the BS2-SX, U1 (SX-28) runs at 50 MHz/12.5 MIPS. Because the BS2's instructions are more complex than those of the BS1, the standard BS2 performs about 4000 BASIC instructions per second; the BS2-SX, about 10,000.

The BS1 and BS2 have 8 and 16 input/output (I/O) pins respectively. These are the pipelines through which information flows between your program and the outside world. Technically, these I/Os are TTL-compatible (with characteristics like the 74HCT series), meaning that they interface with standard digital logic chips without a problem. Moreover, Stamp I/Os have exceptionally high current drive, meaning, for instance, that you can light up an LED indicator without extra circuitry to handle the current. This helps keep Stamp-based hardware simple and uncluttered.

The direction—input or output—of a given (I/O) pin is under the control of your BASIC program. We'll discuss this in more detail later, but this ability to switch between input and output gives rise to many of the Stamps' tricks. When a pin is an output, it can

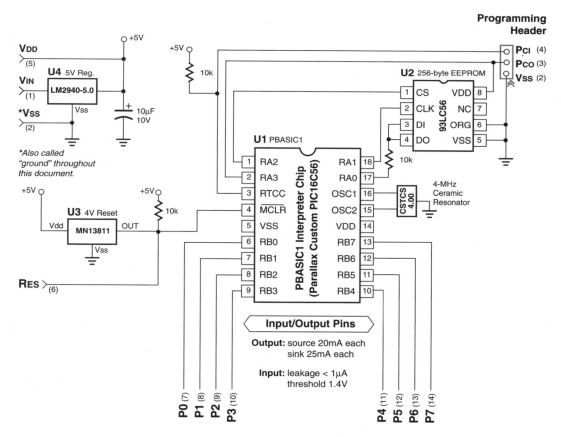

NOTES   1. This diagram depicts the DIP/SOIC version of the PBASIC1 interpreter chip, since users wishing to construct a BS1 from discrete components are most likely to use those parts. Contact Parallax for a schematic depicting the SSOP (ultra-small surface mount) package used in the BS1-IC module.

2. Numbers in parentheses—(#)—are pin numbers on the BS1-IC module and OEM-BS1 kit header.

3. See the OEM-BS1 documentation on the accompanying CD-ROM for more information on components and sources.

Figure 2-3 Schematic diagram of the BS1.

2

INTRODUCING THE BASIC STAMPS

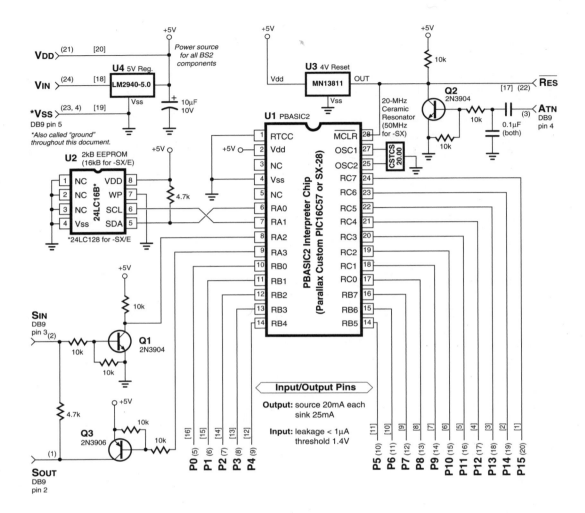

NOTES

1. This diagram depicts the DIP/SOIC version of the PBASIC2 interpreter chip, since users wishing to construct a BS2 from discrete components are most likely to use those parts. Contact Parallax for a schematic depicting the SSOP (ultra-small surface mount) package used in the BS2-IC module.

2. Numbers in parentheses—(#)—are corresponding pin numbers on the BS2-IC module. The BS2-IC has the form factor of a 24-pin, 0.6" DIP. Numbers in brackets—[#]—are pin numbers on the OEM-BS2 module, which has a straight, 20-pin header.

3. See the OEM-BS2 documentation on the accompanying CD-ROM for more information on components and sources.

Figure 2-4 Schematic diagram of the BS2.

signal other electronic devices, light LEDs, etc. When it's an input, it receive signals from other electronics, sense switch settings, etc.

PROGRAM MEMORY (U2)

Since the microcontroller's internal program memory is occupied by Parallax's interpreter firmware, the Stamps need someplace else to store your BASIC programs. That's U2, the electrically erasable, programmable, read-only memory chips (EEPROMs). The BS1 EEP-ROM stores 256 bytes, which works out to about 75 BASIC instructions. The BS2 EEPROM stores 2048 bytes; about 500 BASIC instructions. (we'll discuss the BS2-SX later.) If your programs don't use up all of the EEPROM space (most don't), you can also use the EEP-ROM as long-term storage. This is analogous to the way your computer's disk drives can hold both programs and data files.

EEPROM is great for storing programs and long-term data because it doesn't need power to retain its contents. You can disconnect power from a Stamp for an hour, a year or (in theory) a century, and your program and data will still be there when you reconnect it. You can reprogram EEPROMs only a limited number of times before they wear out, but that limit is around 10 million times, so it's not normally a problem. (If you reprogrammed a Stamp every 10 minutes, 24 hours a day, it would take you over 3 years to approach 10 million programming cycles.)

PROGRAMMING CONNECTION

One of the areas in which the Stamps differ significantly from one another is the programming connection. What they have in common is that you make a temporary connection to a PC running a simple host program in order to type and download your BASIC programs. The way the data gets from the PC and into the Stamp is where they differ.

The BS1 connects to three pins of the PC's 25-pin printer port, also known as the parallel port. The PC host software actually uses the parallel port to communicate serially with the Stamp. (Parallel communication sends whole data bytes simultaneously on eight separate wires, where serial communication breaks each byte into eight individual bits and sends these through one wire.) The Stamp can also send data back to the PC to display the status of a running program via its aptly named Debug instruction. The programming connections on the BS1 are used solely for programming.

The BS2 connects to six pins of the PC's 9- or 25-pin serial port, also known as the comm or modem port. The schematic shows the transistors (Q1 through Q3) that buffer the +12-volt serial-port signals for compatibility with the PIC's standard 0-to-5-volt I/O. The various connections allow the PC to send data to the BS2 (SIN), get data from it (SOUT), and grab its attention for programming (ATN). Two additional connections, not shown in the schematic, help the BS2 host software figure out which serial port the Stamp is connected to.

The BS2 programming port can also be used for serial communication with PCs or other electronic devices, like modems.

POWER SUPPLY (U4)

Every electrical or electronic device operates correctly only within some range of input voltage. Let the power supply voltage get too high, and the electronics may be damaged; too low, and it will work erratically or not at all. The voltage regulator, U4, takes input voltages from approximately 6 to 15 volts (direct current) and converts them to the steady 5 volts that the Stamps require. The same 5-volt supply can also be used to power your circuits, so long as the total current drawn doesn't exceed 50 milliamperes. That's plenty for many small projects.

RESET CIRCUIT (U3)

When power is first connected to the Stamps, or if it falters due to a weak battery, the power supply voltage can fall below the required 5 volts. During such brownouts the PIC (U1) is in a voltage-deprived daze that could make it behave erratically. Reset chip U3 plays the role of adult supervision, forcing the PIC to reset to the beginning of the program and hold until the supply voltage is within acceptable limits.

SPECIAL FEATURES OF THE BS2-SX AND BS2-E

The high-performance branch of the BS2 family, comprising the BS2-SX and BS2-E, has some additional features that I hinted at in the earlier description of the standard Stamps. These features are primarily the result of the use of a Scenix SX-28 microcontroller for U1 and a 16-kB EEPROM for U2. Here's a brief list:

- (-SX and -E) Additional 63 bytes of storage RAM (similar to EEPROM data storage, but erased at power-off, and much faster for write operations)
- (-SX and -E) Eight separate 2-kB program-memory blocks allowing on-the-fly program switching
- (-SX only): 2.5x speedup for up to 10,000 BASIC instructions per second, plus faster serial communications and finer-grained timing instructions.

The -SX and -E variants are completely compatible with their standard BS2 counterparts. Using the new features (storage RAM and program switching) requires learning a couple of new instructions, and the -SX speedup requires adjusting timing parameters to compensate for the 2.5x speed increase, but that's it.

Choosing a Stamp to Start On

If you're just starting out with the Stamps (and you want my advice), start with the standard BS2. True, it's a little more complicated than the BS1, but the additional program-memory space can be very helpful to newbies. The availability of friendly Windows software to program the BS2, rather than the austere DOS-based BS1 stuff, also argues for the BS2. (Windows support for the BS1 is due out about the same time that this book is scheduled to be published.)

I'd advise against purchasing a BS2-SX or BS2-E as your starter Stamp. Yes, they're not that much different from the standard BS2, but these small differences (speed, extra features) can trip up unwary beginners when they try to run programs written for the regular BS2. For example, the BS2 projects in this book are mostly written for the standard model. They would require a bit of educated tweaking before they'd work on the BS2-SX.

The idea is to go as far as you can with the capabilities of the standard BS2 and become a samurai Stamp programmer. Then, when you've exhausted the regular

BS2's capabilities, apply your experience to see whether the BS2-SX or BS2-E can bail you out.

The Future of Stamps

By all accounts, the Stamps have a bright future ahead. At the time of this writing (Fall, 2000), Parallax is at work on another, larger member of the BS2-SX family (tentatively called the BS2-SX+) with more I/O pins and new instructions for interfacing peripherals. There's even an Internet-enabled Java Stamp on the distant horizon.

In the meantime, Stamps have taken classrooms all over the world by storm. Educators looking for a gentle introduction to intelligent electronics and programming have adopted and expanded Parallax's Stamps in Class curriculum.

Stamp-Specific and Stamp-Friendly Kits and Accessories

So, you've got this book full of projects and ideas, but you still can't seem to get off the dime. No problem. A whole cottage industry has sprung up around providing accessories or complete kits for enabling or enhancing Stamp projects.

Here is an admittedly incomplete list of Stamp-related goodies. Contact information for the vendors appears in Appendix E.

ROBOTS

This is probably the largest category of commercial Stamp-based kits. Robots are fun and educational, and the Stamp makes an ideal nerve center for the little critters.

Lynxmotion: grippers and rollers and walkers, oh my!   Lynxmotion manufactures and sells a variety of robot kits that use Stamps (mostly BS2s) for brains. Their line consists of basically three types of robot, with several models of each. See Figure 2-5 for a representative sample.

- Robot arms
  - Stationary-base arms, and arms/claws for attachment to mobile platforms
- Wheeled mobile robots
  - Simple, entertaining carpet-rovers
  - Micromouse maze-runners
- Legged mobile robots (driven by motorized servos)
  - Six-legged hexapods
  - Four-legged quadrupeds
  - Two-legged bipeds (new in late 2000)

All Lynxmotion robots come in kit form with pieces that are computer-cut by mechanical router or laser from strong plastics like PVC, ABS, and acrylic. Assembly instructions

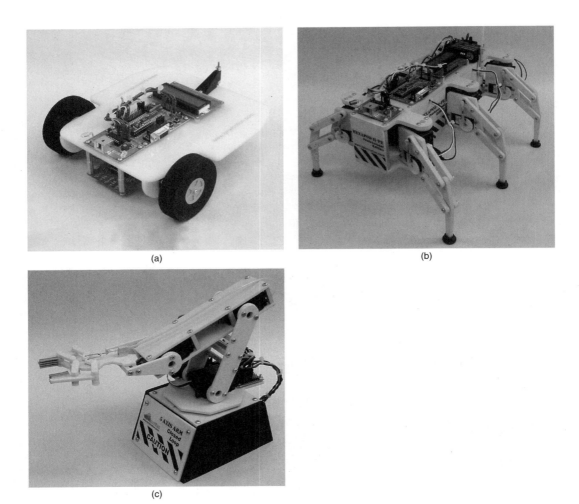

(a)

(b)

(c)

Figure 2-5 Lynxmotion Robots—Carpet Rover (upper left) HexapodII (above) 5-Axis Arm (left).

are extensively illustrated, and Stamp example programs are included. Hardware accessories are available for following lines, avoiding obstacles, etc.

Parallax: quality educational robots   The Stamp's manufacturer, Parallax, makes a couple of dandy wheeled robots that integrate with its hobbyist- and education-oriented products and programs.

Both the Grow-Bot and BOE-Bot (Figure 2-6) share a brushed-aluminum chassis with mounting holes and slots to accommodate your modifications. Where they differ is in the carrier board that holds the Stamp electronics. The Grow-Bot printed circuit board incorporates standardized connectors for accessories call App Mods. The BOE-Bot also accepts App Mods, but its primary mission is as a breadboarding platform for user-designed circuits.

Figure 2-6 BOE-Bot (top) and Grow-Bot (bottom) from Parallax.

Parallax also distributes products from other manufacturers, so they're a good starting point in your search for Stamp-friendly products.

The Robot Store: robots from all over   Mondo-tronics, The Robot Store, is a catalog distributor of robotics products that also manufactures Stamp-friendly robots and motor-control devices. Their Descartes wheeled robot (Figure 2-7) incorporates a coprocessor to run low-level motor-control functions, while a BS2 sits on top, directing the robot's activities. Descartes' navigational prowess is such that it can be fitted with a pen and programmed to draw elaborate spiral patterns on the floor (preferably on a big sheet of paper).

Figure 2-7 The Robot Store's Descartes robot.

Robots need mechanical muscles, and the Robot Store has them. They carry motors, servos, stepper motors, and the appropriate control electronics for interfacing them to a Stamp. They even have a product called Muscle Wire that contracts powerfully when heated by an electrical current. Their Muscle-Wire driven walking 'bots have made frequent TV appearances.

ARobot: Personal robot with a professional pedigree Arrick Robotics manufactures light industrial robotics for serious tasks like picking and placing parts, dispensing adhesives, and positioning sensors. And they make feature-packed mobile robots called Trilobots that are used in academic research.

Though serious, they aren't immune to the fun appeal of the BASIC Stamp, so they created the BS2-based ARobot for hobbyists and educators (Figure 2-8). ARobot (pronounced "A robot") features an aluminum body/frame, easy, no-soldering assembly, motor-control coprocessor to simplify driving and steering, bump-sensing whiskers, wheel encoder for precise distance measurement, and car-style steering.

STAMP PERIPHERALS

PCs get peripherals like monitors, keyboards, and modems; why not Stamps? A variety of companies make Stamp peripherals that provide enhanced capabilities. Here is a sampling of the Stamp-related market:

Solutions Cubed manufactures several Stamp peripherals, including a keypad interface (Memkey), clock module (PocketWatch), and motor-control device (MotorMind).

AWC Electronics and Oak Tree Systems offer a wide range of Stamp coprocessors that provide additional input/output (I/O) pins, measure and generate time delays or pulses, do complex math, interface keyboards, and drive high-current loads.

Figure 2-8 Arobot from Arrick Robotics.

Decade Engineering's BOB-series interfaces let your Stamp display data on a monitor or video screen.

EME Systems makes the OWL-2 data logger for weather and environmental data and offers a wide range of sensors for capturing such diverse data as light and ultraviolet energy, temperature and humidity, and leaf and soil moisture. They also offer a handy device called Stache (pronounced "stash"), a portable Stamp programmer.

Selmaware publishes PC software that enables a Stamp to create real-time data plots on your PC screen.

Scott Edwards Electronics, Inc. (the author's company) makes serial display modules that give the Stamp a small screen for text or graphics. They also offer the Mini SSC II, an interface that controls eight radio-control servos (small positioning motors) under the command of a Stamp or other computer. All products include example programs written in Stamp PBASIC.

ELECTRONIC FOUNDATIONS

This book presents a collection of Stamp-based projects as recipes for you to follow. This is an excellent way for you to gain the skills you will need to design and build your own original projects. Before you can build projects, you will need some fundamental skills:

- Understanding and identifying electronic components.
- Reading and following schematic diagrams—the blueprints of electronic projects.
- Constructing electronic circuits.

If you already possess these skills, great! If not, the sections that follow will help get you started. If you find electronics interesting and fun, you'll probably want to expand your skills through additional reading and education. Appendix F lists books and magazines that can help.

Consider enrolling in an introductory electronics course at your local community college. These classes are a lot of fun, and just one semester will answer all of a beginner's most pressing questions about electronics. Courses emphasize hands-on skills and require only rudimentary math. A growing number of classes even feature BASIC Stamp projects! If you're shy about the competitive aspects of schoolwork, you should know that most colleges will let you "audit" a class on a noncredit basis. No grades, no tests.

Before we pack you up and send you off to college, let's start with a quick seminar on electronics fundamentals.

Electricity

You've probably seen drawings of atoms like Figure 3-1 with tiny electrons orbiting a nucleus consisting of a cluster of protons and neutrons. That drawing is useful for telling the story of electricity. I say "telling the story" because there are many ways to explain electricity that are total fiction, but are useful to understanding the way circuits work. Since I want to help you build circuits, I'm going to tell the story in a way that suits my purpose.

OK, back to the atom. The figure depicts the nucleus as large, solid, and stationary and electrons as small, quick, and flighty. Electrons are attracted to protons; that's what keeps them in orbit around the nucleus. Electrons are repelled (driven away by) other electrons. And electrons can be knocked clear out of orbit by other influences—light, heat, chemical reactions.

There are two ways to look at an atom's loss of an electron: (1) There's a loose electron needing a nucleus to orbit, and (2) there's a nucleus with room in orbit for an electron. Things would balance out if only that electron would come back! Just as good would be if another electron would fill the vacancy in the atom's orbit.

The fate of one electron doesn't mean much in the larger scheme of things, but suppose there was a mass migration of electrons. Over here—lots of electrons. Over there—lots of atoms with empty orbits. If we could create a path to bring the electrons to the atoms that need them, why they would stampede to fill those orbits.

Now we can define some basic electrical terms. A surplus of electrons is called a *negative* charge; a shortage of electrons is a *positive* charge. A path along which electrons can move is called a *conductor*, and a material that blocks electrons is called an *insulator*.

Everyday examples of conductors are metals like copper, silver, gold, and aluminum; insulators include air, glass, rubber, and most common plastics.

Figure 3-1 The atom.

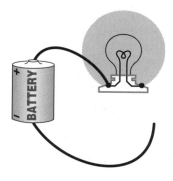

Figure 3-2 With a break in the circuit, the bulb remains dark.

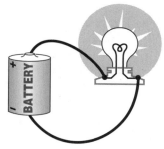

Figure 3-3 A complete circuit lights the bulb.

When a negative charge and a positive charge are joined by a conductor, we call the resulting stampede of electrons electrical *current*. Because that current consists of unruly electrons bouncing off everything in their path, the flow of current in a conductor converts some of the electrical energy into other forms of energy—especially heat.

BATTERIES AND CIRCUITS

The previous discussion goes a long way toward explaining the electrical shorthand seen in everyday life. Batteries are marked with a plus (+) on one hookup and a minus (−) on the other, a quick way of saying positive and negative. The potential energy represented by the difference between these positive and negative charges is written on the label in units of *volts*. This gives us an idea of how much energy would be released by connecting a conductor (for example, a piece of wire) between + and −.

The classic demonstration of the release of electrical energy uses a battery and a light bulb. The battery is connected to the bulb by wires (conductors). In order to light the bulb, there must be a complete path from + through the bulb to −. If there's a break anywhere in the path, the bulb stays dark (Figure 3-2). Once the path is complete, the bulb lights (Figure 3-3).

A complete electrical path from + to − can release energy, like lighting a bulb. All working electrical hookups follow this basic pattern—out from the power source, through some electrical device like the bulb, and back to the power source. You could say that the current flows in a circle, which is exactly why electrical and electronic hookups are called *circuits*.

RESISTANCE

If the wires from the battery are conductors and the innards of the light bulb are conductors, why do the conductors inside the bulb get hot and light up while the wires stay cool and dark? Is there something else going on here?

3

ELECTRONIC FOUNDATIONS

Yes. Not all conductors are the same. Some conductors let current flow easily; others resist the flow of current. This resistance causes electrical energy to be converted into heat. It's not hard to figure out that the filament of the bulb (the curly wire that lights up) has a higher resistance than the other conductors in the circuit. That's why the filament gets hot and the wires don't. See Ohm's Law below for a more complete answer.

ELECTRICAL PROPERTIES, MEASUREMENTS, AND UNITS

In our Stamp projects, we'll discuss measurements like voltage, current, resistance, and power. You can buy an inexpensive tester that will measure these factors, but you can also calculate them using simple math. If you know any two electrical properties, you can calculate the others. Table 3-1 lists the properties we'll be talking about.

Some units are frequently written with the prefix milli-, which means one-thousandth. In other words, 1000 millisomethings is equal to 1 something. We use the milli-versions of units because it's often more convenient to say "three milliamperes" than "point zero zero three amperes."

In the case of ohms, we often deal with large numbers like 47,000 ohms or 1,000,000 ohms. The prefix kilo- stands for one thousand, and lets us say "47 k" or "47 k ohms" when we mean 47,000. Likewise, mega- stands for 1 million, so 1,000,000 ohms becomes "1M" or, in speech, "1 megohm" or even "1 meg."

OHM'S LAW: A QUICK PREVIEW

Appendix C discusses electrical calculations in more detail, but a preview will help explain the bulb-hot/wires-cool phenomenon and provide an insight into working circuits.

Ohm's law defines the relationship between volts, amperes, and ohms with three easy formulas:

$$\text{Volts} = \text{Amperes} \times \text{Ohms}$$

$$\text{Amperes} = \text{Volts/Ohms}$$

$$\text{Ohms} = \text{Volts/Amperes}$$

If you know any two of these values, you can calculate the third. Electronics technicians have a simple trick for remembering the formulas, based on the pictogram in Figure 3-4.

If you cover up the symbol for the value you want to calculate, the other two symbols will show you the correct formula. Cover the horseshoe-shaped ohms symbol (Ω) and you see V over A—the formula for calculating ohms.

TABLE 3-1	ELECTRONIC UNITS OF MEASURE	
PROPERTY	**UNITS OF MEASURE**	**SYMBOLS/ABBREVIATIONS**
Potential	Volts and millivolts	V, mV
Current	Amperes and milliamperes	A, mA
Resistance	Ohms, kilohms, megohms	Ω, k, M
Power	Watts and milliwatts	W, mW

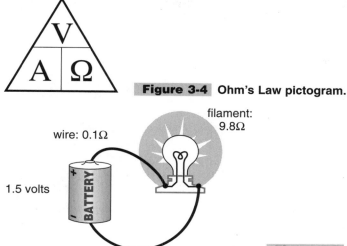

Figure 3-4 Ohm's Law pictogram.

Figure 3-5 Bulb filament has higher resistance than wires.

Figure 3-5 illustrates the situation with the battery and light bulb. The wires have low resistance, while the filament has higher resistance. Since we know the resistance and the battery voltage, we can use Ohm's Law to compute the current through the circuit.

The resistance of the circuit is the total of all resistances in the path between + and −. In this case, it's $0.1 + 9.8 + 0.1 = 10\Omega$. So the current (V/$\Omega$) would be $1.5/10 = 0.15$ A.

One of the effects of resistance is to cause voltage—potential energy—to drop. We can use Ohm's Law and our current calculation to figure voltage drops through various parts of the circuit. For example, each of the wires has 0.15 A of current flowing through 0.1Ω resistance. $V = A \times \Omega$, so each wire has a $0.15 \times 0.1 = 0.015$ V voltage drop across it. That means that the voltage reaching the filament is $1.5 - (0.015 + 0.015) = 1.47$ V.

This partly explains the hot bulb and cool wires; most of the battery voltage is across the filament. Since you've come this far, let's complete the answer using a new concept, power.

Power is a measure of work. In electrical circuits, the most common sort of work is the generation of heat; more power means more heat. Power is expressed in units of watts (W). There are two simple formulas for calculating power:

$$\text{Watts} = \text{Volts} \times \text{Amperes}$$

$$\text{Watts} = (\text{Amperes})^2 \times \text{Ohms}$$

Let's see where the power goes in the battery/bulb circuit. Each of the wires has a 0.015-V drop and 0.15 A of current flowing through it, so each consumes $0.015 \times 0.15 = 0.00225$ W of power. The filament has 1.47 V across it with 0.15 A of current through it, so it's getting $1.47 \times 0.15 = 0.2205$ W of power. That's 100 times as much power (heat-generating potential) as the wires. No wonder the filament is hot!

It may seem like a long way from light bulbs to computers. And to tell the truth, the field of electronics encompasses some fancy physics and mind-boggling math. But digital electronics (the kind we'll be doing with Stamp microcontrollers) can be done at a practical, useful level with little more than Ohm's Law and arithmetic.

3

ELECTRONIC FOUNDATIONS

In fact, if you're content to follow recipes like the ones presented later on in this book, you can build neat projects without any math at all. But you will need to understand, identify, and assemble electronic components.

Electronic Components and Symbols

Electronic circuits are normally drawn not as realistic pictures like our bulb/battery illustration but as skeletal blueprints called *schematic diagrams* or just *schematics*. Using standard symbols for electronic components makes it easy to draw and understand schematics. Unfortunately, beginners find schematics hard to follow, since they don't know what the components might look like. The circuits presented in this book are all in schematic form, so it's important for you to understand schematic symbols and the components they represent. Along the way, I'll even throw in some electronic theory.

BATTERIES AND POWER SUPPLIES

Batteries are represented by the symbol shown in Figure 3-6. The symbol is based on the way early batteries were made from metal plates soaking in acid. You will sometimes see batteries drawn as a single pair of plates (a cell) or more pairs of plates (a high-voltage battery). Don't worry about it. Batteries and schematics are marked with the required voltage rating, so there's no need to count plates.

Note that the wide plate indicates the − end of the battery and the narrow plate the − end.

A more general way of indicating a circuit's power connections is with supply and ground symbols, as shown in Figure 3-7.

For our purposes, supply means the + connection and ground means the − connection. In the more general sense, ground means the common point to which all current in a circuit flows (based on the notion that current flows from + to −).

RESISTORS

Batteries and power supplies provide voltage and current; resistors let us adjust the voltage and current at various points in a circuit. The fact that resistance is the third leg of Ohm's Law—the foundation of all electronics—should go a long way toward explaining resistors' importance.

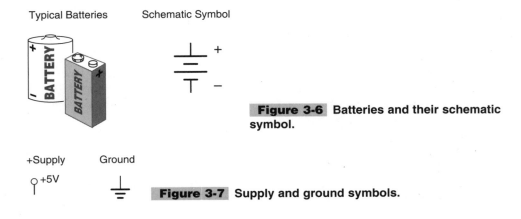

Typical Batteries   Schematic Symbol

Figure 3-6 Batteries and their schematic symbol.

+Supply   Ground

Figure 3-7 Supply and ground symbols.

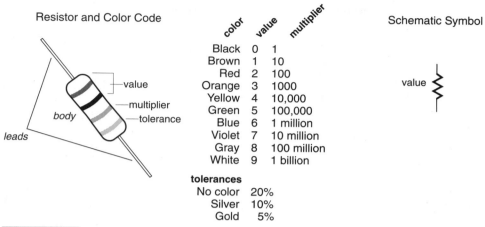

Figure 3-8 Resistors, color code, and schematic symbol.

Most resistors are marked with colored stripes to indicate their value in ohms. Figure 3-8 shows the color code. Figuring resistor values is easy. Suppose a project requires a 47k resistor. We know that k stands for thousand, so that's 47,000 ohms. Looking at the table in Figure 3-8 we see that 4 is yellow, 7 is violet, and thousand is orange. So we want a resistor whose first three bands are yellow-violet-orange.

What about the fourth, tolerance band? It's normal for a resistor's actual value to be a little bit off from its marked or nominal value. The tolerance band tells you how much variation to expect; a 5-percent resistor can range from 95 to 105 percent of its marked value. Many schematics will not specify resistor tolerance, leaving you free to use any tolerance up to 20 percent.

Another resistor characteristic is power rating, given in watts. The more power a resistor must handle, the larger it has to be. Resistors used in the circuits presented in this book can be any wattage from 1/8W on up. If a higher-power resistor is required a note in the schematic will say so.

Resistors, like many components, come with short wires attached for connection and assembly into circuits. These wires are called *leads* (pronounced "leeds").

POTENTIOMETERS ("POTS")

A *potentiometer* (*pot* for short) is a form of adjustable resistor. As the symbol in Figure 3-9 shows, pots consist of a resistor with a sliding connection called the *wiper*. As you turn the control shaft, the resistance between the wiper and one leg of the pot goes up; resistance from the wiper to the other leg goes down. This relationship allows the pot to be used as a voltage divider, a circuit that accepts a steady voltage across the legs and outputs a variable voltage (controlled by turning the shaft) between the wiper and one of the legs. See Appendix C for more on voltage dividers.

CAPACITORS

In its most basic form, a *capacitor* consists of two conductive plates separated by a thin layer of insulation. This setup is a trap for electrical charges. When you connect a voltage

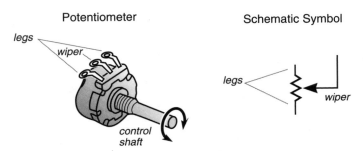

Figure 3-9 A potentiometer is an adjustable resistor.

to a capacitor, + and − charges arrange themselves on opposite plates. The insulator blocks them from coming together, but their attraction holds them in place. The charges are stuck. Trapping charges like this is called *charging* the capacitor.

There is a way to release the charges; connect a conductor between the plates. This causes current to flow until the trapped charges have all come together. This is *discharging* the capacitor.

These properties lead to two main uses of capacitors in the circuits we will build: temporary storage of electricity and timing. The storage aspect is pretty plain. When an external voltage is higher than the charge voltage of the capacitor, the capacitor will charge. When the external voltage is less, the capacitor will discharge. This makes capacitors helpful for smoothing out variations in a power supply.

Timing is more interesting. The process of charging and discharging capacitors takes a predictable amount of time based on the value of the capacitor and the characteristics of the circuit doing the charging or discharging. See Appendix C for examples. The important thing to remember is that a capacitor can help set the timing of a circuit to generate time delays.

Capacitors come in a variety of shapes and sizes. Figure 3-10 shows some examples. Note that electrolytic and tantalum electrolytic capacitors are fussy about *polarity*—the connection of + and −. Make sure to connect them as shown in the schematic or they can be damaged. In power-supply circuits improper connection can make them explode! Fortunately, electrolytic capacitors are marked with a + or − symbol next to one of the leads. As an additional clue, the + lead is usually noticeably longer than the − lead.

All capacitors have a maximum working voltage. Make sure that the capacitor you use has a working voltage equal to or greater than the voltage listed in the schematic. If no working voltage is given, just select a capacitor whose working voltage is more than twice the highest voltage in the circuit.

The basic unit of capacitance is the farad, abbreviated F. However, most common capacitor values are in millionths of a farad (microfarads, abbreviated uF) or trillionths (picofarads, pF). Figure 3-10 shows how to read capacitor markings. To test your understanding, what are the values of the capacitors in the illustration, marked "103M" and "222J?" The formula is value × multiplier, so 103M is 10 × 0.001 = 0.01uF (or 10 × 1000 = 10,000pF). The M tells us that the actual value may be 20 percent higher or lower than the nominal value. For 222J: 22 x 0.0001 = 0.0022uF (22 × 100 = 2200pF), plus or minus 5 percent.

Electrolytic capacitors often use two-digit markings, since their units are always uF or F, never pF.

DIODES AND LEDS

A *diode* is a one-way door for electrical current. It acts like a conductor when + is connected to its anode and − to its cathode, but like an insulator when the connections are reversed. Because of this choosy behavior regarding conduction, diodes are the most basic of a class of components called *semiconductors*. The fancier members of this group are *transistors* and *integrated circuits*.

Most small diodes are cylinders with wire leads, looking somewhat like resistors. They are generally marked with a printed part number, often beginning with "1N" as in "1N4148," a common type of diode. Since it's important to distinguish the anode from the cathode, manufacturers put a stripe on the cathode end of the diode as shown in Figure 3-11. On some diodes the cathode end is tapered to mimic the arrow of the schematic symbol.

A special type of diode lights up when current passes through it. These are *light-emitting diodes* (LEDs). LEDs are often made in a bullet shape so that their package can serve as a simple lens. On such packages, a flat spot on the base marks the cathode, while the longer lead is the anode.

The brightness of an LED is proportional to the amount of current passing through it, but too much current can damage it. Circuits with LEDs usually use a resistor to limit this current. Appendix C shows how to figure the appropriate resistor value.

TRANSISTORS

Transistors let a small current control a large current, like the gas pedal controls a powerful engine. This makes transistors valuable for amplifying weak signals or turning power-hungry

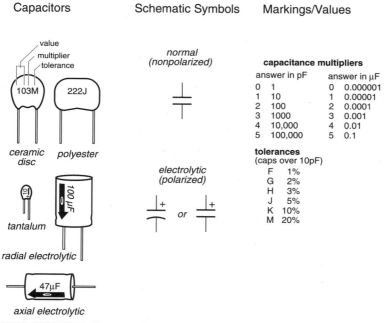

Figure 3-10 Capacitors come in a variety of styles.

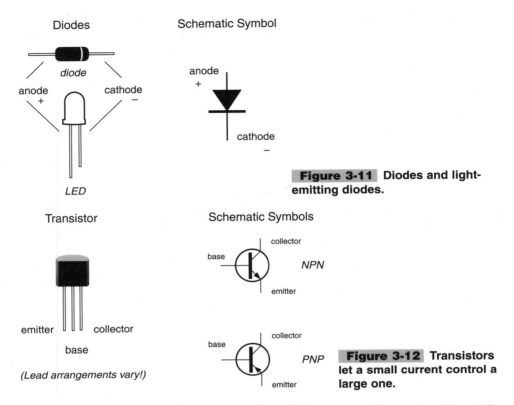

Diodes   Schematic Symbol

diode

anode +

cathode −

LED

Figure 3-11 Diodes and light-emitting diodes.

Transistor   Schematic Symbols

collector

base

NPN

emitter

emitter   collector

base

(Lead arrangements vary!)

collector

base

PNP

emitter

Figure 3-12 Transistors let a small current control a large one.

devices on and off. The control current goes through the transistor's base and emitter, while the larger current being controlled flows through the collector and emitter.

Figure 3-12 shows a typical transistor. I say typical, because there are many different styles of transistors specialized for particular jobs. Each has its own shape and lead arrangement. Don't worry, though; schematics and catalog listings generally include drawings to help you figure it all out.

In this book, the only type of transistors we'll be using are called bipolar transistors. You should be aware that there are other types, like field-effect transistors (FETs) and metal-oxide semiconductor FETs (MOSFETs).

The Stamps can handle enough current to directly drive many small loads like LEDs. Transistors can help them control larger loads. Appendix C includes some basic transistor switching circuits suitable for use with the Stamps.

INTEGRATED CIRCUITS

Integrated means "combined," so an integrated circuit (IC) combines many components into one compact package. When the term was first coined, ICs consisted of only a dozen or so components—mostly transistors. Today's ICs combine thousands, even millions, of transistors and other electronic components onto a single silicon chip.

The term chip is sometimes used interchangeably with IC, but there's a difference. A chip is a minuscule piece of brittle silicon, while an IC is a chip enclosed in a sturdy package made out of plastic, ceramic, or metal.

There are many styles of ICs to suit different methods of electronic assembly, but one of the oldest and easiest to work with is called the *dual-inline package* (DIP). DIPs are relatively large and have stiff leads or pins that can be plugged into sockets for temporary circuit hookups, testing, or prototyping. All of the ICs used in the projects in this book are DIPs.

Figure 3-13 shows how the pins on a DIP are numbered. The number of pins on a DIP depends on the function of the IC and ranges from 6 to 40 or more. However, all DIPs use the same numbering methods. Locate pin 1 by the telltale dot or notch in the IC. The rest of the pins are numbered down that side and up the other as shown in the figure.

All of the components we've examined so far had schematic symbols that hinted at their function. The zigzag line of the resistor certainly gives the impression of a path with more "resistance" than a straight line. But since ICs can include thousands of parts performing very complex functions, a single, standard symbol doesn't cut it. Schematics generally take one of two approaches: depicting the outline of the IC package or using symbols that represent the IC functions with pin numbers next to the connections. Schematics in this book use the representation that makes the most sense for a given circuit.

OTHER COMPONENTS

The symbols used in schematic diagrams are a kind of visual language. Once you get the hang of a few standard symbols, you will find yourself understanding new symbols without much trouble.

Figure 3-14 is an example. It introduces two new symbols: a switch and an electromagnet. A switch is just a break in a wire that can be opened and closed. An electromagnet

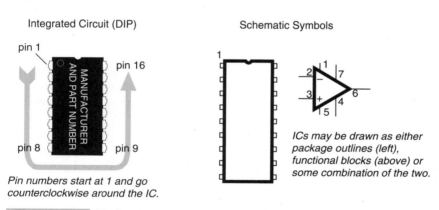

Integrated Circuit (DIP)   Schematic Symbols

Pin numbers start at 1 and go counterclockwise around the IC.

ICs may be drawn as either package outlines (left), functional blocks (above) or some combination of the two.

Figure 3-13 ICs incorporate complex circuits into one small package.

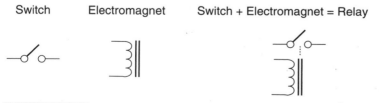

Switch   Electromagnet   Switch + Electromagnet = Relay

Figure 3-14 A relay is a switch controlled by an electromagnet.

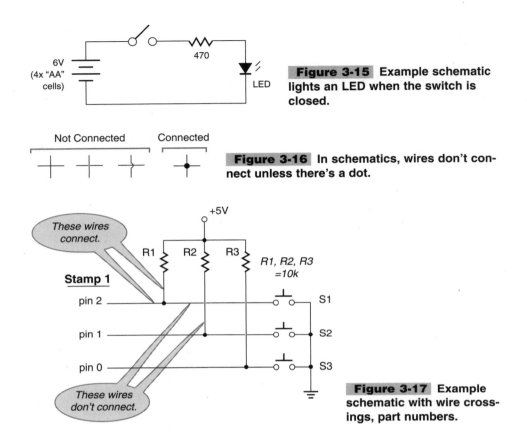

Figure 3-15 Example schematic lights an LED when the switch is closed.

Figure 3-16 In schematics, wires don't connect unless there's a dot.

Figure 3-17 Example schematic with wire crossings, part numbers.

is a coil of wire wrapped around an iron core. When current passes through the coil, the iron becomes temporarily magnetic. If you combine the two as shown in the figure, you have a switch that is controlled by an electromagnet—a component called a *relay*.

Combining Symbols into Schematics

Now that we have seen some of the more important electronic symbols, we can take a look at a schematic diagram. See Figure 3-15. As the figure shows, connections between components—wires—are represented by lines. The schematic includes all the information needed to make the circuit, which lights an LED when the switch is on. The schematic does not specify things that are best left up to you, such as the color of the LED, the exact style of switch, or the method of construction. This is a common characteristic of schematics—saying only what needs to be said.

In more complex schematics, it's common for the lines representing wires to cross one another. Sometimes wires cross without connecting; other times they cross and connect. Figure 3-16 shows how to tell the difference, while Figure 3-17 is an example of a schematic with both kinds of crossed wires.

Figure 3-17 also illustrates a couple of more advanced schematic techniques. Notice the use of part numbers like R1 and S1. These numbers, also known as designators, serve as keys to a separate parts list, which gives component values and other information. Using designators and a parts list keeps the schematic relatively uncluttered but gives you a lot of detailed information about the circuit and its components.

Figure 3-17 also introduces another symbol. S1 through S3 represent a special kind of switch, a pushbutton that closes the circuit when you push it and breaks the circuit when you release it. Note also that the figure does not show the complete schematic of the BASIC Stamp connected to it, but treats it as a sort of supercomponent instead, just calling out the pin numbers to which this extra circuitry will be connected.

The name of the game with schematics is to present the least detail that will let someone build the circuit.

3

ELECTRONIC FOUNDATIONS

BUILDING ELECTRONIC CIRCUITS

CONTENTS AT A GLANCE

There are many ways to assemble circuits. We'll start with the ones that are most suited to beginners' skills and budgets. A couple of words you will encounter in this section are breadboard and prototype. A breadboard is a temporary mockup of a circuit built in order to test basic design ideas. A prototype is the first item or production run of a product that may later be made in quantity. In everyday speech, though, techies use these terms interchangeably to mean the first draft of a circuit.

Plug-in Prototyping Boards

Figure 4-1 is an example of a plug-in prototyping board. These go by a variety of names, such as socket boards, circuit strips, experimenter sockets, and even waffle boards. I like waffle board, because it describes their appearance and their function. They let you change your mind—waffle—over the design of your circuit until you get it just the way you want it!

Waffle boards are easy to use. The perforated plastic conceals a series of metal contact strips inside. When you push wires or component leads into the holes, springy grippers inside grab and hold them. As the drawing shows, the contact strips run vertically in the central part of the board, and horizontally along the edges. Figure 4-2 shows how plugging wires into the board connects them together. (Insulated wire—wire covered with a plastic coating—must have its insulation stripped off the ends before they can make electrical contact.)

Most waffle boards will accept a range of wire sizes, but solid #22 wire works best. This matches the size of most component leads. If wires are to be reused, use wire that has a bright plating on the conductor. Plain, unplated copper wire tarnishes, increasing the resis-

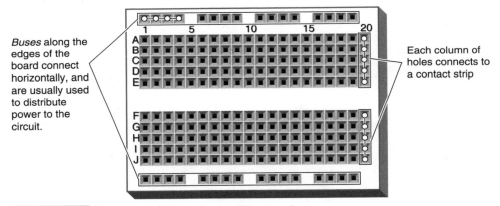

Buses along the edges of the board connect horizontally, and are usually used to distribute power to the circuit.

Each column of holes connects to a contact strip

Figure 4-1 Plug-in protoboard or waffle board.

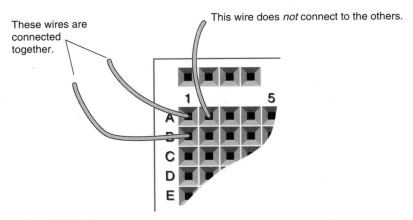

These wires are connected together.

This wire does *not* connect to the others.

Figure 4-2 Wires in the same column are connected.

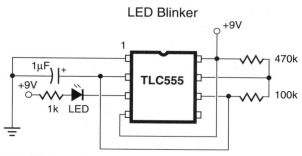

Figure 4-3 Example schematic for waffle-board layout.

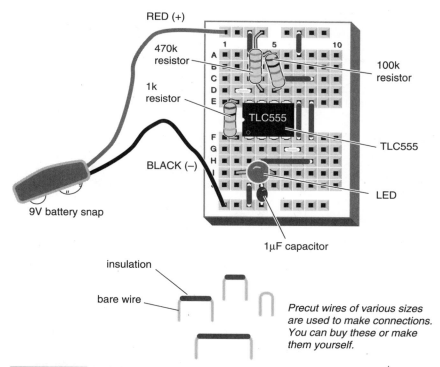

Figure 4-4 Waffle-board layout of circuit from Figure 4-3.

tance of connections or preventing good contact altogether. You can also purchase sets of staple-shaped wire jumpers in various sizes to match the spacing of waffle-board holes. As our example will show, this makes for a neat layout. And the fact that the wires are already stripped and bent is a real time saver.

Let's take a simple schematic and translate it to a waffle-board layout. Figure 4-3 shows a circuit that blinks an LED using a common IC called a TLC555. Figure 4-4 shows how that circuit might be built on a protoboard.

Trace the waffle-board layout and see how it matches the schematic. For example, look at pin 2 of the TLC555. The schematic shows that pin 2 goes to the + end of the 1μF capacitor, and to pin 6. You can see the 1μF capacitor has one lead plugged into the same column of holes as pin 2. And (slightly obscured by the LED), a wire goes from pin 2 to the right. Another wire jumps over the middle of the waffle board. One more wire goes left again, ending at pin 6. In fact, the way this path is laid out on the board looks just like the schematic.

I could have connected pin 2 to pin 6 with a single piece of wire; just bent it into a big hoop and jumped right over the IC. But using the preformed wire staples sold for use with waffle boards makes layouts that are tidy and easy to modify. Imagine a layout with lots of wire arches standing over it like a rib cage—it'd be impossible to change components inside that cage without first removing some of the wires.

Appendix E lists sources for waffle boards, precut jumper wires, and other components mentioned in this section. Parawax, maker of the Stamps, offers several waffle-board kits, such as the Board of Education.

Waffle boards are great for experimenting with circuits. They can be put together and taken apart easily without special tools. Unfortunately, they are not the best for permanent circuits. Parts can work loose if the board is handled too much, and the boards are somewhat bulky compared to other construction methods. When you build a good circuit with a waffle board, record it with a schematic diagram for future assembly using one of the more permanent methods.

CRIMP CONNECTORS

When you need to connect a waffle-board circuit to some other electronic device (like a BASIC Stamp), you may need to make connections to header posts. Header posts are metal pins 0.025" square and about ⅜" tall. Header posts are common because they are inexpensive and can be used with wire-wrapping (see next section) or crimp-on connectors.

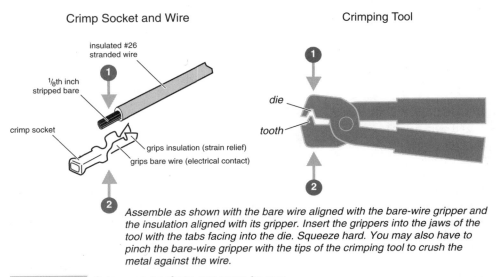

Assemble as shown with the bare wire aligned with the bare-wire gripper and the insulation aligned with its gripper. Insert the grippers into the jaws of the tool with the tabs facing into the die. Squeeze hard. You may also have to pinch the bare-wire gripper with the tips of the crimping tool to crush the metal against the wire.

Figure 4-5 Crimp-on sockets are easy to use.

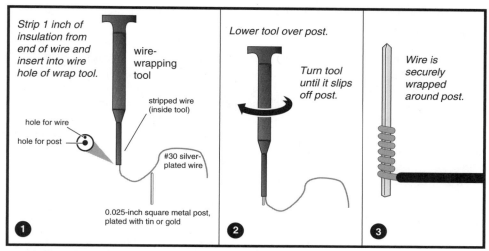

Figure 4-6 **Basic steps in wire wrapping by hand.**

Crimp connectors for headers are tiny sockets that can be squeezed onto the stripped end of a piece of stranded wire. Special pliers let you squeeze—crimp—these connectors with enough pressure to make a reliable electrical connection. Figure 4-5 shows how it's done.

Crimp-on sockets will also fit the leads of some components, like resistors and capacitors. They will not fit the pins of ICs or other small, oddly shaped leads, though.

Wire Wrapping

Wire wrapping is the prototyping technique that helped launch the age of computers. In the 1970s, wire wrapping was the construction method of choice for engineering prototypes, hobby projects, and even limited production runs.

The reason for wire wrapping's bloom of popularity is that it produces a finished circuit that is more-or-less permanent, but can still be modified. Seems like the best of both worlds. Unfortunately, wire wrapping has some significant disadvantages:

- For beginners on a budget, buying an initial set of wire-wrapping tools, panels, sockets, and wire can be overwhelming.
- Wire-wrap posts add a whole inch of thickness to a circuit board.
- Although wire-wrapped connections are in theory more reliable than soldered connections (see upcoming sections), in practice they can be fragile or intermittent.
- Wire wrapping is done from underneath the components, so it is very easy to get confused regarding IC pin arrangements.
- It can be next to impossible to trace connections in a wire-wrapped circuit, with a rat's nest of wires crossing every which-way.

Those drawbacks, along with cheaper and faster printed-circuit-board methods (see upcoming sections), have reduced wire wrapping's popularity.

Although wire wrapping may have passed its prime as a construction method, it's still a useful technique to know. Figure 4-6 shows the basic principle.

The figure shows how wires join to posts, but how do posts join to components? For ICs, there are wire-wrap sockets. The sockets have one post for each pin of an IC. You wrap all your connections to the sockets, then plug in the IC. For other components, like resistors and capacitors, you can mount them in IC-shaped frames called DIP plugs or DIP headers, then plug those into wire-wrap sockets. Or you can use special posts that have little claws designed to grab the component leads. What you should *not* do is try to wire wrap directly to component leads. Leads are round, while wire-wrap posts are square. Wire wrapping works only because the sharp corners of the square posts dig into the relatively soft metal of the wire. Wrapped wires will eventually slip off (or make poor connection with) round component leads.

Wire-wrap posts and sockets require a sturdy support surface. Wire-wrap panels are 1/16-inch fiberglass boards perforated with holes at 0.1-inch intervals—the same as the spacing of pins on standard DIP ICs. Each hole in a panel is surrounded by a metal ring to allow you to solder sockets and pins into place, although a friction fit is sometimes good enough.

If you want my advice, avoid wire wrapping for constructing projects unless you already have the tools and materials. But you should have at least a small spool of wrap wire and an inexpensive wrapping tool. Wire wrapping a couple of connections between header posts can make quick work of small jobs.

Soldering

Waffle boards are great as reusable sketch pads for circuit ideas. And wire wrapping lets you make a more permanent version of a circuit that can still be modified. But circuits that must be built to last are inevitably assembled using solder (pronounced sodder in the United States).

There's so much misunderstanding about soldering that you'd think it was a voodoo ritual rather than a time-tested craft. Even if you have soldered before, please read this section. Soldering is not difficult, but if you don't understand a few basic principles, it is very easy to develop bad habits that make your soldering unreliable.

Solder is an alloy of tin and lead in a ratio of 60/40 or 63/37 (i.e., 60 percent tin and 40 percent lead). Solder melts at a relatively low temperature—less than 400°—compared to other metals commonly used in electronics. For example, a temperature of almost 2000° F is needed to melt copper.

As metals go, the tin/lead solder alloy is not particularly strong. It's soft and easily weakened by repeated bending. What makes it good for securing electronic connections is stickiness. Solder is a sort of metal glue. Most glues are not very strong by themselves. Think of that skin of dried glue that collects on the nozzle of the glue bottle; it's soft and easily broken. Yet a good glued joint can be stronger than the materials it secures.

So we can glue metals together with molten solder. How do we melt the solder? And how do we make sure that the solder sticks properly?

The tool for melting solder is a soldering iron as shown in Figure 4-7. The figure shows some of the features required for electronic soldering: grounded plug, 20- to 30-watt heating element, and a removable cone- or screwdriver-shaped tip. More expensive irons come with a control box that lets you set the temperature, and even adjusts the power to the heating element to hold that temperature steady.

Soldering Iron

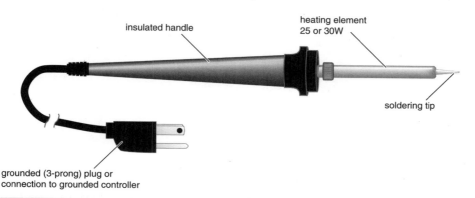

insulated handle

heating element
25 or 30W

soldering tip

grounded (3-prong) plug or
connection to grounded controller

Figure 4-7 **Anatomy of a soldering iron.**

You will also need a stand to hold your soldering iron when it's not in use. Most stands have a pan to hold a moistened sponge, which is essential for cleaning the soldering tip.

The tip of the soldering iron gets hot enough to melt solder, but that's only part of the story. The key to a good solder connection is something called *flux*. The solder sold for electronic purposes looks like wire, but it's actually a tube filled with flux, a fast-acting cleaner. When the solder gets hot, flux gushes out onto the metals being joined. This flux bath removes minor corrosion and contamination that might prevent the solder from sticking. An instant later, the solder itself melts, flows over the freshly cleaned metal, and sticks like crazy. The whole process takes only a couple of seconds.

Enough theory! Here's how to get your soldering off to a running start:

On your mark: Set up your soldering iron stand in a well-ventilated area. Soldering produces fumes that contain lead and other toxic substances. Consider setting up a table-top fan to draw the fumes away from your face. Wear safety glasses to protect your eyes from spatters. Don't eat, drink, or smoke while soldering; you may accidentally ingest lead! Keep soldering equipment and materials out of reach of children, and always stow the hot iron in its stand when you're not using it.

Moisten the sponge with water so that it is damp but not soaking wet. Put the iron in the stand, and plug the iron in. Wait a few minutes for it to heat up. Do not touch the heating element or tip! If you do, painful burns and the stink of burned skin will remind you not to do it again.

Get set: When the iron is hot, take it out of the stand and wipe the tip on the sponge with a striking-a-match motion. Don't leave it in the sponge very long; you just want to wipe the tip of the iron. Quickly apply a little solder to the tip of the iron. If it beads up, wipe the iron on the sponge again and reapply the solder. Repeat this until the solder clings to the tip. This intimate contact between solder and metal is called *wetting*. Wetting the tip with solder is called *tinning*.

GO: Bring the freshly tinned tip of the iron into contact with the wires/metals you want to solder. The tinning of the tip will help transfer heat into the target metals. After a second or so of this preheating, bring the solder into contact with the metals, not the soldering iron itself. If the metals are hot enough, the flux will flow out of the solder and clean

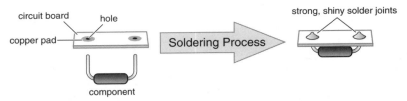

circuit board   hole

copper pad

strong, shiny solder joints

Soldering Process

component

Figure 4-8 Soldering a component to a circuit board.

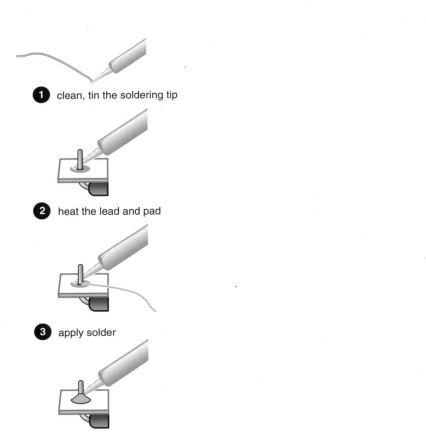

1 clean, tin the soldering tip

2 heat the lead and pad

3 apply solder

4 withdraw the solder, then the iron

5 if necessary, trim excess wire with nippers

Figure 4-9 Steps in producing good solder joints.

the area, then the solder will melt and wet the metals. Remove the solder—pause—then remove the iron without jostling the molten solder left behind.

Figure 4-8 shows the most common sort of soldering job—connecting a component lead to a printed-circuit-board pad. Figure 4-9 illustrates the soldering sequence.

Although it is possible for a good-looking solder joint to be bad, it is nearly impossible for a bad-looking solder joint to be good. Figure 4-10 shows good and bad solder joints. Practice soldering until you consistently produce good-looking solder joints. Even if your first efforts look bad, keep going. Constantly clean and tin the solder tip to keep it clean and shiny. Always apply heat to the joint, not to the solder itself (except when tinning the tip).

It sounds old-fashioned, but good soldering takes discipline. Many beginners go wrong by giving up on good soldering technique before they get the hang of it. They start trying things at random—typically melting big wads of solder on the tip and dropping them on the solder joint. The result may sometimes look OK, but it's not a solder joint! It will work intermittently, or not at all.

Follow the procedures and practice, practice, practice.

Here are some additional hints for good soldering:

■ If you're having trouble tinning a tip, try using a brass-bristle brush to remove stubborn crud or corrosion. There are also tinning compounds and tip cleaners that can help.
■ Once a soldering tip is too old or corroded to tin, replace it with a new one.
■ Hard water in the soldering sponge can leave deposits on the tip. Use distilled or de-mineralized water instead of tap water.
■ Avoid specialty solders—those other than 60/40 or 63/37 tin/lead with a flux (not acid) core. No-lead solders and silver solders can be much more difficult to use without the right training and equipment.
■ Use thin solders, 0.020" to 0.031" diameter, as these melt quickly. Avoid solders that are thicker than 0.031" or thinner than 0.020".
■ Just touch the soldering iron and solder to the joint; never press hard. Pressure damages fragile printed-circuit pads, and can cause them to peel away from the board. This kind of damage is almost impossible to undo. If a joint isn't heating properly, stop. Clean and tin the soldering tip and try again.
■ If you make a bad solder joint, clean and tin the soldering tip, and reheat the joint. You may have to apply a little more solder, primarily to get a fresh application of cleansing flux.
■ To correct really bad solder joints or other mistakes, you may have to remove the solder. See your electronics store for tools and supplies. Spring-loaded solder suckers and desoldering braids are very helpful.

Printed Circuit Boards

Before the printing press was developed, scribes copied books by hand. That's how it was in the early days of electronics, too. Assemblers mounted components on an insulating board and connected each wire by hand. Then, just prior to World War II, engineers realized that the wiring could be manufactured right on the insulating board. The process they came up with resembled printing and increased the speed, ease, and quality of electronic manufacturing in

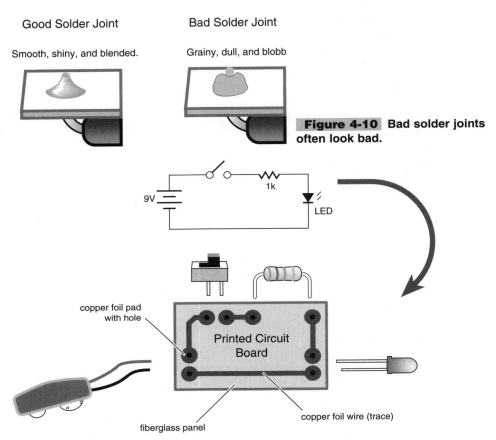

Good Solder Joint

Smooth, shiny, and blended.

Bad Solder Joint

Grainy, dull, and blobb

Figure 4-10 **Bad solder joints often look bad.**

copper foil pad with hole

Printed Circuit Board

fiberglass panel

copper foil wire (trace)

Figure 4-11 How a circuit translates to a circuit board.

the same way printing revolutionized communication. They invented the printed circuit board.

The process is simple. Start with a panel of insulating material—typically fiberglass—and glue a thin sheet of copper foil to it. Print chemical-resistant ink onto the copper in the pattern of the circuit's wiring. Dunk the board into a chemical called *etchant* that dissolves copper. The etchant eats away the copper foil except in the areas protected by the ink. What's left is a perfect reproduction of the printed wiring pattern. You can clean off the ink and solder components to the board.

Figure 4-11 shows how a simple circuit translates to a printed circuit board (pcb). In pcb lingo, foil wires connecting components are called *traces* or *tracks*, and the foil shapes to which the components connect are called *pads*. The pcb in our example is called a through-hole board, because components are mounted by passing their leads through holes in the board. A more modern type of pcb uses surface-mount techniques in which smaller, lead-less components are soldered directly to the pads. Other innovations in pcb technology allow for multiple foil layers separated by thin insulating layers. This allows fabrication of very complicated circuits in which many traces cross without touching.

You can make your own pcbs using materials available from your electronics supplier (Appendix E) and the guidance of a good book (Appendix F). If you'd like to plunge in and start designing your own PCBs, visit www.expresspcb.com. They supply free PCB design software that can automatically upload your design into their manufacturing system. A few days later, your boards are delivered to your door. You can also buy ready-made pcbs for some projects, or generic pcbs that resemble the prototyping boards discussed earlier (Appendix E). Wire-wrapping panels are also generally a form of pcb with pads to support pins and sockets and tracks to distribute power throughout the circuit.

If you're just starting out, the first pcbs you are likely to use are those included with electronic kits and generic boards. Since kits come with their own instructions, let's talk about generic pcbs.

Figure 4-12 depicts a typical generic pcb. I say typical, because there are many different styles available. The type shown is best for the kinds of circuits presented in this book. The row-and-column layout makes it easy to interconnect ICs and other components.

Tools and Test Equipment

Figure 4-13 shows an assortment of tools that you will find useful for electronic assembly and testing. It's by no means a complete set. As you progress, you will find some small jobs that seem harder than they ought to be; the next time you go into an electronics store or browse a catalog, you'll find a tool designed to make that job easy.

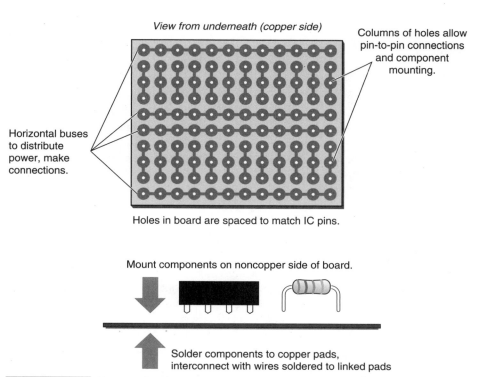

View from underneath (copper side)

Columns of holes allow pin-to-pin connections and component mounting.

Horizontal buses to distribute power, make connections.

Holes in board are spaced to match IC pins.

Mount components on noncopper side of board.

Solder components to copper pads, interconnect with wires soldered to linked pads

Figure 4-12 **Row-and-column arrangement of a generic pcb.**

Tools for Electronic Assembly and Testing

needle-nose pliers　　forceps　　strippers　　nippers

soldering stand w/tip sponge　　20–30W grounded soldering iron　　multitester

Figure 4-13 Basic assortment of hand tools for electronics.

One of the most important tools shown in the figure is the multimeter. These days, you can get a decent unit for less than $100, enabling you to measure volts, ohms, and amps; check for continuity (connection of wires), and test diodes and transistors. Some models include a frequency counter, computer interface, high/low/average sampling, capacitor tester, and a host of other convenient features.

Perhaps the best feature of all is that modern, auto-ranging multitesters are very hard to damage either by physical abuse or incorrect hookup. Buy the best unit you can afford, and read the user manual from cover to cover for practical tips on making common measurements. Build some of the circuits from Appendix C, use Ohm's Law to predict a measurement, then verify your figuring with the meter.

Summary of Construction Techniques

I have only scratched the surface of available electronic construction methods. Cruise through an electronics store or catalog, and you will see many more products, tools, and materials that you may find useful. Manufacturing electronic circuits is advanced technology, but prototyping is a handicraft. Once you learn the basics, you are welcome to improvise your own techniques, or mix and match from existing methods. See what others are doing, through electronics magazines, Internet newsgroups, clubs or classes, and adopt ideas that make sense to you.

GETTING STARTED WITH STAMPS

CONTENTS AT A GLANCE

This chapter will walk you through the process of getting a Stamp up and running for the first time. We'll cover the BASIC Stamp I (BS1) and BASIC Stamp II family (BS2, BS2-SX, BS2-E).

What You'll Need

To get started you will need:

- A PC running Windows '95 or better and CD-ROM drive.[1] For the BS1, an available parallel (printer) port; for the BS2 family, available serial (modem/comm) port.
- BASIC Stamp controller (BS1 or BS2).
- Carrier board (e.g., Parallax Board of Education is highly recommended, but even a plug-in breadboard/waffle board will do).
- Stamp programming cable.
- CD-ROM from this book.

Since the Stamps have no keyboard or display of their own, the PC allows you to type, edit, and download programs as well as view messages from the Stamp.

You can buy a Stamp, carrier board, and a programming package containing the cable and host software directly from Parallax or one of its dealers; see Appendix E for information. Or you can buy just the Stamp; copy the host software from the CD-ROM that accompanies this book, and build your own programming cable. Figure 5-1 shows how to wire programming cables for the BS1 and BS2. The technique for crimping pin sockets onto the BS1 cable is covered in Chapter 4, Figure 4-5.

I've mentioned carrier boards—just what are they? Stamps have been miniaturized to the extent that they are no larger than some ICs. Like ICs, they must be supported by a prototyping board (e.g., waffle board) for connection to power and programming inputs, or placed in a custom carrier board. The carrier board has convenient hookups for programming, snaps for a 9V battery, and a prototyping area that can be used to construct small circuits.

Mounting your Stamp in a carrier board takes most of the effort out of getting it hooked up and running on the first try—highly recommended.

Connecting the Stamp Hardware

With your PC turned off and positioned for easy access to the back panel, connect the Stamp programming cable to the appropriate port. The BS1 uses the 25-pin-female parallel port, while the BS2 uses the 9-pin male serial port. If the port is already in use by another peripheral—a printer or modem—temporarily disconnect it. Later, you can pop down to the computer store and buy an A-B box that will allow you to switch between the Stamp and the peripheral.

---

1. Note that Windows isn't absolutely required for working with Stamps. Alternatives exist for MS-DOS and the Macintosh (using Windows compatibility software). But owing to the near universality of Windows, and the fact that the free CD-ROM that accompanies this book is designed for it, that's what we'll discuss here. For support on other operating systems, check with Parallax, www.parallaxinc.com.

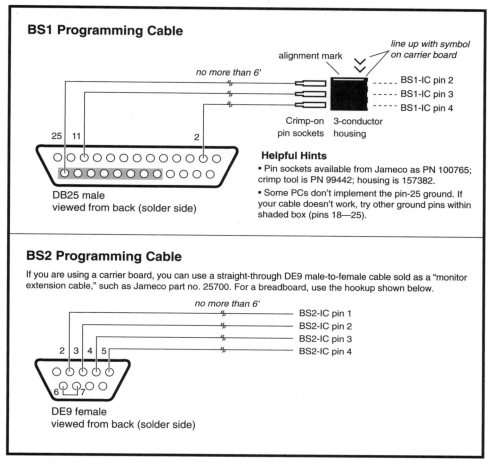

Figure 5-1 Wiring programming cables for the BS1 and BS2.

Installing and Starting the Stamp Software

The CD-ROM that accompanies this book is a customized version of the one that Parallax uses to distribute software, documentation, and application notes. It couldn't be easier to use—just pop it into your CD-ROM drive and wait a few seconds. The screen in Figure 5-2 will appear, ready for you to browse, preview, and install software.

If this screen does not appear, it probably means that your PC does not support the CD-ROM auto-play feature, or that some smarty pants has turned this feature off. Solution: Double-click the My Computer icon to view your disk drives. Locate the Parallax CD in the list and right-click it. A menu similar to that shown in Figure 5-3 will appear; pick AutoPlay.

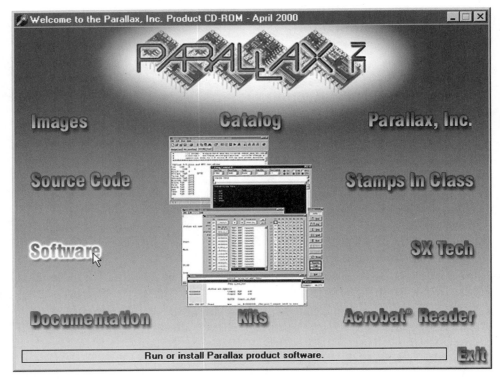

Figure 5-2 Startup screen of the CD-ROM.

Once you have the main screen (Figure 5-2), click on Software. A list of available categories will appear as folders; pick Windows 95/98/NT as shown in Figure 5-4. (This list may appear slightly different. The figure is taken from an earlier release of the software.) Pick the version of the software that lists the Stamp(s) you want to program (BS1, BS2, or BS2/BS2-SX) and click the Install button.

The installation program will lead you from here. Install the software anywhere that is convenient on your hard drive. If you wish, you may use the software without installing it at all. It will run directly from the CD-ROM; just click Run instead of install.

CONFIGURING THE STAMP SOFTWARE

Go ahead and run the Stamp software. In most respects, the Stamp software looks and acts like any other Windows program. However, in order to communicate with the Stamp, it needs to be configured for the appropriate port. Pull down the Edit menu and select Preferences. Click the Editor Operation tab so that the window looks like Figure 5-5.

Although the software can often figure out for itself what port the Stamp is connected to, it's much more reliable if you simply tell it.

Pull down the menu next to Default Parallel Port (for BS1) or Default Serial Port (for BS2 and -SX) and select the correct port number. If you are selecting the com port and the port number you want isn't listed, click the button labeled "..." next to the list and add the desired port number. Then return to the list and select that port.

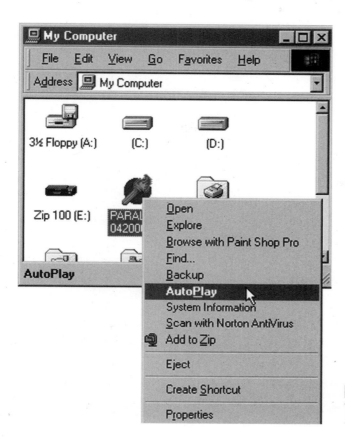

Figure 5-3 If the CD-ROM fails to start up, right-click it and pick AutoPlay.

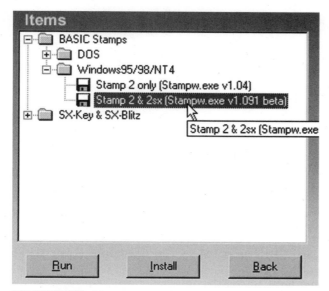

Figure 5-4 Select the software you want and click install.

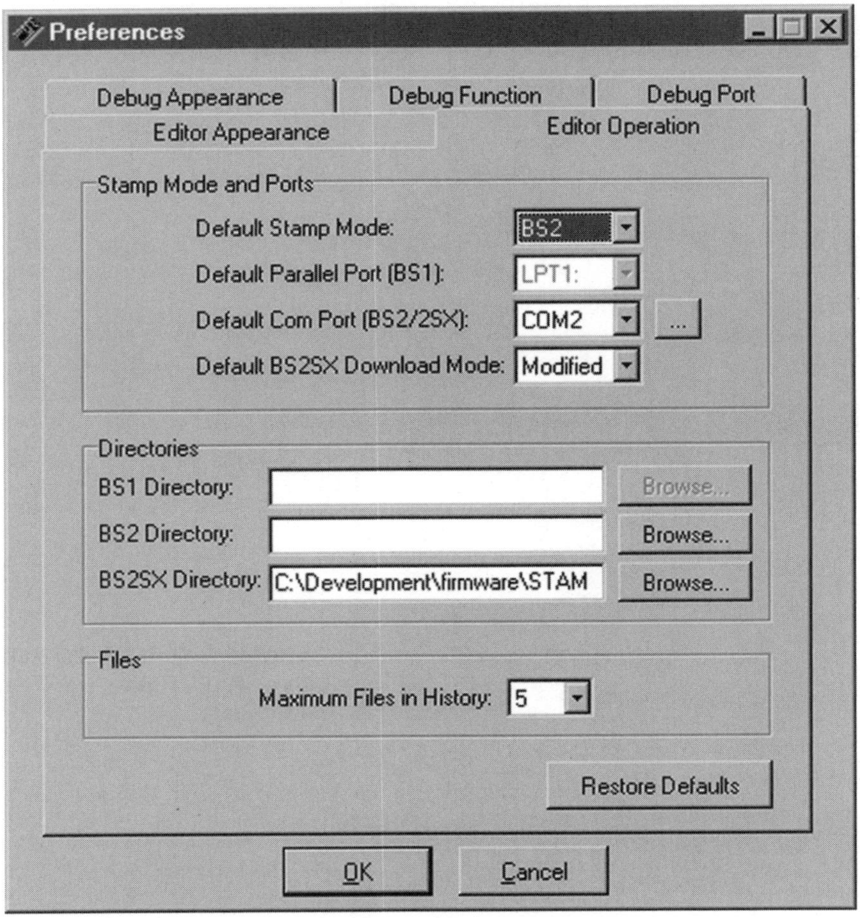

Figure 5-5 Configure the Stamp software to your system setup.

For now, you can safely ignore the other options in the Preferences dialog. Just click OK to dismiss the Preferences window.

INITIAL CHECKOUT: HELLO WORLD!

Earlier you connected your programming cable to the PC; now connect the other end to the Stamp as shown in Figure 5-1. If you are using a Parallax carrier board, make the hookup in accordance with the instructions that came with it. With the BS1, be careful to align the mark on the connector with the arrows on the board. The BS2 serial connector won't allow you to connect it the wrong way.

Apply power to the Stamp, usually by snapping on a 9V battery. Nothing will happen at this point because the Stamp is unprogrammed. Type the following one-line program into the Stamp software on your PC:

```
debug "Hello World!"
```

When that's typed in, hold down the control ("Ctrl") key and press "R" for "Run." A progress bar will appear briefly, then another window will pop up and print the result of your first Stamp program: "Hello World!" Figure 5-6 shows what you should see (more or less, depending on your Windows screen settings).

Maybe you got an error message instead of "Hello World!" Don't panic. Table 5-1 lists some common error messages, what they mean, and how to eliminate them. Try the fixes from the table before calling tech support. You may save the embarrassment of having to confess to a world-weary technician that you forgot to connect the battery.

If all else fails and you cannot get your program to download, contact the vendor who sold you your Stamp hardware. They can offer additional troubleshooting help.

One additional error that I have seen from time to time does not produce a formal error message. It involves characters being dropped from BS2-family Debug messages; for example, "Hello World!" comes out "ello Worl" on the Debug screen. The problem has to do with serial-port buffering. Here's how to fix it:

- From the Windows Start menu, select Settings - Control Panel.
- Double-click the System icon and select the Device Manager tab.

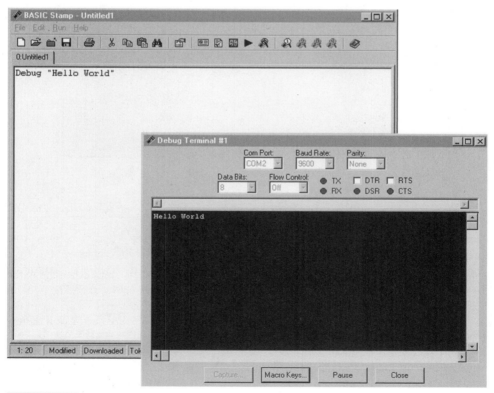

Figure 5-6 Successful test run.

MESSAGE	MEANING	TROUBLESHOOTING/FIXES
Hardware not found	The software could not establish communication with the Stamp.	■ Test 9V battery and replace if necessary. ■ Check cable connection. BS1: Is mark aligned with arrow on carrier board? ■ Try a different PC port. Use lowest numbered port (1 or 2) if possible.
EEPROM verify failed	A problem occurred when the software was checking the downloaded BASIC program.	■ Test 9V battery and replace if necessary. ■ The cable may be loose—reconnect and try again.
Various error messages accompanied by highlighted text in your program	The software doesn't understand something you typed.	■ Double-check the highlighted text against the program example you're trying to run. Correct any errors and try again. ■ Check for typos in the text preceding the highlighted point. Fix errors and try again.

■ Click Ports, then double-click the com port to which your Stamp is connected.
■ Select the Port Settings tab and click the Advanced button.
■ Uncheck the box labeled Use FIFO Buffers…
■ Click the OK buttons on all of the windows opened by the steps above.

Installing the Program Examples

The CD-ROM contains ready-to-run copies of all of the major program listings from this book. I say "major" because many of the little program snippets, especially from the Boot Camp chapters, are not included. Why? Because it's critically important that you type and run these examples in order to begin programming your own mental computer!

To copy the program listings to your hard drive, go back to the CD-ROM main screen (by pressing the Back buttons until you see the screen shown in Figure 5-2 again). Now click on Source Code. You'll get another installation screen with a list like the one shown in Figure 5-7. Click on PCBSC Book Source Code and work your way through the chapter folders that appear until you locate the program listing you wish to copy. Click Install to copy the listing to your hard drive, or View to take a peek.

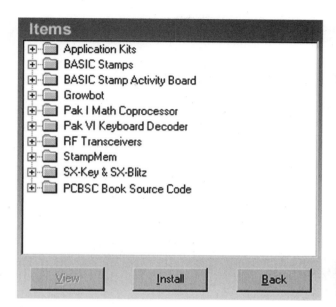

Figure 5-7 View or install programs from this book via the "Source Code" window.

Understanding the Programming Process

When you press Ctrl-R to run a Stamp program, you start the process outlined in Figure 5-8. If there's an error at any point in the procedure—a typo in your program or a bad cable connection—the software stops the process and presents an error message.

I've provided this background information as a way of preparing you for some of the minor errors you may encounter while working with Stamps. If you know what to expect, you'll feel free to experiment and won't take error messages too seriously.

Where Do I Go from Here?

This chapter helped you install and check out the Stamp software. It is meant to be a quick-start, not an all-inclusive guide. You will want to print and read the Parallax documentation included on the CD-ROM if you wish to be a power user of this software. On the other hand, if you've been around the block with Windows programs, you'll probably pick up everything you need to know by osmosis and experimentation.

You can now download programs to the Stamp, so you could start building and programming the projects presented later in this book. And if you are experienced in programming and electronics, you may want to do just that. If you're not, I recommend that you go through BASIC Stamp Boot Camp, my high-intensity introduction to PBASIC programming.

When you run a Stamp program...

The PC:
• Searches for a Stamp connected to a COM or LPT port
• Checks your program for errors
• Converts it to compressed data "tokens"
• Downloads tokens to EEPROM & displays bargraph
• Verifies downloaded data
• If the program contains "debug," displays data window

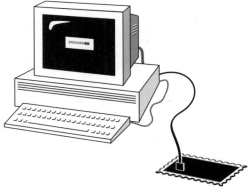

The Stamp:
• Immediately stops any program currently running
• Accepts the downloaded data into its EEPROM
• Cooperates with verification of downloaded data
> Resets to beginning of the BASIC program in EEPROM
• Reads and decodes token in EEPROM
• Carries out instruction represented by token
• If instruction is "debug," transmits data to PC via cable
• Continues until power down or END instruction
• Upon power up, performs steps beginning with > above.

Figure 5-8 **The download-and-run process.**

BASIC STAMP BOOT CAMP, PART 1

The army sends all new recruits to boot camp, where they learn the skills every soldier needs. That's the idea of BASIC Stamp Boot Camp: to introduce essential concepts before you go on to build the projects presented later or design your own. I've divided boot camp into two sections; part 1 covers fundamental programming concepts, while part 2 introduces basic electronic circuits like switches, dials, lights, and speakers.

To get the most out of boot camp, you should try the examples on your own Stamp. If you left the Stamp set up after the checkout described in the previous chapter, you're ready to go. If not, connect the Stamp to the PC and to its battery, turn on your PC, and boot the Stamp host software.

The material we'll cover applies to both the BS1 and BS2. There are some differences in the two Stamps' dialects of BASIC. Where necessary, I'll point out those differences and present examples tailored for each of the Stamps.

Variables: Storage Spaces for Numbers

Every computer or microcontroller is a number-cruncher at heart, so it's fitting that we start by seeing how the Stamps handle numbers. We'll demonstrate the use of BASIC variables, memory locations used to store and retrieve numbers.

Start your Stamp host software with a blank screen (or delete whatever is on the screen) and type in the program below. Make sure to use the version for the kind of Stamp you have (BS1 or BS2). When you're done, check your typing, make sure the Stamp is connected and powered by its battery, and press *Run* (Control-R for Windows, ALT-R for DOS). This will be the procedure for all of the examples that follow. To save space, I won't repeat it. In some cases, you won't need to clear the whole screen, but just to make a few changes to what's already there.

```
BS1:                    BS2:
b0 = 15                 b0 = 15
b1 = 22                 b1 = 22
b2 = b0 + b1            b2 = b0 + b1
debug b0, b1, b2        debug ? b0, ? b1, ? b2
```

The Stamp displays:

```
b0 = 15
b1 = 22
b2 = 37
```

That program stores two numbers (15 and 22) into variables b0 and b1. It then adds the contents of b0 and b1 and stores the result into variable b2. Finally, the Debug instruction displays the contents of the three variables.

Notice that adding variables b0 and b1 does not affect their contents. They still contain the numbers originally stored in them (15, 22) after the addition. The result of the addition is stored in b2.

We don't have to use all three variables in this program. Change the line

```
b2 = b0 + b1
```

to

```
b0 = b0 + b1
```

and rerun it. The Stamp displays:

```
b0 = 37
b1 = 22
b2 = 0
```

This time, the result of adding b0 and b1 is stored into b0. The old value of b0 (15) is lost, written over by the result of the addition. And b2, which never had a value stored into it, contains 0. Before a program runs, the Stamp clears all variables to 0.

These examples demonstrate that:

- Programs store numbers in variables.
- When variables are used in calculations like

```
b2 = b0 + b1
```

the program uses the numbers stored in the variables to calculate the result.

VARIABLES HAVE FIXED SIZES

Let's try another example.

```
BS1:                  BS2:
b0 = 155              b0 = 155
b1 = 220              b1 = 220
b2 = b0 + b1          b2 = b0 + b1
debug  b0, b1, b2     debug ? b0, ? b1, ? b2
```

The Stamp displays:

```
b0 = 155
b1 = 220
b2 = 119
```

Whoa! That's wrong. Variable b2 is supposed to contain 155 + 220, which is 375, not 119. Maybe the Stamp added wrong. Let's try this:

```
BS1:            BS2:
b0 = 375        b0 = 375
debug b0        debug ? b0
```

The Stamp displays

```
b0 = 119.
```

So apparently the Stamp thinks that 375 is 119. What's happening here?

The Stamps have several different sizes of variables. Each has a fixed range of values that it can hold. If you try to store a number that is larger than the variable can hold, it overflows, and a portion of the number is lost. The variables we've been using (b0, b1, b2) are byte variables. The biggest number they can hold is 255. We can fix our problem by substituting a word variable, which maxes out at 65,535:

```
BS1:            BS2:
w0 = 375        w0 = 375
debug w0        debug ? w0
```

Now we get the correct result:

```
w0 = 375
```

It's important to use variables that are large enough to hold the results of a program's calculations without overflowing. But it's equally important to avoid using variables that are larger than necessary, because they use up the Stamp's limited memory more quickly. Table 6-1 lists the available variable sizes. Table 6-2 shows how variables are organized in the Stamps' memory.

TABLE 6-1 VARIABLE TYPES, SIZES, AND RANGES OF VALUES

TYPE	SIZE	RANGE OF VALUES
bit	1 bit	0 or 1
byte	8 bits	0–255
word	16 bits	0–65,535

Note: The BS2 has an additional variable type, called a nibble, consisting of 4 bits and having a range of 0–15. See the BS2 instruction summary.

TABLE 6-2 ORGANIZATION OF STAMP VARIABLE MEMORY

WORDS	BYTES	BITS (SEE NOTE)
w0	b0, b1	bit0–bit7
w1	b1, b2	bit8–bit15
w2	b3, b4	
w3	b5, b6	
(continues through w6 for BS1, w12 for BS2)	(continues through b13 for BS1, b25 for BS2)	

Note: The BS2 does not have named bit variables like bit0. See the BS2 instruction summary.

As you can see from the tables, word variables can hold larger values than bytes, but take up twice as much memory space. The BS1 has enough memory to hold 14 byte variables, but only 7 word variables; the BS2, 26 bytes, or 13 words. That's why it's important to use the smallest variable that will do a particular job.

NAME THOSE VARIABLES

The idea of variables was borrowed from mathematics, which has a custom of giving variables names, as in the formula area = length × width. But the Stamp variables we've dealt with so far have had generic names, like w0 or b2, that don't tell us much about the numbers they contain other than their maximum size (word or byte).

We can change that. Let's use the Stamp to calculate the area of a 27-×-13-foot room:

```
BS1:
SYMBOL length = b0
SYMBOL width  = b1
SYMBOL area   = w1
length = 27
width = 13
area = length * width
debug "Area= ", # area, " sq ft."
```

```
BS2:
length  VAR   byte
width   VAR   byte
area    VAR   word
length = 27
width = 13
area = length * width
debug "Area= ", DEC area, " sq ft."
```

The Stamp displays:

```
Area = 351 sq ft.
```

Note that BASIC uses the asterisk (*) instead of × for multiplication. Otherwise there might be confusion between × for multiplication and × as a possible variable name.

BS1 users: Naming a variable just means substituting a symbol like "length" for the built-in name of the variable, like b0. When the Stamp program runs, it substitutes b0 wherever the name "length" appears.

With the BS1, the responsibility for picking variables falls to you, the programmer. You must be careful not to assign more than one name to the same variable, or pick overlapping variables.[1] For example, what if you assigned both length and width to variable b0? The program would store 27 to b0, then 13 to b0, then calculate b0 times b0, which is $13 \times 13 = 169$—the wrong answer for 27 times 13.

As for overlapping variables, see Table 6-2. The 16-bit variable w0 occupies the same space in memory as the two 8-bit variables b0 and b1. So changes in b0 or b1 affect the contents of w0 and vice versa. If we had given the name area to w0 in the program example, the result would have been correct, but b0 and b1 would no longer contain 27 and 13 after the calculation

```
area = length * width
```

If you needed those values for some future calculation you'd get the wrong results and have a hard time figuring out why!

BS2 users: The VAR directive tells the Stamp software to set aside space in memory for a variable of a given size (e.g., byte or word) and gives it a name like length or area. This is slightly different from using the prenamed variables like b0 and w1, since the Stamp software makes sure that the memory space it assigns does not overlap with any other variable. This is a very good reason to use VAR to assign variables—it helps prevent program bugs.

In more complicated programs, assigning names to variables is vitally important. Good variable names make a BASIC program easy to read and understand.

CONSTANTS ARE UNCHANGING VALUES

We've seen how you can name variables; you can also name constants. What are constants? We've been using them all along to assign values to variables; they are unchanging numbers. For example, in the line

```
length = 27
```

the number 27 is a constant.

You can name constants in much the same way you name variables. Example:

```
BS1:                          BS2:
SYMBOL DogYears = 7           DogYears CON   7
SYMBOL Age     = 15           Age      CON   15
SYMBOL dogAge  = b1           dogAge   VAR   byte
dogAge = Age * DogYears       dogAge = Age * DogYears
debug # dogAge, "dog years"   debug DEC dogAge, " dog years"
```

The Stamp displays:

```
105 dog years.
```

1. There are programming situations in which you might want to assign one variable two different names, or take advantage of the way that byte and word variables overlap. The Stamp software will let you do those things. But most of the time, these things are a mistake.

BS1 users: Naming a constant works the same as naming a variable.

BS2 users: Naming a constant is similar to naming a variable, but you must use the directive CON instead of VAR.

SUMMARY

Variables are locations in memory that your program can use to store and recall numbers. The Stamps allow you to use variables of different sizes to get the maximum use out of available memory. You must be careful to pick variables that are large enough to hold the range of values you expect, but not larger than necessary. In order to make your programs read more like plain English, you should assign clear names to variables. You can also assign names to constants, the numbers that don't change while the program is running.

If you have been typing and running the program examples, take a break and let this information sink in. Don't fret if you don't understand everything yet. The BASIC programming language is like an intricate machine, and I've given you a close-up look at an important component. But you won't really grasp it completely until you have seen all of the parts and how they work together.

Labels, Gotos, and Loops

All of the programs we've written so far have run in a straight line. They started with the first instruction and went line-by-line through each subsequent instruction. But most of the interesting things that programs can do come from changing the order in which instructions are carried out. Let's see how that might work.

```
BS1:                      BS2:
debug "Ready..."          debug "Ready..."
debug "Set.."             debug "Set.."
debug "Go!"               debug "Go!"
```

By now, you can probably predict the outcome of that program; the Stamp displays

```
Ready...Set..Go!.
```

We can change that:

```
BS1:                      BS2:
debug "Ready..."          debug "Ready..."
goto Start                goto Start
debug "Set.."             debug "Set.."
Start:                    Start:
debug "Go!", cr           debug "Go!", cr
```

Now we get

```
Ready...Go!.
```

The Stamp skipped over the instruction that would have printed

```
Set..
```

on the screen. The Stamp saw the instruction

```
goto Start,
```

looked for a label called

```
Start,
```

found it, and continued the program there. In the process, it skipped over the instruction

```
debug "Set..".
```

This illustrates the simplest way to change the course of a program—with a label and a goto instruction. Labels are names ending with a colon (:) that you place before an instruction that will be the destination of a goto.

The most common use of the goto/label pair is to create a loop, a repeating instruction or sequence of instructions. Move the goto in the example above to match this listing:

```
BS1:                    BS2:
debug "Ready..."        debug "Ready..."
debug "Set.."           debug "Set.."
Start:                  Start:
debug "Go!"             debug "Go!"
goto Start              goto Start
```

If you ran that program, you see what I mean about repeating instructions, with the Stamp displaying

```
Ready...Set..Go!Go!Go!
```

with the Go! repeating endlessly. (You may close the debug screen if all that Go!ing makes you nervous. The program will continue, but the Stamp host software on your PC will ignore it.)

Repeating sequences of instructions are called loops, and loops are among the most useful programming tricks. As a result, there's more than one way to loop.

For...Next, Another Kind of Loop

Suppose we wanted the previous example to print

```
Go!
```

just three times. Or 10 times. Or some number of times to be decided later. How would we do that? The goto/label pair doesn't let us pick the number of times it loops, it just loops. But there's another pair of instructions that gives us complete control.

```
BS1:                      BS2:
symbol loops = b0         loops var byte
FOR loops = 1 to 3        FOR loops = 1 to 3
 debug "Go!",cr              debug "Go!",cr
NEXT                      NEXT
debug "Done."             debug "Done."
```

There we are,

```
Go! Go! Go! Done.
```

The For...Next pair uses a variable to count the number of trips through the loop. The For instruction sets the starting value of the variable:

```
FOR loops = 1... .
```

The rest of the For instruction,

```
...to 3
```

is actually a message for the Next instruction. Next adds 1 to the variable (loops) and compares the result to the ending value sent to it by For (3 in this case). If the variable is less than or equal to the ending value, Next makes the program loop back to the instruction right after For. If the variable is greater than the ending value, Next ends the loop by letting the program continue with the instruction following Next.

In short, For...Next keep executing the instructions between them and increasing the variable until the variable exceeds the ending value.

Try changing the starting and ending values in the example and see how the number of loops changes. Try ...1 to 5 and ...1 to 10. Just for kicks, try ...1 to 1. The loop executes at least once, no matter what. Can you figure out why?[2]

Here's a program that shows the inner workings of For...Next by letting you watch the value of the variable change as the loop executes:

```
BS1:
symbol loops = b0
FOR loops = 0 to 5
  debug "looping: ", # loops, cr
NEXT
debug "Done: ", # loops
```

```
BS2:
loops var byte
FOR loops = 0 to 5
  debug "looping: ", DEC loops, cr
NEXT
debug "Done: ", DEC loops
```

You can clearly see the For...Next machinery at work, with the Stamp displaying

```
looping: 0
```

through

```
looping: 5
```

and finishing with

```
Done: 6.
```

GETTING IN STEP

If you want to count up by twos (2, 4, 6...) or down (6, 5, 4...), or some combination of the two (6, 4, 2...), For...Next has an option called Step that will interest you:

```
BS1:
symbol loops = b0
FOR loops = 0 to 6 STEP 2
  debug "up by 2s: ", # loops, cr
NEXT
debug cr
FOR loops = 3 to 0 STEP -1
  debug "down by 1s: ", # loops, cr
NEXT
debug cr
FOR loops = 6 to 0 STEP -2
  debug "down by 2s: ", # loops, cr
NEXT
```

```
BS2:
loops var byte
FOR loops = 0 to 6 STEP 2
  debug "up by 2s: ", dec loops, cr
NEXT
debug cr
FOR loops = 3 to 0
  debug "down by 1s: ", dec loops, cr
NEXT
debug cr
FOR loops = 6 to 0 STEP 2
  debug "down by 2s: ", dec loops, cr
NEXT
```

2. The instructions between For and Next always execute at least once, because it is Next that decides whether or not the loop should repeat. In the process of getting to Next, the program executes the instructions inside the loop.

The Stamp displays count up by twos, down by ones, and down by twos, thanks to the versatile Step option. There are slight differences in BS1 and BS2 versions of Step:

BS1: When you want the variable to increase, give Step a positive value; to decrease, a negative value.

BS2: The BS2 compares the start and end values and decides for itself whether the value after Step should be positive or negative. Always give Step a positive value and let the BS2 do the rest.

Both Stamps: The For...Next examples above all use constants for the start, end, and step values. But BASIC will also let you use variables instead. For instance, in a gumball-vending machine, your program could count the number of pennies deposited and store that value into a variable. Then it would dispense the correct number of gumballs within a loop that looked something like this:

```
FOR gumballs = 1 to pennies ... Next
```

[That example assumes that the program has already defined variables named gumballs and pennies. The ellipses (...) represent the program steps needed to dispense one gumball.]

Making Decisions with If...Then

You may have noticed that part of the For...Next mechanism involves deciding whether a variable is greater than the end value. Making decisions is the soul of programming. In BASIC, the primary decision-making tool is the instruction If...Then. Here's how it works:

```
BS1:
SYMBOL x = b0
x= 99
IF x < 100 THEN saySo
debug "x is 100 or more"
end
saySo:
debug "x is less than 100"
```

```
BS2:
x var byte
x= 99
IF x < 100 THEN saySo
debug "x is 100 or more"
end
saySo:
debug "x is less than 100"
```

The program will announce that x, which holds 99, is less than 100. Change

```
x = 99
```

to give x a value of 100 or more (not more than 255, the maximum value for a byte) and rerun the program. Now the program will display the message "x is 100 or more."[3]

There are two parts to If...Then. The first is the condition—a statement about the relationship of two values, like "x < 100" (read "x is less than 100"). If that condition is true, the second part comes into play, the action. The action is to go to a program label, just like the Goto instruction we saw earlier. If the condition is not true, If...Then does nothing. It just allows the program to continue with the next instruction.

Table 6-3 lists the comparisons used to set up conditions for If...Then. If you want some practice with them, try modifying the example program with varying conditions and values of x and running it to see the results.

3. In the If...Then examples an End instruction stops the program after printing a Debug message. This prevents the program from continuing on and printing both Debug messages—corresponding to true and false conditions—in the event that a condition is false.

TABLE 6-3   COMPARISON OPERATORS USED WITH IF...THEN

SYMBOL	MEANING
=	is equal to
<	is less than
>	is greater than
<>	is not equal to
<=	is less than or equal to
>=	is greater than or equal to

I mentioned that For...Next has a built-in decision of the sort that If...Then makes. To improve our understanding of both instructions, let's rewrite the second For...Next example using If...Then:

```
BS1:
symbol loops = b0
loops = 0
repeat:
 debug "looping: ", # loops, cr
 loops = loops + 1
IF loops <= 5 THEN repeat
debug "Done: ", # loops
```

```
BS2:
loops var byte
loops = 0
repeat:
 debug "looping: ", DEC loops, cr
  loops = loops + 1
IF loops <= 5 THEN repeat
debug "Done: ", DEC loops
```

Running the program gives the same results as the For...Next version. The For...Next program is generally better because it uses fewer instructions and is easier to understand. But If...Then is more flexible, allowing loops to repeat or not based on complex conditions.

COMBINING IF...THEN CONDITIONS

If...Then sends the program to a label if a condition is true. You could describe its operation this way: IF (true) THEN (goto label). That's similar to the way we use if in everyday speech; "If it is dark then turn on the headlights." If the statement "it is dark" is true, we turn on the headlights.

In speech, we often combine conditions, as in:

- "If it is dark or it is raining, turn on the headlights."
- "If light is green and intersection is clear, step on the gas."

So we use the words *or* and *and* to say what we'll do based on whether or not multiple conditions are true. You can do the same in your programs. The general forms are

- IF (true) OR (true) THEN (goto label)
- IF (true) AND (true) THEN (goto label)

An example will show how this works:

```
BS1:
SYMBOL x = b0
x= 150
IF x >= 100 AND x <= 200 THEN inRange
```

```
BS2:
x var byte
x= 150
IF x >= 100 AND x <= 200 THEN inRange
```

```
debug "x is not between 100 and 200"      debug "x is not between 100 and 200"
end                                       end
inRange:                                  inRange:
debug "x is between 100 and 200"          debug "x is between 100 and 200"
```

Since x holds 150, it is greater than or equal to 100 and less than or equal to 200. So If...Then makes the program go to the label inRange: and print the message. Change the value stored to x to 50 or 250 and rerun the program to see the other "not between 100 and 200" message.

If...Then processes multiple conditions by combining them into a single true/false result. Tables 6-4 and 6-5 show how this works.

If...Then is not limited to two conditions; you can add more conditions using AND and OR to your heart's content. The Stamp will start with the leftmost condition and perform the necessary comparisons from left to right until it ends up with a single answer of true or false. That last remaining true or false will determine whether or not the program goes to the label specified by Then.

I have only scratched the surface of this kind of decision making, called Boolean logic. See Appendix B for a more complete discussion.

TABLE 6-4 COMBINING CONDITIONS WITH OR

CONDITION 1		CONDITION 2	RESULT
false	OR	false	false
false	OR	true	true
true	OR	false	true
true	OR	true	true

TABLE 6-5 COMBINING CONDITIONS WITH AND

CONDITION 1		CONDITION 2	RESULT
false	AND	false	false
false	AND	true	false
true	AND	false	false
true	AND	true	true

Multiple Possibilities with Branch

Sometimes a program has to recognize and respond to one of a number of choices. One way to handle this is with a series of If...Then instructions:

```
BS1:                      BS2:
SYMBOL n = b0             n var byte
```

```
for n = 0 to 2              for n = 0 to 2
  IF n = 0 then Zero          IF n = 0 then Zero
  IF n = 1 then One           IF n = 1 then One
  IF n = 2 then Two           IF n = 2 then Two
nextN:                      nextN:
next                        next
end                         end
Zero:                       Zero:
debug "zero",cr             debug "zero",cr
goto nextN                  goto nextN
One:                        One:
debug "one",cr              debug "one",cr
goto nextN                  goto nextN
Two:                        Two:
debug "two",cr              debug "two",cr
goto nextN                  goto nextN
```

The For...Next loop cycles the value of the variable n through values of 0, 1, and 2. Three If...Then instructions are set up to recognize one of these values and send the program to one of three labeled points in the program. The result is that the program displays the messages, "zero one two."

The Stamps' Branch instruction can replace the multiple If...Then instructions in this kind of program. Branch takes a list of labels and assigns an index number to each. Index numbers start with 0 and increase by 1 for each label: 0,1,2... Branch accepts a value (like the variable n in the example), selects the label whose index matches this value, and makes the program go to that label.

If the value given to Branch is greater than the highest index value in its list of labels, Branch does nothing. The program continues with the next instruction.

It's probably easier to see how Branch works than to describe it.

```
BS1:                        BS2:
symbol n = b0               n var byte
for n = 0 to 2              for n = 0 to 2
  BRANCH n, (zero,one,two)    BRANCH n, [zero,one,two]
nextN:                      nextN:
next                        next
end                         end
Zero:                       Zero:
debug "zero",cr             debug "N is zero",cr
goto nextN                  goto nextN
One:                        One:
debug "one",cr             debug "N is one",cr
goto nextN                  goto nextN
Two:                        Two:
debug "two",cr             debug "N is two",cr
goto nextN                  goto nextN
```

The effect is the same as the multiple If...Thens, but the program is shorter and neater. It's also easier to read and modify.

Programmers spend a lot of time fixing program mistakes—bugs—and adding new features or changing old ones. Smart programmers design their programs to be easy to fix or change. So although there's nothing earth-shattering about Branch, it's often a better choice than a gauntlet of If...Thens.

Looking Up Values from a List

We have seen how Branch uses an index value to pick from a list of labels for the program to go to. The Lookup instruction is similar, but instead of labels Lookup's list contains numbers. Lookup indexes the items in its list starting with 0 and increasing by 1. Lookup accepts a value (like the variable n in the example), selects the listed item whose index matches this value, and stores that item into a variable (result in the example).

```
BS1:                            BS2:
symbol n = b0                   n      var byte
symbol result = b1              result var byte
for n = 0 to 4                  for n = 0 to 4
 LOOKUP n,(83,84,65,77,80), result   LOOKUP n,[83,84,65,77,80], result
 debug result                    debug ? result
next                            next
```

The program displays the numbers from the Lookup list in order. An interesting wrinkle to Lookup is how it deals with an out-of-range index value. Change the 4 in

```
for n = 0 to 4
```

to 5 and rerun the program. If you number the items in the list starting with 0, the maximum index number is 4. What happens when Lookup is asked to find the nonexistent item 5? Nothing. This means that the last value stored in the variable result stays there, so the last number in the list, 80, is displayed twice.

By the way, are you wondering what the significance is of the numbers I put in the Lookup list? I couldn't resist hiding a message in the example. To view my message, change the For... instruction back to

```
for n = 0 to 4
```

and change the debug instruction as follows:

```
BS1:                BS2:
debug @ result      debug result
```

These changes tell debug to look at the value of result as ASCII codes—numbers that are interpreted as character symbols like those that appear on your computer's keyboard and screen.[4] To learn more about ASCII codes, see Appendix B.

Another Look at a List: Lookdown

Lookup accepts an index number and returns the corresponding item from a list. The Stamps have another instruction, Lookdown, that does the reverse. Given a number, it

---

4. To a computer, everything is a number. Beginners have a hard time grasping this, but even letters are numbers! The program decides whether a number will be used as a character on the screen, a value in a calculation, or a pattern of dots on a display. So what do the numbers 83,84,65,77,80 represent? STAMP.

searches through a list for a match, and stores the index number in a variable. If the list does not contain a match,[5] Lookdown does nothing. An example:

```
BS1:
symbol n = b0
symbol result = b1
for n = 80 to 85
 result = 255
 LOOKDOWN n,(83,84,65,77,80), result
 debug result
next
```

```
BS2:
n      var byte
result var byte
for n = 80 to 85
 result = 255
 LOOKDOWN n,[83,84,65,77,80], result
 debug ? result
next
```

The program cycles through values from 80 to 85, asking Lookdown to search for a match in its table. It displays 4, 255, 255, 0, 1, 255. Those numbers represent the result of Lookdown's search through the list for numbers 80 through 85. It finds 80, 83, and 84 in the list at positions 4, 0, and 1. It does not find 81, 82, and 85 in the list, so it leaves the value 255 in result.

Try removing the line

```
result = 255
```

and rerunning the program.[6] Now the program displays 4, 4, 4, 0, 1, 1. That's because in the no-match situations Lookdown does not change the value of the variable result. This should help you understand the purpose of loading 255 into result in the original example—it serves as a flag to indicate that Lookdown didn't find a match. I picked 255, because the highest index number in the list is 4, so Lookdown could never return 255 for a match.[7]

I offer this as an example of the way programmers think; they consider all possibilities and plan for them. They try to make programs complete without making them overly complicated. The simple act of preloading a value into result prior to Lookdown would give a program a simple test to determine whether or not Lookdown found a value in its list (If result <> 255 Then foundMatch).

Create Your Own Instructions with Gosub

The Stamps' BASIC instructions are a toolbox for building programs. And that toolbox is small, just 32 instructions for the BS1, 36 for the BS2. As you master the tools BASIC

5. The BS1's Lookdown instruction requires an exact match, but the BS2 can optionally search for items using any of the If...Then comparison operators. For example, it can find the first item in a list that is less than or equal to a given value, and return its index number. See the BS2 instruction summary for more information.

6. Programmers have a trick for temporarily taking lines out of a program. They put the tick mark (') at the beginning of the line to be removed. This causes the Stamp to regard the line as a comment and ignore it. The technique is called commenting out a piece of code. It's a handy trick for temporarily deactivating a program line or lines. If you later want to reactivate the line, just delete the tick mark. Beats retyping a deleted line! We'll discuss comments and program formatting in more detail later on.

7. You could never create a 256-item list in a Lookdown table because the instruction would exceed the size of the BS1's program memory. The BS2 has enough program memory, but won't accept a single program line that long. So 255 is an entirely safe no-match indicator for any Lookdown table.

gives you, you'll begin to make your own tools by combining instructions in ways that solve particular programming problems. In some cases, a group of instructions will be so useful that it will be used more than once in the course of a single program. That's where the instructions Gosub...Return come in.

You can think of Gosub as a variation of Goto that marks its place in the program before going to a labeled location. The Return instruction finds the mark and goes back to that location in the program. The net effect is that you can use a group of instructions between a label and a Return from many locations within a program. Groups of instructions that are used this way are called subroutines, which is where Gosub got its name. An example:

```
BS1:                                BS2:
symbol loops = b0                   loops var byte
debug "Subroutine Example"          debug "Subroutine Example"
GOSUB mySub                         GOSUB mySub
for loops = 0 to 5                  for loops = 0 to 5
 debug # loops                       debug DEC loops
 GOSUB mySub                         GOSUB mySub
next                                next
end                                 end
mySub:                              mySub:
pause 500                           pause 500
debug cls, ">"                      debug cls, ">"
RETURN                              RETURN
```

The example subroutine mySub pauses the Stamp program for ½ second (500/1000ths of a second), clears the Debug screen, and prints a pointer character >. When you run the program, the subroutine does its stuff at the beginning of the program, and in each trip through the For...Next loop.

Using mySub is the same as inserting those Pause and Debug instructions at various points in the program. Using the subroutine makes the program shorter by substituting one instruction, Gosub, for several. The longer the subroutine and the more often it is used, the greater the savings.

PASSING DATA TO SUBROUTINES

Our subroutine example above lacks one feature that almost all BASIC instructions have. That's the ability to accept parameters—values that control the way that the instruction works.

For instance, the Pause instruction in the example above accepts a value that determines the length of time in thousandths of a second (milliseconds) that the program is supposed to pause (wait before continuing to the next instruction). The subroutine mySub gives Pause a constant value of 500. What if we wanted that time delay to change under various conditions? How would we pass this information to the subroutine?

Easy. Write the subroutine so that it gets its data from a variable. Then make sure that the main program puts the appropriate data into the variable before executing Gosub. We can modify the previous example:

```
BS1:                        BS2:
symbol loops = b0           loops var byte
symbol time = w1            time  var word
time = 500                  time = 500
GOSUB mySub                 GOSUB mySub
```

```
time = 100                time = 100
for loops = 0 to 5        for loops = 0 to 5
 debug # loops             debug DEC loops
 GOSUB mySub                GOSUB mySub
next                      next
end                       end
mySub:                    mySub:
pause time                pause time
debug cls, ">"            debug cls, ">"
RETURN                    RETURN
```

Now the Pause instruction in mySub[8] gets the length of its delay from the variable time. Whenever the program needs to change the length of the delay, it stores a new value in time.[9]

At Ease! End of Boot Camp Part 1

That's it for part 1 of our BASIC Stamp Boot Camp. By now, you should be able to read some of the programs that accompany the projects and recognize fundamental program structures, like

■ Variables for data storage and math calculations
■ Loops with Goto or For...Next
■ Decisions using If...Then
■ Operations on lists using Lookup and Lookdown
■ Subroutines

In part 2 of Boot Camp, we'll turn these ideas loose on the real world by adding electronic hardware to the Stamps.

8. The natural-born efficiency experts out there may have noticed that mySub no longer makes the program shorter. It would actually be better to insert the Pause and instructions at each point that

```
Gosub mySub
```

appears. That's exactly the sort of analysis you should do in order to decide whether or not a subroutine is appropriate.

9. Note that the variable time is a word variable rather than a byte. Remember that bytes only hold numbers up to 255; words can store up to 65,535.

BASIC STAMP BOOT CAMP, PART 2

CONTENTS AT A GLANCE

In part 2 of BASIC Stamp Boot Camp, we will apply the programming fundamentals of part 1 to control simple electronic circuits. In the process, we'll discover some of the Stamp's most interesting capabilities.

Seeing the World through an Input

A lot of sayings go, "There are two kinds of _____ in the world, those that _____, and those that don't."

If a Stamp could fill in the blanks, it would say, "There are two kinds of voltages in the world, those that exceed 1.5 volts and those that don't."

That's not very profound. But to a Stamp, it's everything. That's because a Stamp looks at the world through its inputs, and those inputs recognize only two states: 1 (above 1.5 volts) and 0 (below 1.5 volts).

Let's hook up a Stamp and see how this works. Wire the circuit shown in Figure 7-1. If you don't have a switch or pushbutton handy you can fake it by substituting a pair of bare wires. Separate them to open the switch and hold them together to close the switch. Run the appropriate example program below.

```
BS1:
again:
 debug cls, "Pin is: ", # PIN0
 pause 250
goto again
```

```
BS2:
again:
 debug cls, "Pin is: ", dec IN0
 pause 250
goto again
```

When you run the program, you will get a Debug screen saying "Pin is: 1." Don't worry if the screen flashes; the program is clearing and rewriting the message four times a second. Now close the switch. The Debug screen says "Pin is: 0." Open the switch and the pin changes back to 1.

That's the essence of digital input. When the switch is open, the resistor places a 5-volt level on the pin, which the Stamp sees as a 1. When the switch is closed, the pin is connected through the switch to 0 volts (ground), which the Stamp sees as a 0.

Remove the resistor and switch from pin 0 (P0) and connect a piece of wire to the pin instead. Run the program again, and move the wire between +5 volts (Vdd) and 0 volts (ground or Vss). The Debug screen registers 1 for 5V and 0 for ground. What happens when the wire is not connected to anything? Watch the screen. It may show 1 or 0, or flip back and forth between the two.

This condition is called a *floating input*, and it is something to avoid. The Stamp does not get valid data from a floating input. Many folks with a little electrical or electronic background assume that a disconnected input pin will see 0 volts. After all, a volt meter that's not connected to anything reads 0 volts. But digital inputs work differently than

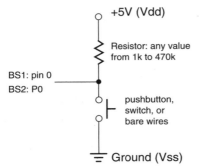

+5V (Vdd)

Resistor: any value from 1k to 470k

BS1: pin 0
BS2: P0

pushbutton, switch, or bare wires

Ground (Vss)

Figure 7-1 Switch input.

meter inputs, and behave in unpredictable ways when left floating. That's why our switch circuit includes a resistor to put the input into a predictable state when the switch is open. If it weren't for the resistor, the input would float.

WHERE DOES INPUT COME FROM?

In the example program, we got the state of pin 0 from the variable PIN0 (BS1) or IN0 (BS2). These are specialized variables that work a little differently than the ones we've seen before. These variables are windows into part of the Stamps' input/output (I/O) circuitry. The voltages present at the Stamp's pins determine their contents. Whenever a Stamp program needs to know the state of a pin or pins, it looks at these variables.

In our example, we looked at one pin through a bit variable—PIN0 (BS1) or IN0 (BS2). A program can also look at all of the pins at once, through the byte variable PINS (BS1) or word variable INS (BS2). We'll get into more specifics of the I/O structure in a moment. First, we need to look at the other side of the coin, output.

Changing the World through an Output

It's a little dramatic to say that lighting an LED is "changing the world," but it's an important step in understanding Stamp I/O. And that could change the world. Reconnect the switch circuit of Figure 7-1 and hook up the LED circuit of Figure 7-2. Run the following program (or modify the previous program to match this listing):

```
BS1:                                    BS2:
OUTPUT 7                                OUTPUT 7
again:                                  again:
 debug cls, "Pin is: ", #PIN0            debug cls, "Pin is: ", dec IN0
 pause 250                                 pause 250
 PIN7 = PIN0          .                   OUT7 = IN0
goto again                              goto again
```

Now open and close the switch, and watch the LED. The LED will light when the switch is closed and go out when it's open. There is a slight delay in this response because the instruction

```
pause 250
```

ensures that the loop only copies the status of the switch to the LED every quarter second.

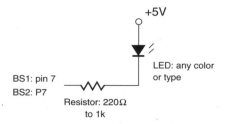

Figure 7-2 LED output.

It's interesting to see how this program differs from the previous one. The first instruction,

```
OUTPUT 7
```

makes pin 7 an output. There is no corresponding

```
INPUT 0
```

instruction in either program, but pin 0 is an input. That's because Stamp pins are inputs by default. Unless your program says otherwise, all pins are inputs. So the instruction

```
INPUT 0
```

is not strictly necessary.

The only other change from the original program is the line

```
BS1:            BS2:
PIN7 = PIN0     OUT7 = IN0
```

That copies the state of pin 0, the switch input, to pin 7, the LED output. We have the switch wired to report a 0 when it is closed, and the LED wired to light when a 0 is output on pin 7 (completing the circuit from +5 volts, through the LED and resistor, to ground).

Just for kicks, comment out the line

```
OUTPUT 7.
```

You know how to do that; just put a tick (') at the beginning of the line to make the Stamp ignore it. Rerun the program. This time, pin 7 is an input (its default setting), so the LED won't light no matter what the state of the switch is.

Deeper Meanings of "Input" and "Output"

From the demo programs you can see that:

- Stamp pins default to input.
- Inputs can sense the state (0 or 1) of a voltage.
- A program cannot control the state of a pin that's an input.
- Pins can be set to output, allowing the program to control them.

That's what engineers call a black-box perspective. That is, without knowing exactly what's going on inside the Stamp (the black box), we can develop a working knowledge of input and output. But to really understand these ideas, we have to pry open the box.

Figure 7-3 is a simplified view of the input/output structure of one Stamp pin. As the diagram shows, there are three bits that determine the complete behavior of each pin: the output bit, the direction bit, and the input bit. Let's look at each of these in detail.

OUTPUT BIT

The output bit consists of a pair of electronic switches (known as MOSFETs) controlled by the Stamp's internal circuitry. The electronic switches are wired in a way that ensures that

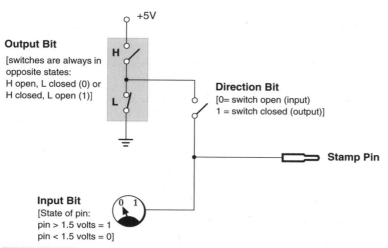

Figure 7-3 Input/output circuitry of a Stamp pin.

they are always in opposite states: one closed, the other open. When you store a 1 to the output bit, switch H (for High) closes and switch L (for Low) opens; a 0 opens H and closes L.

The BS1 accesses the output bit for a given pin whenever you write to the corresponding pin variable. For example, the instruction

```
PIN0 = 1
```

stores a 1 to pin 0's output bit. This happens regardless of the setting of pin 0's direction bit. The BS1 does not provide a direct way to read back the contents of output bits when the direction is set to input. More on this later.

The BS2 accesses the output bit for a given pin through the variable OUTS. For example, the instruction

```
OUT0 = 1
```

puts a 1 into pin 0's output bit. This happens regardless of the setting of pin 0's direction bit. The BS2 can read the contents of output bits even when the direction bit is set to input.

DIRECTION BIT

The direction bit is an electronic MOSFET switch. The switch is closed when the direction bit = 1 and open for direction = 0. Closing this switch connects the output bit to the corresponding pin of the Stamp. This means that the output bit controls the state of the pin—what we've been calling output mode. Opening the direction bit disconnects the output bit from the pin. Now the pin's state is controlled by external circuitry—input mode.

INPUT BIT

As the figure shows, the input bit is always connected to the Stamp pin. Its state always matches the state of the Stamp pin. When the direction bit is 1 (output mode) the input bit will also match the output bit. When the direction bit is 0 (input mode) the input bit will reflect the state of the pin, which may not match the output bit.

SUMMING UP THE I/O TRIO

At this point in your Stamp career it's not necessary to dwell on the subtler points of input and output.[1] Most Stamp instructions automatically manage the I/O trio for their particular needs. But knowing a little about I/O structure can help you understand the way certain programs and circuits work.

For instance, in the previous example, did you notice that the LED came on when the program first started, even if the switch was open? That's because all Stamp variables start off containing 0s until your program changes them. So the output bit for pin 7 initially contained 0, which turned on the LED as soon as the direction bit was set to 1 by the instruction

OUTPUT 7.

Variations on an LED Blinker

The Stamp instructions High and Low write to both the output and direction bits at the same time[2] in order to make a given pin output a 1 or a 0, respectively. They're commonly used to turn something on or off in a single instruction. Table 7-1 summarizes their effects.

Let's demonstrate High and Low using the LED circuit of Figure 7-2. If you still have the switch circuit of Figure 7-1 connected, don't remove it. Type and run this program.

```
BS1:            BS2:
again:          again:
 LOW 7           LOW 7
 pause 50          pause 50
 HIGH 7          HIGH 7
 pause 200       pause 200
goto again      goto again
```

The program turns the LED on, waits 50 milliseconds (ms), turns the LED off, waits 200 ms, then repeats. Notice that the program never explicitly makes pin 7 an output—High and Low take care of that.

You can fiddle with the two Pause instructions to change the on and off time of the LED. For example, change the 50 in pause 50 to 200, and the LED will be on and off for 200 ms each.

There's another instruction called Toggle that's perfect for writing an equal-on/off-time flasher. Try this:

```
BS1:            BS2:
again:          again:
 TOGGLE 7         TOGGLE 7
```

[1.] The BS1 and BS2 instruction summaries go into much more detail on the I/O variables. You'll find that the BS2's I/O variables exactly match the structure of the I/O pins, with separate variables for direction, output, and input. The BS1 reduces it to two variables, direction and pins. Data stored to a pin's variable goes into the output bit(s). Data read from a pin's variable actually comes from the input bit(s). You can see this from the program line in the example: PIN7 = PIN0, which copies the input bit of pin 0 to the output bit of pin 7. The compromise involved in this approach is that you cannot read the state of the output bit(s) of a pin that is set to input.

[2.] Actually, High and Low write to the output bit first, then a couple of microseconds later, to the direction bit. There are not many things that microcontrollers can do truly simultaneously.

```
pause 200        pause 200
goto again       goto again
```

Toggle sets the pin to output and reverses the state of the output bit (i.e., changes 1 to 0 or 0 to 1). Each time through the loop, the LED switches to the opposite state: on...off...on...off...

TABLE 7-1 EFFECTS OF HIGH AND LOW ON I/O BITS

EMPTY	DIRECTION	OUTPUT
High	1	1
Low	1	0

Controlling the LED Blinker

Let's put the LED blinker under the control of the switch. Make sure that both the switch circuit of Figure 7-1 and the LED circuit of 7-2 are connected to the Stamp.

```
BS1:                    BS2:
again:                  again:
 if pin0 = 1 then again  if in0 = 1 then again
 toggle 7                toggle 7
goto again              goto again
```

When you run the program, a flaw in its logic becomes apparent. The If...Then instruction jumps back to itself as long as the switch is open (1). As soon as the switch is closed (0), the program continues with the next instruction, Toggle. Then it loops back to the beginning to look at the switch again.

All of that is OK. What's not OK is that when you hold the switch closed, the LED seems to be on continuously. That's because the Stamp is toggling it on and off hundreds of times a second. The resulting flashes are so fast that they blend together. It's also almost impossible to simply change the state of the LED. In theory, you should be able to close the switch briefly and toggle the LED from lit to dark or dark to lit. In practice, though, the LED toggles several times during one switch closure and may end up back in its original state.

You can fix this by adding a Pause instruction after Toggle, like so:

```
BS1:                    BS2:
again:                  again:
 if pin0 = 1 then again  if in0 = 1 then again
 toggle 7                toggle 7
 PAUSE 200               PAUSE 200
goto again              goto again
```

That cures the problems of the previous version, but creates a new one—the program can't respond to two quick switch closures in a row. During the Pause instruction, the Stamp is no longer monitoring—polling—the switch. So it may miss fast switch actions.

Juggling the various actions that go with switch or pushbutton input can become difficult in a more complex program, so the Stamps have an instruction called Button to do most of the work. I'm not going to repeat all the details of Button's operation here, since it's well documented in the instruction summary and in the manufacturer's documentation (included on the CD-ROM that accompanies this book). But the following program should give you a taste of how Button works.

```
BS1:
symbol btn = b0
again:
 BUTTON 0,0,200,100,btn,0,again
 toggle 7
goto again
```

```
BS2:
btn var byte
again:
 BUTTON 0,0,200,100,btn,0,again
 toggle 7
goto again
```

Button's long list of parameters should give you some idea of how much it is doing for you. Reading from left to right, that Button instruction says, "There's a switch on pin 0 that reads as 0 when closed. When it is closed, give a delay of 200 loops before automatically repeating the pushed-button action. Once that delay is over, make the action repeat every 100 loops. Use the byte variable btn to keep track of these values. When the button is not pressed (0), go to the label named again."

Button gives you a smooth response to a switch action with all of the professional touches thrown in.

One of Button's additional virtues is that it debounces the switch. Switch contacts are springy pieces of metal that can bounce open and closed many times in the fraction of a second after they are initially closed. Stamps respond so quickly that they may see this bouncing as a series of switch closures and act accordingly. That's not good, because the human user of the device thinks he or she pressed the button or switch once, and is expecting one response. Button waits for the switch contacts to settle down before taking action.

Final Example: Dial-Controlled Timer

You've almost earned your stripes in BASIC Stamp Boot Camp. I want to present one last example that introduces another instruction, reinforces many of the concepts we've covered, and sets the stage for the format of the projects in the remainder of this book.

Our final example will be a dial-controlled timer with a buzzer, switch, and LED. We'll work out a rudimentary user interface that makes it easy to use the timer, even in the dark.

The program listings for this project appear in Figures 7-4 (BS1) and 7-5 (BS2). As you can see, these listings are a lot longer than the previous examples. The extra text consists mostly of comments, indispensable notes of explanation that help make a program as clear to humans as it is to the Stamp.

COMMENTS

In a couple of the previous examples, you deactivated an instruction by placing a tick mark (') at the beginning of the line. This caused the Stamp to ignore that line and go on to the next one. This is called *commenting out* an instruction, and it's a handy way of temporarily deleting an instruction when you're modifying or debugging a program.

But the primary purpose of the tick mark is to allow you to embed comments in your programs. Most programs contain lots of comments (see Figure 7-6). When you're writing a program, comments help you clarify your thinking about what you plan to do. If you need more than one sitting to write a program, comments help you pick up where you left off. And when you return months later to modify or expand a program, comments can help jog your memory about how the program worked in the first place.

```
'Program: TIMER.BAS
'This program makes a 10-second countdown timer. See the schematic
'for circuit details. The timer works like this: The user sets the
'duration by adjusting the pot. The buzzer beeps for each unit change
'in the time setting. This makes it easy to set the dial in the
'dark—just twist the dial fully left, then adjust right while
'counting beeps. Once the desired time is set, pushing the button
'turns on the LED and starts the countdown. The buzzer beeps
'once a second during the timing interval. When time is up,
'the LED turns off, and the buzzer lets out a final long beep.
'Pressing the button again starts another timing cycle of the
'same length; adjusting the dial changes the time setting.

'========Constants: pin assignments
SYMBOL LED      =       7       ' LED pin, 0=ON.
SYMBOL buzz     =       6       ' Buzzer pin, 0=ON.
SYMBOL potIn    =       1       ' Pot pin.
SYMBOL start    =       pin0    ' Start button, 0=press.
'========Constants: timing, etc.
SYMBOL beep     =       2000    ' Length of short pulse to buzzer (20ms).
SYMBOL longBeep =       50000   ' Length of long pulse to buzzer (500ms).
SYMBOL shortSec =       973     ' Length of second, minus overhead (973ms).
SYMBOL pScale   =       220     ' Scale value for Pot instruction.
'========Variables
SYMBOL old      =       b10     ' Result of previous Pot instruction.
SYMBOL new      =       b9      ' Result of current Pot instruction
SYMBOL time     =       b8      ' Number of seconds to count down.


'========Program Start
high buzz                 ' Make buzzer pin output and off (0=ON)
high LED                  ' Same for the LED pin.

set:                      ' Set the timer.
 gosub checkDial          ' Check the pot via subroutine below.
 if start=1 then set:     ' Keep checking until button pressed.
runTimer:                 ' Button pressed—start countdown.
 low LED                  ' Turn on the LED
 for time = 0 to new      ' Loop for # of seconds set by dial.
  pulsout buzz,beep       '   Beep to mark the second.
  pause shortSec          '   Wait 1 sec (minus time for instructions).
 next                     ' End loop.
done:                     ' Timing interval is over.
 high LED                 ' Turn LED off.
 pulsout buzz,longBeep    ' Make a long beep.
goto set                  ' Start over

'========Pot subroutine "checkDial"
'This subroutine checks the setting of a pot. Each time it checks
'the pot, it scales the setting to the range 0-10 by dividing the
'Pot result by 25 [(0 to 255) divided by 25 is (0 to 10)]. It adds
'this value to the old value and takes the average by dividing
'the result by 2. This yields a number from 0-9, because the old
'value starts at 0 and increases toward the new value. As a result,
'even if the new value is 10, the old value never gets larger than 9.
'The average of the two is 19/2 = 9 in the Stamp's integer math.
'The purpose of the running average is to prevent the setting from
'jittering when set to some in-between value, like 6-1/2. If the
'setting fluctuated, the extra beeps would spoil the user's
'ability to judge the setting by counting beeps.
checkDial:
 pot potIn,pScale,new       ' Take a pot reading.
 new = new/25+old/2         ' Scale it and average with old setting.
 if new = old then skipBeep ' If no change, then no beep.
 pulsout buzz,beep          ' Setting changed—beep!
skipBeep:
 old = new                  ' Update old.
return                      ' Back to main program.
```

Figure 7-4 Timer program listing for BS1.

```
'Program: TIMER.BS2
'This program makes a 10-second countdown timer. See the schematic
'for circuit details. The timer works like this: The user sets the
'duration by adjusting the pot. The buzzer beeps for each unit change
'in the time setting. This makes it easy to set the dial in the
'dark—just twist the dial fully left, then adjust right while
'counting beeps. Once the desired time is set, pushing the button
'turns on the LED and starts the countdown. The buzzer beeps
'once a second during the timing interval. When time is up,
'the LED turns off, and the buzzer lets out a final long beep.
'Pressing the button again starts another timing cycle of the
'same length; adjusting the dial changes the time setting.

'=======Constants: pin assignments
LED       con     7         ' LED pin, 0=ON.
buzz      con     6         ' Buzzer pin, 0=ON.
potIn     con     1         ' Pot pin.
start     var     IN0       ' Start button, 0=press.
'=======Constants: timing, etc.
beep      con     10000     ' Length of short-beep pulse (20ms).
longBeep  con     500       ' Length of long-beep Pause (500ms).
shortSec  con     973       ' Length of second, minus overhead (973ms).
'=======Variables
old       var     byte      ' Result of previous RCtime instruction.
new       var     word      ' Result of current RCtime instruction
time      var     byte      ' Number of seconds to count down.

'=======Program Start
high buzz                   ' Make buzzer pin output and off (0=ON)
high LED                    ' Same for the LED pin.

set:                        ' Set the timer.
  gosub checkDial           ' Check the pot via subroutine below.
  if start=1 then set:      ' Keep checking until button pressed.
runTimer:                   ' Button pressed—start countdown.
  low LED                   ' Turn on the LED
  for time = 0 to new       ' Loop for # of seconds set by dial.
    pulsout buzz,beep       '    Beep to mark the second.
    pause shortSec          '    Wait 1 sec (minus time for instructions).
  next                      ' End loop.
done:                       ' Timing interval is over.
  high LED                  ' Turn LED off.
  low buzz                  ' Turn buzzer on.
  pause longBeep            ' Wait long-beep duration.
  high buzz                 ' Turn the buzzer off.
goto set                    ' Start over

'=======Pot subroutine "checkDial"
'This subroutine checks the setting of a pot. Each time it checks
'the pot, it scales the setting to the range 0-10 by dividing the
'RCtime result by 25 [(0 to about 255) divided by 25 is (0 to 10)].
'It adds this value to the old value and takes the average by dividing
'the result by 2. This yields a number from 0-9, because the old
'value starts at 0 and increases toward the new value. As a result,
'even if the new value is 10, the old value never gets larger than 9.
'The average of the two is 19/2 = 9 in the Stamp's integer math.
'The purpose of the running average is to prevent the setting from
'jittering when set to some in-between value, like 6-1/2. If the
'setting fluctuated, the extra beeps would spoil the user's
'ability to judge the setting by counting beeps.
checkDial:
  high potIn
  pause 1
  RCtime potIn,1,new                ' Take a pot reading.
  new = ((new/25)+old)/2            ' Scale it and average with old setting.
  if new = old then skipBeep        ' If no change, then no beep.
  pulsout buzz,beep                 ' Setting changed—beep!
skipBeep:
  old = new                         ' Update old.
return                              ' Back to main program.
```

Figure 7-5 Timer program listing for BS2.

Up to now, I've omitted comments from short program examples. I used separate text to explain how the programs worked because I didn't want you to feel obliged to type the programs and the comments. But all full-scale programs contain lots of comments. Most programs begin with comments explaining what the program does, what external circuitry it requires, and generally how it works. After that, each major section of the program is introduced by its own comments. Most programmers also comment each line of a program to explain how a particular instruction fits into the larger scheme of the program.

FORMATTING

Comments clarify the purpose and meaning of a program and its parts; formatting clarifies its structure. The most common kinds of formatting are grouping, alignment, and indention.

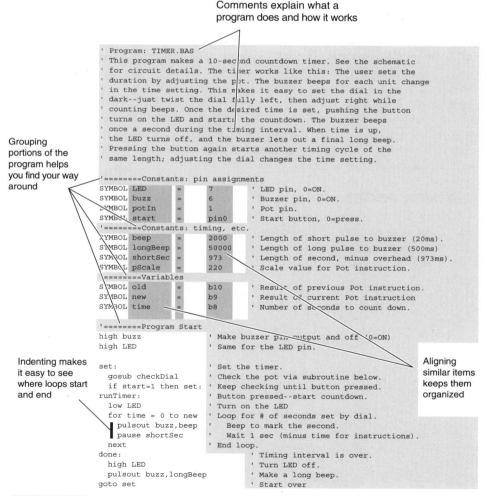

Figure 7-6 Formatting makes programs easier to read.

In the program listings, you can see that the parts of the programs that name constants and variables are logically grouped and labeled with comments. Equal signs (=) are used to draw lines in order to set these areas apart from one another. There's nothing magical about the use of the = sign; some programmers use asterisks (*), hyphens (-), or just about anything else.

Within the sections, similar items are aligned; for example, the names and values of constants are set in orderly columns. This is easy to do: Just use the Tab key to automatically jump to the next eight-character column in the Stamp program editor screen.

Within loops, code is indented slightly from the left margin. In the For...Next loop, you can see that instructions are indented a couple of spaces. This makes it easier to locate the beginning and end of loops. Since the label

```
set:
```

and the instruction

```
goto set
```

amount to one big loop, all of the code between them is indented two spaces. Spaces are used instead of tabs in order to save room. Some programs have loops within loops within loops... Indenting by a whole tab (eight spaces) for each loop could get ridiculous as the instructions marched toward the righthand edge of the screen.

Some employers and programming courses have more complicated formatting rules than these. But commonsense use of grouping, alignment, and indention will make your programs tidy and readable without excessive effort. As you progress, you will probably develop your own style. As long as you don't stray too far from the basics, your stuff will be perfectly understandable to others.

THE TIMER PROGRAMS

We've spent a lot of time talking about how the timer programs look, now let's see what they do. These programs create a simple timer that activates an output (the LED) for a period of time ranging from 1 to 10 seconds. Figures 7-7 and 7-8 show the circuits that the timer programs use. A dial (the pot) sets the time, and the timing interval starts when the user presses the button.

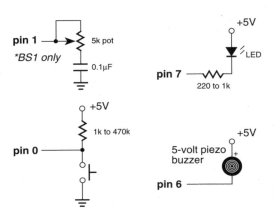

Figure 7-7 Schematic for timer programs.

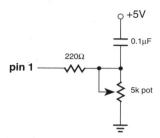

Figure 7-8 Pot connection for BS2 timer program.

Everything described so far could be done with an inexpensive timer IC, similar to the circuit used to illustrate circuit prototyping a couple of chapters ago. In order to show off the real genius of the Stamps, I added one more requirement to the list of features: the timer had to be usable in the dark!

If we were building an ordinary IC timer, this requirement would force us to go back to the drawing board. With the Stamp, however, I just added a buzzer[3] to an unused I/O pin and wrote the program so that it let out a beep when the dial is turned from one setting to another. To set the timer in the dark, you'd turn the dial all the way to the left (the minimum, 1-second setting), then dial upward while counting beeps (2 seconds, 3, 4...) until you reached the correct setting.

As a bonus, the buzzer lets us hear the progress of the timer countdown, beeping with each passing second, then letting out a longer buzz to signal the end of the timing period.

Another new feature of this program are the instructions that read the position of a potentiometer (pot) to produce a proportional value. The BS1 and BS2 use different instructions for this purpose, Pot and RCtime, respectively. Both rely on the same basic principle—that the time required to charge a capacitor depends on the resistance in series with it. The higher the resistance, the longer the time. The Pot and RCtime instructions measure this time to determine the resistance, and therefore the setting, of the pot.

Appendix C goes into more detail on this kind of circuit.

The subroutine checkDial not only reads the pot, but also averages each reading with a previous one. This helps to counteract the effects of noise that might otherwise cause the pot reading to flip back and forth between two settings. This would make the buzzer chatter like crazy, and cause the user to lose track of the dial setting. A feature that doesn't work properly is no help at all, so the programs use a running average to smooth out noisy pot readings.

BOOT CAMP GRADUATION

The timer programs are simple, but they have a lot to show you. Enter and run them. Read the comments in the listings and try to work out what each instruction does. Make small changes and see how they affect the program. Look up each of the instructions in the quick-reference guides or the manufacturer's documentation and see whether you can state the meaning and purpose of each instruction and each value given to each instruction.

[3.] Radio Shack part nos. 273-059, 273-060, or 273-065, or Jameco nos. 76064 or 76021 are suitable buzzers. If you don't have a buzzer, or prefer visible output, just connect another LED to pin 6 in exactly the same way as the one connected to pin 7.

This process of program dissection is the most effective way to learn about programming. Once you understand how an instruction is used in one program, you can apply that knowledge to your own programs.

You've graduated from BASIC Stamp Boot Camp; congratulations! I hope you continue your microcontroller education by dissecting and customizing the projects that make up the rest of this book, designing your own projects, and by participating in the exciting community of Stamp users.

MAGIC MESSAGE MACHINE

BS1

CONTENTS AT A GLANCE

This project is a just-for-kicks contraption that displays a short message that appears to float in thin air. It demonstrates a scientific principle called persistence of vision, and shows how a simple program can achieve an impressive result. It's also an opportunity to haul out a tub of Tinkertoys and play with them—who could resist that?

Persistence of Vision (POV)

Our brains receive and process huge amounts of raw data from our senses. Sometimes this data is incomplete or confusing, so the brain, in partnership with the senses, fills in the gaps and straightens out the jumbles. We normally don't notice this process; we take for granted that the world we perceive moves smoothly and behaves logically.

One of the easiest perceptual tricks to demonstrate is called persistence of vision. The mind takes a brief glimpse of something and holds onto it for awhile. If the scene changes abruptly, the mind seems to smooth over the transition. In school, you may have made flip-books with slightly different drawings on each page. When you riffled through the pages, the drawings sprang to life, like little animated movies.

And of course movies are based on the same principle. By flashing 24 or more still pictures per second in front of our eyes, we get a convincing illusion of smooth motion. The usefulness of persistence of vision is huge; the images on TVs and computer monitors are created by a single, rapidly moving dot. In fact, if not for persistence of vision, it would be almost impossible to transmit and reproduce anything but the crudest image electronically. Instead of a single moving dot our display devices would have to have an electronically controlled dot for each picture element (pixel) to be displayed. A modest computer monitor of 640 x 480 pixels would require 307,200 individual electronic dots.

Putting POV to Work

In microcontroller applications, it is common to take advantage of POV to reduce the amount of circuitry required to drive a display device. Consider a four-digit numeric display, the kind you might see on a simple timer. Each digit is made up of seven cleverly arranged segments, plus dots for decimal points. Figuring eight outputs per digit, times four digits, it would take 32 output pins for a microcontroller to drive a four-digit display.

By taking advantage of POV, a microcontroller can drive a four-digit display with just 12 output pins. Here's how: Assume that the display digits are made up of eight LEDs. Each digit has all of the cathodes (– ends) tied together and all of the anodes (+ ends) separate. You can turn on any segment or combination of segments by applying + to the anode(s) and – to the common cathode connection.

Now suppose you have four digits wired just like that. You connect together the corresponding anodes of each digit. Figure 8-1 shows the hookup. You can turn on any combination of segments of any particular digit by applying + to the shared anodes and – to one digit's common cathode. The only catch is that you can't light up different combinations of segments on two or more digits at the same time.

With POV, that's not a problem. Let's say you want to display "1234." Your micro applies + to the segments that make up the 1 and – to the common cathode of the first digit. A fraction of a second later, it turns off – to the first digit, changes the + to the pattern for 2, and applies – to the second digit. It repeats this process for digits 3 and 4, then starts over.

Because the microcontroller rapidly scans all four digits many times (30+) per second, your eye and brain merge the momentary images together, and you see a solid "1234" on the display.

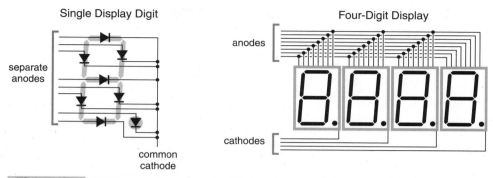

Single Display Digit

Four-Digit Display

Figure 8-1 Displays reduce connections using persistence of vision.

A Different Spin on POV

We've looked at the practical, workaday aspect of POV—how about using it to have a little fun? Here's the general idea: We use the BS1 to drive seven LEDs. The BS1's program rapidly flips through a series of patterns of lit and unlit LEDs. If we move the LEDs rapidly, the POV effect lets us see a larger picture formed by the patterns.

We'll actually make two versions of this project. The first will display simple geometric patterns when waved through the air by hand. The second will use a Tinkertoy pendulum and a magnetic trigger to provide better control over the motion of the LEDs. This version can actually display short text messages.

Initial Experiments with POV

Figure 8-2 is the hardware and 8-3 is the program for our first experiment with POV graphics. As the figure shows, the hardware just consists of seven LEDs connected to pins 1 through 7 of the Stamp. If you use individual (discrete) LEDs, mount them about 1/4" apart on a piece of perfboard or prototyping board. Alternatively, you can use the LED bargraph assembly from the parts list at the end of this chapter. Either way, make the hookup sturdy so that the assembly will survive being picked up and shaken around.

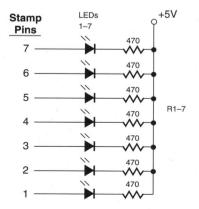

Figure 8-2 Circuit for POVPAT.BAS.

```
'Program: POVPAT.BAS (persistence of vision pattern-maker)
'This program demonstrates how a rapidly changing pattern of
'lights on an LED array can be perceived as a single, solid
'picture, due to the persistence-of-vision effect. A bit
'pattern stored in EEPROM is scanned across a line array of
'7 LEDs. If you sweep this array in front of your eyes in a
'darkened room, you will see a sort of 'chain-link' pattern.
'See the accompanying text for information on computing other
'EEPROM values to display other patterns, or even text.

'========Constants
SYMBOL  colDelay        =       4       ' Delay (ms) between columns.
SYMBOL  maxEEaddress    =       6       ' Highest EEPROM address.

'========Variables
SYMBOL  column          =       b2      ' Current column address.

'========EEPROM byte definitions
'The EEPROM directive tells the STAMP host software to write the
'values listed into the unused portion of EEPROM (the part not
'occupied by the program).
EEPROM  0,($7C,$BA,$D6,$EE,$D6,$BA,$7C)

'========Setup
'Set output latches to 1s, turning the LEDs off, then set the
'directions for pins 1-7 to output to drive the LEDs.
pins = $FF      ' Set all pins to 1s (LEDs off)
dirs = $FE      ' All outputs except pin 0. (Not used here).

'========Main Loop
Display:
  for column = 0 to maxEEaddress        ' For each of the columns...
    read column,pins                    ' ..put EEPROM data onto pins,
    pause colDelay                      ' ..and leave it for a few ms.
  next
goto Display                            ' Do it again.
```

Figure 8-3 Program listing for POV pattern display.

When you download the program of Figure 8-3 to the Stamp, the LEDs will flicker visibly. If you dim the room lights and wave the LEDs in front of your eyes, you will see a chain-link-fence pattern—sort of like XXXXXX—that appears to hang in mid air.

The program of Figure 8-3 creates this illusion by rapidly outputting a series of patterns to the LED outputs. The patterns are stored in the Stamp's EEPROM program memory by the EEPROM directive at the beginning of the program.

EEPROM is called a directive rather than an instruction. The difference is that directives are performed by the Stamp host software on the PC while instructions are performed by the Stamp itself. Another directive we have used often is Symbol, which allows us to give names to constants and variables.

EEPROM takes an address in memory and stores a list of bytes starting at that address and working upward (e.g., 0,1,2,3,4,5...). If you want EEPROM to start at 0, the default, you can leave the address out. Data stored by EEPROM shares space with the Stamp program (which starts at EEPROM address 255 and works downward). If you have a lot of data or a large program, or both, there's a possibility that the two might overlap. The

Stamp host software will detect this error and refuse to download to the Stamp until you correct it.

In this program, I specify the bytes stored by EEPROM in hexadecimal (hex) format. Once you get used to it, hex is convenient shorthand for expressing patterns of bits (see Appendix B). I could have used the Stamp's binary format, but instead of compact numbers like $3A (hex), I'd have to type monsters like %000001111010 (binary). No thank you.

Once the data is stored, the program itself is simple. The Stamp sets pins 1 through 7 to output and writes 1s to them in order to turn off the LEDs. Then for addresses from 0 to the maximum address of the bytes stored in EEPROM, the Stamp Reads the EEPROM directly into the Pins variable. Then it loops back to the beginning and does it all over again.

Figure 8-4 shows how I arrived at the bytes that make up the patterns of lights. I encourage you to devise your own patterns, calculate corresponding hex values, and modify the program to see the results. Remember that if you store a different number of bytes, you will have to change the values of the constant EEmax to match the highest address of your data. Simply count the bytes stored by your EEPROM directive starting with 0. The highest number you reach is the right value for EEmax.

Displaying a Text Message

If you modified the first program to display different patterns, you probably noticed a couple of flaws in the display scheme. First of all, the lights flicker continuously, so you get

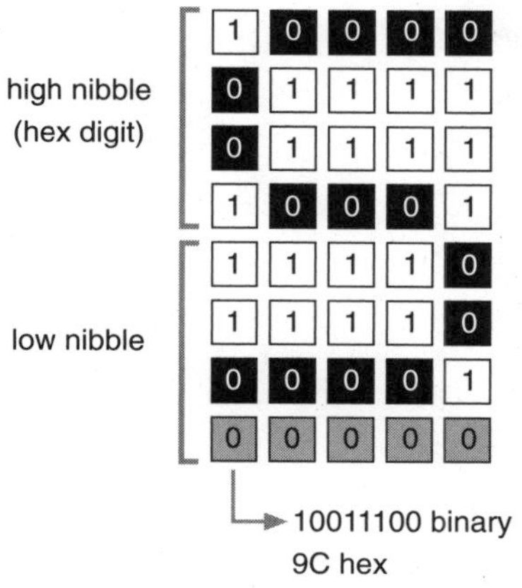

Binary nibble	Hex digit
0000	0
0001	1
0010	2
0011	3
0100	4
0101	5
0110	6
0111	7
1000	8
1001	9
1010	A
1011	B
1100	C
1101	D
1110	E
1111	F

high nibble (hex digit)

low nibble

10011100 binary
9C hex

Each column of the LED picture is defined by a byte stored in EEPROM. The programs use hexadecimal (hex) notation to represent these bytes. To convert bit patterns to hex numbers, follow the example above. Note that the lowest bit (colored gray) is not used in this application.

Figure 8-4 Figuring bit patterns is an opportunity to use hex numbers.

the same image repeated in paper-doll fashion as long as you wave the lights. Second, if you wave the lights back and forth, the images have to be symmetrical horizontally or they'll appear to flip when you change direction. Finally, the images stretch and squash horizontally depending on the speed with which you wave the lights.

If we want to display text, we have to eliminate these flaws. My solution is partly mechanical and partly electronic.

The mechanical setup, shown in Figure 8-5, mounts the array of LEDs on a pendulum constructed of Tinkertoy parts. A pendulum is the simplest way to make the waving motion of the display fairly consistent. Although the pendulum gradually runs down, the message is readable for the first several swings.

The pendulum provides a place to mount our electronic enhancement—a magnetically activated Hall-effect switch shown in Figure 8-6. This switch, mounted on the swinging part of the pendulum, pulls Stamp pin 0 low when it passes a magnet mounted on the stationary part. The new program, shown in Figure 8-7, waits for this signal before displaying the patterns of lights.

Nothing about the construction of the pendulum is particularly critical. Use your ingenuity and whatever materials you have on hand. I used rubber bands to attach the electronics and batteries to the Tinkertoy framework. The magnet and Hall-effect switch I stuck

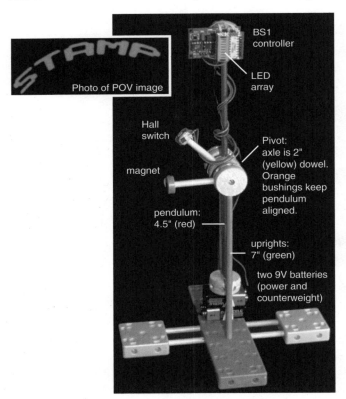

Figure 8-5 Pendulum arrangement lets the Stamp display text.

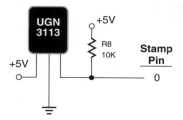

Figure 8-6 Hall-effect switch senses magnet, triggers display.

in place with blobs of putty sold for temporary mounting of posters. Chewing gum would have worked as well. I avoided using glue because I wanted to be able to adjust the position of all the components.

There are two 9V batteries rubber-banded to the bottom end of the pendulum arm. Only one is needed to power the Stamp; the other just lends additional weight. You could use some other weight, like a fishing sinker or blob of modeling clay, if you prefer.

Once you have built the pendulum and downloaded the program to the Stamp, you must adjust the alignment between the magnet and the Hall switch. The switch must pass within ⅛" of the magnet in order to work. But it should not hit the magnet as it passes. These are pretty tight tolerances for a Tinkertoy contraption, but not impossible. Adjust the spacing between the plastic bushing disks in the pivot to bring the Hall switch as close as possible to the magnet without touching it.

Connect one of the 9V batteries to the Stamp and move the pendulum arm so that the switch passes the magnet. At that moment, the LEDs should flash. If they don't, double-check your wiring and alignment. Some Hall switches trigger on only a particular pole of the magnet; if you can't get the LEDs to flash, try turning the magnet around, or substitute a stronger magnet.

Once you have the kinks worked out, you're ready for a test run. I found that the pendulum worked best when clamped to the table top. Connect the battery to the Stamp and lift the battery end of the pendulum arm until it's horizontal. Press the Stamp's reset button. (If your Stamp doesn't have a reset button, wait until the arm is horizontal before connecting the battery.) Have someone turn out the lights, and let go of the arm. Presto! You will see the word STAMP in ghostly, floating lights.

Modifications and Photography

I used an LED bar-graph assembly that I had on hand for the display portion of this project. The LEDs in this component are wider than they are tall, so they produced the stretched-out letters you see in the photo of Figure 8-5. After the photos were taken, I located another LED bar graph with tall, narrow LEDs that makes a better-looking display. Of course, you can use other LEDs in any shape that strikes your fancy. Scan the catalogs; you can get round, square, rectangular, or even triangular LEDs.

If you change the spacing between the LEDs, you may also have to adjust the timing of the program. Comments in the program listing show which constant to adjust.

The display effect of this project is fleeting. It appears for a fraction of a second, and then it's gone. If you want to record it for posterity, try my setup. I loaded my 35mm camera with 100-speed film, and mounted it on a tripod a few feet from the pendulum. With

```
'Program: POVSIGN.BAS (persistence of vision sign)
'This program displays a one-word message ("STAMP") using just
'seven LEDs. To view the message, the LEDs must be mounted on
'a pendulum arm and swung through the air. The accompanying
'text and figures show how to build the pendulum from Tinkertoys(R).
'The program waits for a pulse on a Hall-effect switch mounted
'on the pendulum. This pulse indicates that the pendulum is
'swinging toward the top of its arc from left to right. The
'program then goes into a loop, reading bit patterns from EEPROM
'and lighting the LEDs accordingly. As the pendulum moves, the
'changing patterns of lights paint the complete message. Your
'eyes' persistence of vision makes this message appear solid.

'========Constants
SYMBOL  colDelay      =       4      ' Delay (ms) between columns.
SYMBOL  magSwitch     =       pin0   ' Input from magnetic switch.
SYMBOL  maxEEaddress  =       28     ' Highest EEPROM address.

'========Variables
SYMBOL  column        =       b2     ' Current column address.

'========EEPROM byte definitions
'The EEPROM directive tells the STAMP host software to write the
'values listed into the unused portion of EEPROM (the part not
'occupied by the program). In order to break the list between
'two lines, I've started the second EEPROM directive at next
'address after the one at which the first EEPROM left off.
EEPROM  0,($9C,$6C,$6C,$6C,$72,$FE,$7E,$7E,$00,$7E,$7E,$FE,$80,$76)
EEPROM 14,($76,$76,$80,$FE,$00,$BE,$CE,$BE,$00,$FE,$00,$6E,$6E,$6E,$9E)

'========Setup
'Set output latches to 1s, turning the LEDs off, then set the
'directions for pins 1-7 to output to drive the LEDs. Pin 0
'is an input to sense the state of the Hall-effect (magnetic)
'switch.
pins = $FF      ' Set all pins to 1s (LEDs off)
dirs = $FE      ' All outputs except pin 0 (Hall switch).

'========Main Loop
Display:
  if magSwitch = 1 then Display     ' Wait till magnet passes switch.
  for column = 0 to maxEEaddress    ' For each of the 29 columns...
    read column,pins                '  ..put EEPROM data onto pins,
    pause colDelay                  '  ..and leave it for 4ms or so.
  next
  pins = $FF                        ' Message done: Turn LEDs off
hold:
  if pin0 = 1 then hold             ' Wait for pendulum to swing back.
  pause 50                          ' Give it time to swing past.
goto Display                        ' Then start the display again.
```

Figure 8-7 Program listing for POV text display.

the room lights on, I focused sharply on the LEDs. I set the shutter to B, the setting that stays open as long as the button is pressed, and adjusted the aperture to f5.6. The f-stop setting was purely a guess based on past experience shooting LEDs. I attached a cable-release to the shutter button. This is a flexible cable about 18 inches long that lets you press and hold the shutter without touching (and possibly shaking) the camera.

Taking the picture was purely reflex. A friend shut off the lights and worked the pendulum; I pressed and held the shutter release as the pendulum approached the Hall switch, releasing it as soon as the display ended.

Shoot a whole roll of film, varying the f-stop between wide-open and f8, and you're almost sure to get a few decent shots. The LEDs should stand out vividly against a nearly-black background.

PARTS LIST

(See Appendix E for contact information on parts sources.)

LED1–LED7—Any color, shape or package style LED. Should be spaced 0.1 to 0.25 inches apart. Digi-Key carries 7-LED bar graphs, P531-ND, P532-ND, or P533-ND (red, green, and yellow, respectively), that are particularly good.

R1–R7—any 470-ohm resistors, 1/8W or greater.

R8—any 10k resistor, 1/8W or greater.

Hall Switch—UGN3113U, Newark/Farnell Electronics or HAL115UA-C-ND, DIGI-KEY.

Misc.—Tinkertoys or other mechanical-building set; button magnet (sold by craft stores for refrigerator magnets); rubber bands; sticky putty; batteries; etc.

8

MAGIC MESSAGE MACHINE

INTELLIGENT TRAFFIC SIGNAL SIMULATION

BS1

CONTENTS AT A GLANCE

Traffic School

Stoplight Circuit

Construction and Modifications
PARTS LIST

Long before there were microcontrollers, engineers had to develop reliable control systems for sequencing traffic lights. Their creations were like the guts of a player piano, with motors and cams and switches and relays. Now traffic signals are largely controlled by electronics, in some cases even networked to large computers that analyze citywide traffic flow and alter the timing of lights at critical intersections to avert gridlock.

This project simulates a traffic light controlling an intersection. It sequences two sets of lights through the normal green-yellow-red pattern with preset timing for each step of the sequence. Our Stamp-based stoplight also looks at a couple of other factors—a car waiting on a sidestreet and whether it's day or night—and alters the stoplight's behavior accordingly.

With a steady hand and small LEDs, this project could probably be made to fit the stoplights in a diorama or model-train layout.

Traffic School

To design this project, I had to sit down and think about the way traffic signals work. Figure 9-1 shows what I came up with. An ordinary intersection might have two signals, one to control east-west traffic and the other north-south traffic. These lights cycle through six states, as shown in the figure. The length of time the lights stay in a particular state varies.

So each state of the lights has two properties—a pattern of lights and a length of time to wait before changing to the next pattern of lights. This suggested the structure of the program's main loop:

```
for state = 0 to 5
 lookup state,(state0,state1,state2,state3,state4,state5),patternAndTime
 ' Extract the pattern bits and use them to turn lights on and off.
 ' Extract the time bits and use them to set a time delay (pause).
next
```

That's not a real program, just a sketch to show you my first-draft thinking. As the variable state increments from 0 to 5, get the corresponding 16-bit value from the table. Take six bits of the table value and use them to set the states of the six lights of the traffic signal. Then take the remaining 10 bits and use them to determine a length of time to leave the lights in that state. A 10-bit number can represent values from 0 to 1023, which is more than enough to produce appropriate time delays for a traffic signal. The actual program counts time by units of ¼ second, so the length of time a light is on can range from ¼ second to 4 minutes, 15 seconds. That seemed like enough to me, although I swear I've been at red lights that were twice that long! Well, in the Stamp world, I'm the king, and I say that no one should have to endure a red light of more than 4 minutes or so.

With six pins being enough to control the stoplight, there are two Stamp pins uncommitted. That left an opening to add two other features: a change-on-demand function that shortens the green light on the main, east-west road when cars are waiting on the north-

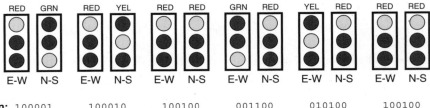

Bit pattern: 100001   100010   100100   001100   010100   100100

(Bit patterns are based on lights being wired so that the lowest three bits control N-S light, upper three bits control E-W light, and a light is on when its bit =1.)

Figure 9-1 Stoplight sequence boils down to six states.

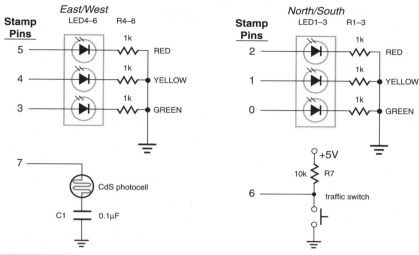

Figure 9-2 Circuit for stoplight simulation.

south street, and the ability to tell the difference between day and night, and use that information to alter the timing of the change-on-demand feature.

Stoplight Circuit

The circuit of Figure 9-2 is quite straightforward. Red, yellow, and green LEDs for each of the signals are controlled by Stamp pins 0 through 5. The switch, which would be closed by the presence of a car at the north-south sidestreet, is connected so that Stamp pin 6 sees a 1 when the switch is open and a 0 when it is closed.

To distinguish day from night, a photocell and capacitor connect to pin 7 of the Stamp. A photocell is a light-sensitive resistor. In bright room light, its resistance might be around 2k; in the dark, it rises to 1M or more. The stoplight program uses the Stamp's Pot instruction to take an approximate reading of this resistance and decide when it's dark enough for the change-on-demand feature to go into its nighttime schedule.

Since the Stamp doesn't really need to know exactly how light or dark it is, I could have set up a resistive voltage divider with the photocell to ground, a fixed resistor to +5V, and the Stamp pin connected in the middle. (See Appendix C for more about voltage dividers.) I'd have to select the resistor so that the pin would see less than 1.5V (0) by day and more than 1.5V (1) at night. This isn't difficult, but it would require some fiddling around. The Pot instruction reduces any adjustment to changing a number in the program, so I opted to use it.

Figure 9-3 is the program that runs the intersection. In a way, it is similar to the LED persistence-of-vision project presented earlier. The Stamp's primary job is to retrieve a series of values and act on them. In this case, the values are 16 bits in length, so it is simpler to use the Stamp's Lookup instruction to retrieve them. The Read instruction used in the previous project only retrieves 8 bits (one byte) at a time, so we would have to Read twice to get each value. Lookup is easier.

```
'Program: STOPLT.BAS (Sequence a stoplight.)
'This program generates proper green-yellow-red sequencing for a
'pair of traffic signals controlling an intersection. I refer
'to one street as "EW" (east-west) and the other as "NS" (north-
'south). Pins are connected to LEDs as follows:

'         pin5   EW/red       pin2   NS/red
'         pin4   EW/yellow    pin1   NS/yellow
'         pin3   EW/green     pin0   NS/green

'The program automatically runs through the normal traffic-light
'sequence at fixed timing intervals. However, it has two inputs
'that can changes its operation: a traffic switch that indicates
'a car waiting at the NS intersection, and a photocell, which
'indicates day or night. When a car is waiting at NS, the EW
'green light is shortened. If it's day time, the minimum length
'of the EW green light is longer than if it's night (when
'traffic is assumed to be less).


'====Constants: Stoplight patterns and times===
'The program uses six 16-bit constants to represent the states
'of the lights (lower 6 bits) and the length of time to leave
'the lights in those states (upper 10 bits). The program divides
'that value by 64, effectively shifting the upper 10 bits down
'to the lower 10 bits. So our time can range from 0 to 1023 units.
'Each unit is approximately 1/4 second in length, so times can
'range from less than a second to 255 seconds (4+ minutes).
'Here's how the constants are organized:
'            Duration         Pattern of lights
'                     \         /
'             |=========|====|
SYMBOL NSgo   = %0000010000100001    ' NS green/EW red, 8 secs.
SYMBOL NSyel  = %0000001000100010    ' NS yellow/EW red, 2 secs.
SYMBOL allRed = %0000000010100100    ' NS red/EW red, 1/2 sec
SYMBOL EWgo   = %0001000000001100    ' NS red/EW green, 16 secs.
SYMBOL EWyel  = %0000001000010100    ' NS red/EW yellow, 2 secs


'====Other Constants===
SYMBOL day    = 8       ' Minimum day EW green time in seconds.
SYMBOL night  = 4       ' Minimum night EW green time in seconds.
SYMBOL NScar  = pin6    ' Switch: 0=car at NS light.
SYMBOL phCell = 7       ' Photocell.


'===Variables===
SYMBOL seq = b11        ' Current state (0-5) of stoplight sequence.
SYMBOL qSecs= b10       ' Number of quarter-seconds to remain in state.
SYMBOL lkup= w4         ' Lookup table entry w/ time and light pattern.
SYMBOL time = w4        ' Alternative name for lkup for timing.
SYMBOL minTime=w3       ' Minimum wait time for NS light to change.
SYMBOL dark = b5        ' Photocell reading.


'===Program===
dirs = %00111111        ' Set lower six pins to output.
again:                  ' Endless loop.
gosub dayNite           ' Check to see whether it's dark.
for seq = 0 to 5        ' For each of six stored patterns/times..
  lookup seq,(NSgo,NSyel,allRed,EWgo,EWyel,allRed),lkup  ' Get bits.
  pins = lkup & %00111111    ' Copy lower 6 bits to pins.
  time = lkup /64            ' Shift upper 10 bits to lower 10 bits.
```

Figure 9-3 Stoplight program.

```
   for qSecs = 1 to time          ' Count off 1/4-second intervals.
     pause 250                     ' 1/4-sec pause (250 ms).
 'The IF...THEN below means, "If the EW light is not red, or if
 'there's no car at the NS intersection, continue without looking
 'at the minimum timing.
     if seq <> 3 OR NScar = 1 then continue
 'To get here, the EW light must be red, and there's a car waiting at
 'the EW intersection. This IF...THEN looks at the value of qSecs.
 'If the EW light has been red long enough to meet the minimum time
 'requirement, then the light will change. Otherwise, the EW light
 'will stay red until the minimum time is met.
     if qSecs > minTime then changeLight
 continue:
   next                           ' Next qSecs.
 changeLight:
 next                             ' Next entry from the table.
 goto again                       ' Repeat endlessly.

 '===Subroutine "dayNite"===
 'Based on a photocell reading, this routine stores the minimum 'go'
 'time for the EW intersection into the variable minTime. If it's
 'dark, minTime gets the night setting; light, the day setting.
 'You may have to adjust the dark threshold (the constant 230 below)
 'to match the characteristics of your photocell.
 dayNite:
   pot 7,255,dark                 ' Read the photocell.
   minTime = night * 4            ' If it's dark (photocell/pot input > 230)
   if dark > 230 then skip        ' ..minTime = night seconds.
   minTime = day * 4              ' If it's light, minTime = day seconds.
 skip:
 return                           ' Return.
```

Figure 9-3 Stoplight program. (*Continued*)

Construction and Modifications

You can adjust the timing of virtually every aspect of the traffic light's operation. In the process of fiddling with the values (constants NSgo, NSyel, etc.), you may inadvertently pick up a better understanding of the relationship between binary and decimal number systems. For a more formal approach to the subject, see Appendix B.

If you build dioramas or model-train layouts, you've probably already started figuring how to build tiny stoplights with LEDs. In all likelihood, the wiring for the LEDs will be the hardest to construct to scale. Since only small currents are involved, you may find that hair-thin magnet wire will do the trick. This wire is extra thin because it is insulated with a coating of lacquer instead of a plastic sheath.

PARTS LIST

R1–R6—1k resistors, 1/8W or greater

R7—10k resistor, 1/8W or greater

C1—0.1μF capacitor, any type

CdS photocell—Jameco 136047 or similar (adjust the value in dayNite subroutine to match others)

Misc.—switch for traffic-sensing on NS street; model traffic light, etc.

ROBOTIC BUG WITH WHISKER SENSORS

BS1

Small-scale robotics is one of the most popular hobby applications for the Stamps. A relatively simple PBASIC program and inexpensive hardware can produce impressive results.

As a result, there are lots of commercial kits and noncommercial plans and application notes available for Stamp-based robotics. In this chapter, we'll take a commercial robot chassis kit and outfit it with a pair of homemade feelers that help the Stamp explore its environment.

Motors, Ready-to-roll

Although the experts disagree on the precise definition of "robot," it would be hard to find a definition that didn't include movement, and movement requires motors. We're going to concentrate on a particularly versatile and Stamp-friendly sort of motor—the radio-control (R/C) servo.

Robots use two basic categories of motors: positioning and propulsion. An example of the positioning type would be the motors that control the joints in a robotic arm. A propulsion motor would drive the wheels that move the robot.

R/C servos are positioning motors used in radio-controlled cars, boats, and planes. They accept a command signal and precisely adjust the angle of a shaft to the nearest fraction of a degree. Their range of motion is typically 90 to 180 degrees. Figure 10-1 depicts an R/C servo and its control signal.

In an R/C application, moving a joystick on the transmitter causes a servo to adjust its angle to match. This lets you control the movements of a model plane as though you were a (tiny) pilot in the cockpit.

The Stamp can easily mimic the R/C servo control signal. All it must do is generate pulses ranging from 1 to 2 ms in width at intervals of approximately 10 to 20 ms. (Servos are very forgiving about the repetition rate of the pulses, but they do need a continuous stream of pulses in order to maintain their position.)

Figure 10-2 shows how to connect the Stamp to a servo; the listing below shows how simple Stamp-based servo control can be. Note that the servo is powered by an external

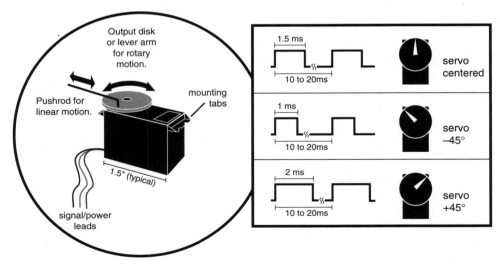

Figure 10-1 R/C servo principles of operation.

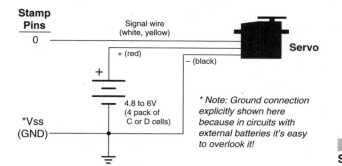

Figure 10-2 Basic Stamp-to-servo hookup.

battery pack. Do not try to power the servo from the Stamp's built-in 5V supply! It won't work, and might damage the Stamp.

```
again:
for b2 = 100 to 200     ' Pulse widths from 1ms to 2ms
  for b3 = 1 to 5       ' 5 pulses of each width
    pulsout 0, b2       ' Pulse pin 0.
    pause 15            ' Wait 15 ms.
  next                  ' Next pulse.
next                    ' Next pulse width.
goto again              ' Repeat forever.
```

That program moves the servo slowly clockwise as the pulse widths increase. When it has reached its clockwise limit (2-ms pulses), the program jumps back to the label "again:" and repeats the process. This causes the pulse width to abruptly change from 2 ms to 1 ms, making the servo turn back counterclockwise as quickly as it can.

This demonstrates a couple of servos' properties. A servo moves to a new position as rapidly as it can, depending on its motor and gear train. If you want it to move more slowly, you must guide it through a series of intermediate positions, the way that the outer for/next loop does.

An additional property that the program can't make apparent is that the servo's internal control electronics vary the speed of the motor in proportion to the difference between the pulse-commanded position and the actual position. This makes sense; it would be very hard on the servo's gears if the motor went from full-bore to full-stop the instant the final position was reached.

That last servo characteristic opens up an opportunity to convert servos from positioning motors to propulsion motors.

MODIFYING SERVOS TO DRIVE WHEELS

Servos make great positioners, but our project needs wheels. Instead of a 90-degree range of positioning, we require 360-degree, variable-speed, reversible rotation. With slight modifications a servo can do this job, too.

Servos are similar in principle but differ in construction, so I'm going to explain the theory behind the servo modification. That way, you can apply your knowledge to whatever servo you happen to have. (If you purchase a robot kit, you will receive instructions specific to the servos included with that kit. There's also a servo-mod tutorial on line at www.rdrop.com/users/marvin/explore/servhack.htm.)

Figure 10-3 is a generalized sketch of what's inside the servo casing. Modifying a servo for continuous rotation involves two basic steps—removing the mechanical stops that limit rotation, and substituting a fixed resistance for the position-feedback pot. Figure 10-4 shows the effects of the modification.

There are two places that you will find mechanical stops: on the final gear that drives the output shaft, and inside the feedback pot. Most servos have plastic gears, so you can easily cut the stop tab off the gear with a pair of sharp diagonal cutters. Avoid filing or grinding this tab, as you will create grit that will stick to the grease on the gears and foul up the gear train.

To eliminate the stop(s) inside the pot, you have to open the pot casing. First, however, mark the center (wiper) wire to the pot, and cut the wires off at the pot. Next, use a multimeter set to ohms and measure the resistance across the two outside (leg) terminals of the pot. Write this reading down. Most servo feedback pots are 5k or 10k.

Opening the pot usually involves prying up three soft metal tabs that hold the casing together. Inside the pot, you'll find a circular track and metal fingers that ride on it. Remove

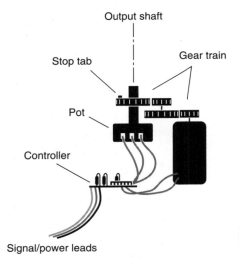

Figure 10-3 Major components of a typical servo.

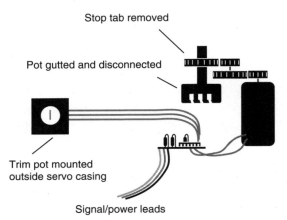

Figure 10-4 Modifications to use a servo as a variable-speed drive motor.

the track and the fingers, but leave the pot shaft and bushing intact. You may find that the back shell of the pot casing has been cleverly formed to create stop tabs—bend these out of the way. When you are done, the pot shaft should turn freely. Reassemble the pot.

With the mechanical stops gone, you can turn your attention to the electronics. The wires you disconnected from the pot must be connected to a new trim pot whose value matches (as closely as practical) the resistance you measured from the original pot. For example, if the original pot measured 5.7k, you would use the nearest standard value—5k.

You may have to extend the pot wires in order to mount the trim pot outside the servo casing. The best way is to run new, unbroken lengths of stranded hookup wire straight from the controller board to the trim pot. If the board is too small for you to comfortably solder, you may connect to the old wires instead. Just be sure to cover the connections with heat-shrinkable tubing to prevent them from shorting together when you reassemble the servo.

When you connect the trim pot, make sure that the center/wiper wire you marked goes to the center/wiper connection on the trimmer. The leg connections aren't critical.

When you're done with these modifications, carefully put the servo back together. Adjust the trim pot to its halfway position, hook the servo back up as shown in Figure 10-2, and rerun the example program. Now instead of turning back and forth, the servo will rotate in one direction, slow down, stop, then rotate in the other direction.

When you prepare a pair of servos for use in the mobile robot, make one more modification to one of the servos: Reverse the motor wires. Since the servos will be mounted back-to-back in the chassis, you want them to rotate in opposite directions for a given input. Reversing the motor wires on one servo will do the trick.

Building and Programming the Robot

Those of you with patience and skill with hand tools can probably roll your own robot chassis based on nothing more than Figure 10-5. The principles are very simple: Mount 3" diameter wheels on the output shafts of two modified servos. Attach them to a small platform with a swivel-caster wheel at the other end. Voila! A very capable wheeled robot.

I'm not patient, and somewhat dangerous with hand tools, so I opted for the commercial chassis kit listed in the parts list. All major parts are precut, and the kit comes with pages of detailed instructions and drawings.

A nice feature of the commercial kit is a pair of power switches, one for the Stamp and the other for servo power. When you program the Stamp, it's a good idea to leave the servos turned off so that the newly programmed robot doesn't run off the edge of the table as soon as the download is complete. Instead, you can place the robot in the middle of the floor, then turn on power to the motors. Of course, you can accomplish the same thing by temporarily removing one of the AA batteries from its holder, but the switch solution is convenient.

Before you start writing programs to control your robot, you must adjust the trim pots on the modified servos for proper speed and direction control. Connect the servos and Stamp as shown in Figure 10-6 and run the following:

```
again:
 pulsout 0,150
 pulsout 1,150
 pause 15
goto again
```

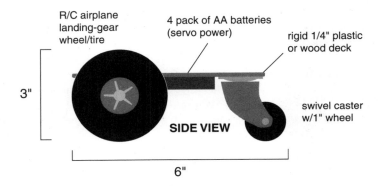

R/C airplane landing-gear wheel/tire

4 pack of AA batteries (servo power)

rigid 1/4" plastic or wood deck

swivel caster w/1" wheel

3"

SIDE VIEW

6"

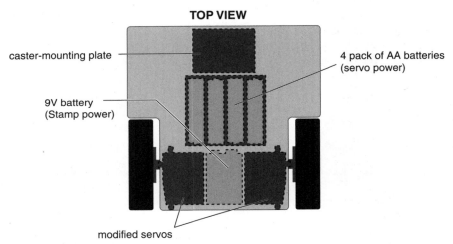

TOP VIEW

caster-mounting plate

4 pack of AA batteries (servo power)

9V battery (Stamp power)

modified servos

Figure 10-5 Simple robot chassis using servos to drive wheels.

With that program running and the servos powered up, adjust the trimmers so that the motors stop turning. Your programs will now be able to use the pulsout value of 150 to mean stop; above 150, forward; and below 150, reverse.

With the servos adjusted, we can now write our first robot-behavior program—one that simply makes the robot take a random cruise around the floor. This program (Figure 10-7) should only be run under close human supervision, since the robot has no sensors at this point, and could lodge up against an obstacle and stall its motors. This would quickly deplete the batteries, and might damage the motors. If your robot gets into trouble, pick it up and move it out of harm's way.

ADDING COLLISION-DETECTION SENSORS

It gets pretty tiresome chasing the robot around to rescue it from collisions with walls and chair legs, so the next step in robot evolution is naturally a set of collision-avoidance sensors. The most basic sensors consist of nothing more than switches connected to flexible feelers (whiskers) positioned to touch an obstacle a moment before the robot would.

Switches are cheap and easy to implement, but you need one I/O pin for each switch. You can quickly use up all of the available BS1 pins that way. I decided to try something a little different.

Figure 10-8 shows my variation on the whisker idea. A pot acts as a pivot for a ⅟₁₆ " thick piece of music wire—the whisker. A second, thinner piece of music wire serves as a spring to return the pot to center. The Stamp can read the pot's resistance through a single I/O pin and determine the whisker's position. That way, one I/O pin gives us the equivalent of two switches, plus the ability to tune the sensitivity of the whisker by changing numbers in the program rather than making mechanical adjustments to a switch.

About the only drawback of the pot/whiskers is that they must be read with the Pot instruction, which takes longer to execute than just reading a switch input. Pot charges a capacitor through the unknown resistance, then measures the time required for the cap to discharge. Since the resistance is unknown, Pot charges the cap for 10ms (long enough to fully charge a 0.1μF cap through 20k of resistance; see Appendix C for more on RC timing) before beginning the discharge cycle that actually produces the measurement. The higher the resistance, the longer the measurement cycle will be. In our robot application, we have to send fresh pulses to the servos every 20 ms. Taking two Pot measurements consumes all of that time and more, if the resistance being read is high. For instance, at 8k with a 0.1μF capacitor, a Pot instruction takes about 25ms.

However, I found that we can relax the schedule for pulsing the servos as long as we send most of the pulses on time, then occasionally interrupt with a longer delay. Still, it makes sense to keep the Pot routines as short as possible. With two pots to measure, a high resistance would produce a 50-ms interruption in pulses, which would make the servos hiccup, and cause the robot to lurch instead of rolling smoothly.

To keep the Pot instructions short, use a meter to adjust the pots to approximately 500 ohms when the whisker is centered. Loosen the setscrew on the knob that supports the whisker, adjust the pot, then reinstall the whisker without moving the pot. If you don't have

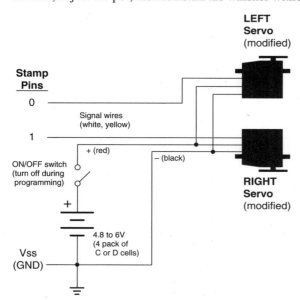

Figure 10-6 Servo hookup for robot programs.

```
'Program: RND_ROBO.BAS (Random robot behavior)
'This program guides the modified-servo robot through a random cruise.

'==Variables and Constants
SYMBOL fate = w5          ' Random 'what-to-do-next' variable
SYMBOL choice = b9        ' Selector for Branch
SYMBOL ticks = b8         ' Amount of time to continue a course.
SYMBOL mot_L = b7         ' Left-motor speed
SYMBOL mot_R = b6         ' Right-motor speed
SYMBOL Lp = 0             ' Left-motor pin.
SYMBOL Rp = 1             ' Right-motor pin.
SYMBOL hafSec = 25        ' Approximate number of ticks in 1/2 second.

'==Main Program Loop
again:                         ' Repeat forever.
 random fate                   ' Pick a random number.
 choice = fate & %1111 ' Limit choice to 0-15.
'The most sensible random drives are made up of mostly forward motion,
'with occasional stops and turns. Since choice falls in a range of
'0-15, we want most choices to be straight ahead. To do this, we
'use a branch instruction with 5 choices for left/right/stop; the
'remaining 11 choices drop through to drive forward.
 branch choice,(left,right,stop,right,left)
 mot_L = 200: mot_R = 200      ' Go forward.
run:
 for ticks = 1 to hafSec       ' Run motors for 1/2 second.
  pulsout Lp,mot_L             ' Output control pulses.
  pulsout Rp,mot_R
  pause 15                     ' Wait 15ms.
 next                          ' Complete 1/2-sec cycle.
goto again                     ' Repeat main loop.

left:                          ' Set for left turn.
 mot_L = 100: mot_R = 200
goto run

right:                         ' Set for right turn.
 mot_L = 200: mot_R = 100
goto run

stop:                          ' Set for full stop.
 mot_L = 150: mot_R = 150
goto run
```

Figure 10-7 Program listing for random robot behavior.

a meter, connect the pots to the Stamp as shown in Figure 10-9, connect the Stamp for programming, and press ALT-P in the STAMP.EXE program. This will give you an opportunity to watch the actual Pot readings while you make the adjustments. Set the pots so that the readings are close to 32 (plus or minus a couple of units; the program is forgiving of a little slop).

Figure 10-10 shows how the whisker readings are interpreted by the whisker-robot program of Figure 10-11. Basically, the Stamp encodes the two whisker readings into a single 4-bit number. It uses that number to retrieve appropriate motor-control values from a table, then runs the motors for a half-second or so at those settings. That simple scheme is enough to get the robot out of many common predicaments, although there's plenty of room for improvement.

Modifications and Improvements

An easy way to improve the performance of the whisker-robot is to study the 16 sensor outputs of Figure 10-10 and apply some strategic thinking. For example, the response to the "both-out" situation—what you would expect when the robot ran straight into a wall—is clearly inadequate. The robot simply backs up a short distance. When the sensors no longer

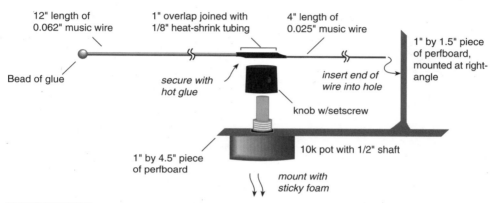

Figure 10-8 Construction of sensor whisker.

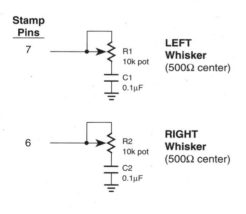

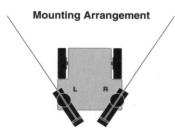

Mounting Arrangement

Figure 10-9 Connection and mounting of sensor whiskers.

variable: *sense*

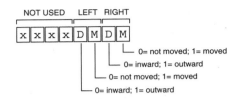

sense bits	value	meaning	action
0000	0	no obstacle	forward
0001	1	left-in	pivot right
0010	2	no obstacle	forward
0011	3	left-out	back right
0100	4	right-in	pivot left
0101	5	left-in, right-in	back
0110	6	right-in	pivot left
0111	7	left-out, right-in	turn left
1000	8	no obstacle	forward
1001	9	left-in	pivot right
1010	10	no obstacle	forward
1011	11	left-out	back right
1100	12	right-out	back left
1101	13	left-in, right-out	turn right
1110	14	right-out	back left
1111	15	left-out, right-out	back

"IN"

IN means toward
the centerline of
the robot

"OUT"

OUT means away
from the centerline
of the robot

Figure 10-10 How the robot interprets sensor inputs.

```
'Program: WHSKR1.BAS (robot control with whisker-sensors)
'This program reads the states of a pair of resistive-whisker sensors
'to detect and respond to obstacles in its path.
'==Variables and constants
SYMBOL temp = b3          ' Temporary storage for pot reading.
SYMBOL sense = b0         ' Outcome of whisker readings.
SYMBOL detectL = bit0     ' 0=no detect; 1=left whisker bumped
SYMBOL directL = bit1     ' 0=whisker pushed inward; 1=outward
SYMBOL detectR = bit2     ' As above, for right whisker.
SYMBOL directR = bit3     ' "   "          "
SYMBOL mot_ctl = w2       ' Motor direction.
SYMBOL mot_L = b5         ' Left-motor speed
SYMBOL mot_R = b4         ' Right-motor speed
SYMBOL fate = w3
SYMBOL ticks = b6         ' Timer for motor movements.
SYMBOL scale = 255        ' Pot scale value.
SYMBOL CNTR = 32          ' Whisker center position.
SYMBOL LO = 22            ' Low reading for whisker center.
SYMBOL HI = 42            ' High reading for whisker center.
SYMBOL LW = 7             ' Left whisker on pin 7.
SYMBOL RW = 6             ' Right whisker on pin 6.
SYMBOL Lpin = 0           ' Left-motor pin.
SYMBOL Rpin = 1           ' Right-motor pin.
SYMBOL fulSec = 50        ' Approximate number of ticks in 1 second.
'==Motor-speed-control constants
SYMBOL F = $C8C8          ' Both motors forward
SYMBOL R = $C896          ' Right turn (left fwd, right stop)
SYMBOL RP = $C864         ' Right pivot (left fwd, right rev)
SYMBOL L = $96C8          ' Left turn (left stop, right fwd)
```

Figure 10-11 Program listing for whisker sensors.

```
SYMBOL LP = $64C8          ' Left pivot (left rev, right fwd)
SYMBOL B = $6464           ' Both motors reversed (back)
SYMBOL BR = $6496          ' Back right (left rev, right stop)
SYMBOL BL= $9664           ' Back left (left stop, right back)
'==Main Program Loop
'The program's primary job is to fill in the lower four bits of
'the variable sense in accordance with the states of the whiskers.
'It then uses sense to lookup the appropriate motor speed and
'direction settings from a table.
again:
 sense = 0                      ' Clear sense variable
 pot LW,scale,temp              ' Get left pot reading.
 if temp < HI AND temp > LO then cont1 ' Centered—no detection.
  detectL = 1                   ' Left whisker touched something.
 if temp < CNTR then cont1      ' Whisker direction: If inward (low)
  directL = 1                   ' then direction = 0; if outward, 1.
cont1:
 pot RW,scale,temp              ' Get right pot reading.
 if temp < HI AND temp > LO then cont2 ' Centered—no detection.
  detectR = 1                   ' Right whisker touched something.
 if temp > CNTR then cont2      ' Whisker direction: If inward (low)
  directR = 1                   ' then direction = 0; if outward, 1.
cont2:
'Motor speed/direction settings are packed into 16-bit values in the
'following table. The lower 8 bits of the values are used to generate
'pulses for the right motor; the upper 8 bits for the left. The way
'this works is that mot_ctl is a 16-bit variable (w2), mot_L is b5
'(upper byte of w2), and mot_R is b4 (lower byte of w2). If you
'select other variables, make sure that you preserve this relationship.
 lookup sense,(F,RP,F,BR,LP,B,LP,L,F,RP,F,BR,BL,R,BL,B),mot_ctl
for ticks = 1 to fulSec        ' Run the motors for a second
 pulsout LPin,mot_L            ' before taking another set of
 pulsout RPin,mot_R            ' sensor readings.
 pause 15
next
goto again                     ' Main loop and another sensor check.
```

Figure 10-11 (*Continued.*)

feel a wall in the way, the program responds by driving the robot forward again. In the unlucky circumstance that the robot is exactly perpendicular to the wall, it will butt its feelers against the wall until the batteries run out!

The program as written gives the robot only the simplest kind of reflexes. Faced with a particular stimulus, it performs one, short-lived action. You could make it perform more complex actions by eliminating the current Lookup instruction and substituting a Branch, like the one used in the earlier random-behavior program. For a given set of sensor stimuli, the program could Branch to a multistep response. For example, to solve the "both-feelers-out" problem discussed above, the robot might branch to a routine that would first back up, then pivot the robot away from the wall.

On the hardware side, the robot could benefit from at least one bump switch on the front edge of the chassis between the feelers. That would make it less vulnerable to getting snagged on chair legs, which slip unnoticed past the feelers.

This is the spirit of hobby robotics—the urge to tinker, enhance, and evolve a project. Robotics is a great way to get involved in creative problem solving and to develop an appreciation for the complexities of the everyday world.

PARTS LIST

R1, R2—10k linear potentiometer with 0.25"-diameter shaft, 0.5" long

C1,C2—0.1µF capacitor, any type

servos—Tower Hobbies TS-51 or any standard R/C servo

trimpots (for servo modification)—see text for value

misc.—0.062" music wire; 0.025" music wire; small control knobs for 0.25" shaft, with setscrew (e.g., Radio Shack 274-403); swivel caster; wood or plastic for robot deck; 9V battery holder; 4xAA battery holder; R/C plane landing-gear wheels/tires; glue; screws; sticky foam; woodworking tools; etc.

Note: a kit of all parts required to build the robot chassis is available from Lynxmotion; see Appendix E for contact information.

TIME/TEMPERATURE DISPLAY

BS1

CONTENTS AT A GLANCE

Keeping Time	A 2x16 serial LCD module
Measuring Temperature	All Together Now
ADC0831 ANALOG-TO-DIGITAL CONVERTER	PARTS LIST

Despite the fact that you can buy a digital clock for less than $5 from a discount store, clocks remain very popular as electronics projects. The principles of timekeeping are familiar but sufficiently intricate to be an interesting project. And you can always use another clock.

But this project isn't just another clock. It also measures and displays the current temperature. Think of it as a miniature version of those time-and-temperature signs that grace downtown banks.

From a technical standpoint, this project has plenty of educational value. It uses an alphanumeric liquid-crystal display (LCD) with a serial interface, a precise oscillator for tracking the passage of time, and an analog-to-digital converter to measure the output of a temperature sensor.

Keeping Time

Measuring time, in one form or another, is a microcontroller's most common job. Many of the Stamp's PBASIC instructions have something to do with time: Pause, Nap, Sleep, Pulsin, and Pulsout all generate or measure time delays. Many other instructions use time indirectly: Serout and Serin send and receive serial data using timing (the baud rate) to distinguish one bit from the next, Pot measures the time required for a capacitor to discharge in order to determine an unknown resistance, and Sound toggles a pin high and low at preset intervals to generate tones.

All of these instructions derive their timing from the Stamp's internal clock, which runs at 4 million cycles per second (4 MHz). Every fourth cycle, the Stamp's PIC microcontroller executes one machine-language instruction. Because of the predictable relationship between instructions and time, most of the Stamp's timing functions base their notion of time on the number of instructions they execute. For instance, Pulsin and Pulsout both reckon time in units of 10 microseconds (μs), meaning that the routines that make up these PBASIC instructions execute in exactly 10 machine-language instructions.[1]

That approach doesn't work very well in higher-level languages like PBASIC. PBASIC instructions take varying amounts of time to execute. Even if instructions did execute in a fixed amount of time, you would have to structure your program so that it always executed the same number of instructions, regardless of the outcome of any If...Then or other decision instructions. This might be difficult or impossible, and would certainly make the program larger than necessary.

The clock portion of this application works around these limitations quite neatly. It uses an IC oscillator/divider to provide an external source of timing. All the Stamp program has to do is periodically check the state of an input pin and maintain a counter.

Here's how it works. An oscillator generates a signal that toggles on and off 32,768 times a second (Hz). The rate of oscillation is controlled by a crystal, making it very stable. The signal is fed into a series of elements called flip-flops. Each flip-flop divides the frequency by two; the first flip-flop divides the 32,768-Hz signal down to 16,384; the second takes 16,384 and outputs 8192, and so on until the 14th flip-flop, which outputs a 2-Hz signal derived from the accurate 32,768-Hz crystal oscillator.

Figure 11-1 shows the oscillator/divider circuit. Note that the majority of the IC's pins are not used. Most are outputs that tap into the chain of flip-flops at various intermediate points. For example, pin 1 connects to the output of the 12th flip-flop, so the signal that appears there is the oscillator frequency divided by 2^{12} (4096).

In our application, we use the output from the 14th flip-flop, so we get 32,768 Hz divided by 2^{14} (16,384), which is 2 Hz. The timekeeping program adds one to a counter every time the 2-Hz signal changes (transitions) from 1-to-0 or 0-to-1. Every fourth transition, the program adds one to the clock's count of seconds. It takes four transitions of a 2-Hz signal to equal one second because, as Figure 11-2 shows, it takes two transitions to make a cycle.

Using changes in an external signal to keep track of time imposes one small obligation on our Stamp program—it must check the state of its input from the oscillator/divider

---

1. If you are knowledgeable about PIC micros, you know that this is an oversimplification. Most instructions execute in four clock cycles; goto instructions take eight. Since the timing routine for Pulsin/Pulsout is obviously a loop, there has to be at least one goto.

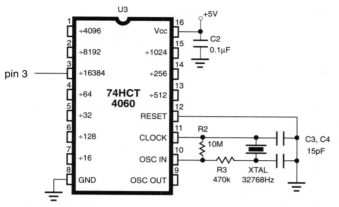

Figure 11-1 Timebase circuit produces accurate 2-Hz output.

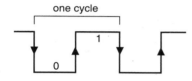

Figure 11-2 One cycle consists of two transitions; high-low and low-high.

more than four times a second. If it checks less frequently, it will miss transitions, and therefore lose time. This is an easy requirement to meet. The Stamp executes about 2000 PBASIC instructions per second, so it can execute almost 500 instructions between checks of the oscillator/divider. We'll use some of those instructions to implement the thermometer portion of our project.

Measuring Temperature

Temperature is extremely easy to sense electronically. In fact, one of the problems faced by electrical engineers is how to keep the effects of temperature from showing up in the performance of their designs. Many electronic components have a specified temperature coefficient (tempco)—the degree to which their values change with changes in temperature.

Usually, manufacturers try to minimize tempco, since it's a nuisance to designers trying to build circuits that operate over a wide temperature range. The exception, of course, is a temperature sensor, whose response to temperature should be as large and as predictable as possible.

There are many kinds of temperature sensors, but some of the most common are specialized kinds of diodes. Temperature-sensing diodes are constructed in such a way that passing a small current through them produces a temperature-dependent output voltage.

A popular diode-temperature sensor is the LM34/35 series from National Semiconductor. You can connect these to a power source and a voltmeter and read the temperature directly (see Figure 11-3); if the meter reads 0.65 volts, then the temperature

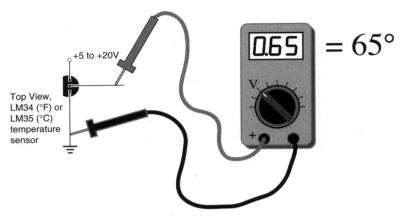

Figure 11-3 LM34/35 sensors' output is proportional to temperature.

is 65 degrees. In the case of the LM34, that would be 65 degrees Fahrenheit—for LM35, 65 degrees Celsius. The basic relationship is 0.01 volts (10 millivolts) per degree.

Just like regular thermometers, the LM34 and 35 come in varying temperature ranges and accuracy ratings. In general, you pay more for better accuracy and/or wider measurement range. Table 11-1 lists the part numbers of the various types.

ADC0831 ANALOG-TO-DIGITAL CONVERTER

The ADC0831 is an analog-to-digital converter (ADC). It measures a voltage and reports it as an 8-bit number (byte). Because its output is meant to be interpreted by a microcontroller or microprocessor, it has some features that an ordinary voltmeter does not. Figure 11-4 shows an ADC0831 connected to an LM34 temperature sensor and to the Stamp.

Let's look first at the ADC's inputs: Vin+, Vin−, and Vref. The Vin pins are inputs that accept the voltage to be measured. The input voltage must fall within the ADC's power-supply range of 0 to 5V. If it does not, the ADC can be damaged.

The ADC measures the difference between the voltages present at Vin+ and Vin−, with the requirement that Vin+ be higher (more positive) than Vin−. In our circuit, Vin− is connected to ground (0V), making it a sure bet that the temperature-sensor voltage at Vin+ will indeed be more positive.[2]

Vin− is the low end of the ADC voltage-measurement range. An input equal to Vin− will cause the ADC to output 0 (%00000000 binary). Vref sets the high end of the range—the voltage that will cause the ADC to output the highest possible byte value, 255 (%11111111 binary). Now there's a limit to how close together you can set Vref and Vin−; for practical purposes, there should be no less than 1V difference between the two. That means that each unit of the ADC output from 0 to 255 can represent a voltage difference of as little as 1/255 = 0.0039 = 3.9 millivolts (mV).

2. Some circuits have dual-polarity power supplies with three (or more) outputs—positive, ground, and negative. None of the circuits in this book make use of a negative supply, so you might get the impression that ground (0V) is always the lowest voltage in a circuit. Not so, if there's a negative supply voltage in there somewhere. Bear this in mind when grafting analog circuits to digital ones. Many analog components require negative supply voltages, and most digital components can be damaged by inputs outside the power-supply (0–5V) range.

Let's consider the voltage we want to measure, the temperature-sensor output. The temperature sensor outputs a voltage of 10mV per degree. If we set Vref to 2.55V then each ADC unit will be 2.55/255 = 0.01V = 10mV. Therefore, each ADC unit will equal 1 degree. The measurement range will be from 0 to 255 degrees. Admittedly, the upper end of that range is wasted, since it exceeds the maximum operating temperature of the Stamp and the LCD display (about 122° F). But the trade-off is that the Stamp can accept the output of the ADC as the temperature, without doing any math. This saves program memory, and makes the application straightforward.

Now that we know how to set up the ADC to make the measurement, how do we get the results out of the ADC and into the Stamp? As Figure 11-4 shows, the ADC connects to the Stamp through just three pins: CS, Clk, and DO.

CS stands for chip select. When this pin is high, the ADC is turned off. It ignores its inputs and turns off its data output (DO) pin. Putting a low on CS signals the ADC to wake up and do some work.

TABLE 11-1   CHARACTERISTICS OF LM/35 TEMPERATURE SENSORS

PART	RANGE	ACCURACY	OUTPUT
LM34A	−50 to +300°F	±2.0°F	
LM34	−50 to +300°F	±3.0°F	10mV
LM34CA	−40 to +230°F	±2.0°F	per °F
LM34C	−40 to +230°F	±3.0°F	
LM34D	+32 to +212°F	±4.0°F	
LM35A	−55 to +150°C	±1.0°C	
LM35	−55 to +150°C	±1.5°C	10mV
LM35CA	−40 to +110°C	±1.0°C	per °C
LM35C	−40 to +110°C	±1.5°C	
LM35D	0 to +100°C	±2.0°C	

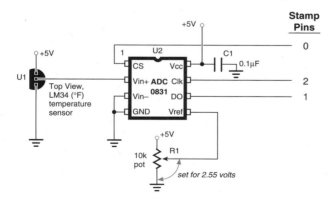

Figure 11-4 Analog-to-digital converter lets Stamp measure voltage.

11

TIME/TEMPERATURE DISPLAY

Once the ADC is selected (CS low), it is ready to take a measurement and output it one bit at a time through DO. The Stamp controls this process through the clock (Clk) pin. Clk is initially low (0). The Stamp pulses Clk by outputting a 1 to it briefly, then returning it to 0. The pulse signals the ADC to place a bit on DO. The Stamp grabs that bit and adds it to a byte variable in which it will accumulate all eight bits. Each time the Stamp acquires a new bit, it shifts the old bits left by one place to make room. In this way, it acquires a bit at a time until it has all eight bits aligned in a row in a byte variable. This process is called synchronous serial communication. The ADC outputs bits only when told to by the Stamp pulsing the clock line. Synchronous serial is very reliable at a wide range of operating speeds, making it a popular means for microcontrollers to communicate with other ICs.

A 2x16 Serial LCD Module

The final specialized component in this project is a serial LCD module as shown in Figure 11-5. This unit accepts data transmitted by the Stamp's Serout instruction and displays it on a 2-line by 16-character display. I'm very proud of this device—I designed and started manufacturing it in response to the introduction of the original BASIC Stamp. The serial LCD makes up for the Stamp's lack of a Print instruction. In other BASICs, Print allows you to see the results of your program on a display screen. The Stamp has no screen, and therefore no Print.

However, the Stamp's Serout instruction, which outputs asynchronous serial data (like the com/modem port on a PC[3]), has some of the same features traditionally included in Print. The serial LCD module allows you to display text, numbers, and block-graphic symbols on an LCD screen using simple Serout instructions.

If you purchase one of these modules, you'll get a comprehensive instruction manual.[4] But to help you understand what's going on in the program, I'll hit the highlights here.

Any data that you send serially to the serial LCD at the correct baud rate appears on the display screen. The following instruction

```
Serout 0,N2400,("Hello")
```

would print the word "Hello" on the LCD screen (assuming that the serial LCD is connected to pin 0 and set for 2400 baud).

The serial LCD also offers a way to pass instructions to the LCD controller. Just precede the instruction with the code 254 ($FE hex or %11111110 binary) and it will be recognized as an instruction. Table 11-2 lists commonly used LCD instructions.

The time/temperature program uses just the clear-screen and position-cursor instructions. Clear-screen is simple enough; the following instruction erases the contents of the LCD and resets the printing position to the upper-lefthand corner:

```
Serout 0,N2400,(254,1)   ' Clear the LCD.
```

3. Asynchronous means "without a clock." Synchronous serial, like that between the Stamp and the ADC0831, relies on a series of clock pulses to allow sender and receiver to distinguish one bit from the next. Asynchronous serial relies on the sender and receiver agreeing on a fixed timing between bits in order to eliminate the clock line.

4. See the listing in Appendix E for Scott Edwards Electronics, Inc.

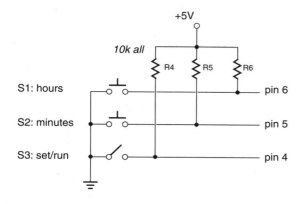

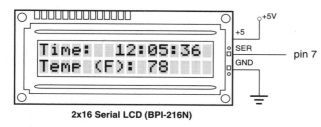

2x16 Serial LCD (BPI-216N)

Figure 11-5 Pushbutton input and serial display output.

TABLE 11-2 COMMONLY USED LCD INSTRUCTIONS	
INSTRUCTION/ACTION	**CODE**
Clear screen	1
Home cursor (undo scrolling, movement)	2
Scroll screen one character left	24
Scroll screen one character right	28
Move cursor one character left	16
Move cursor one character right	20
Blank display (retaining contents)	8
Turn on underline cursor (and unblank display)	14
Turn on blinking-block cursor (and unblank display)	13
Turn off cursor (and unblank display)	12
Set display (DD) RAM address (set cursor position)	128 + addr
Set character-generator (CG) RAM address	64 + addr

Positioning the cursor is also easy, but requires a little inside knowledge about the LCDs themselves. The LCD screen is a window into 80 bytes of internal memory called data-display RAM (DD RAM). When you print to the LCD, you are actually writing data into this memory. If the memory lines up with the screen, you can see the data as text. If it doesn't, it's not lost, but it's not visible either. So it's important to send data only to the locations that are visible.

The first line of the LCD always begins at address 0, the second line at 64. As you can see from Table 11-2, you set the printing position by adding 128 to the address and sending it as an instruction. For example, to move the cursor to the beginning of the second line of the display—

```
Serout 0,N2400,(254,192)        ' Go to start of line 2.
```

In the time/temperature program, I've defined symbols for particular locations on the screen that the program visits frequently. In fact, I've defined symbols for all of the parameters that control the display. This makes the Serout instructions that talk to the display almost self-explanatory.

All Together Now

The program listing of Figure 11-6 brings together all of the elements we've discussed— timing, temperature sensor, ADC, and serial LCD module—to create the time and temperature display. Assemble each of the circuits shown in Figures 11-1, -4, and -5, connect them to the specified Stamp pins, and run the program.

```
'Program: TIMETEMP.BAS (time-and-temperature display)
'This program implements a complete time-and-temperature display
'using a BS1, a precise time base, an analog-to-digital converter
'and a serial LCD module. The program is written for a 2x16 LCD;
'the time appears on the top line and the temperature in degrees
'F on the bottom line.


'=================================
'Variables and constants
'=================================
SYMBOL ticks = b0         ' Count of time-base pulses.
SYMBOL count = b1         ' Value to be displayed.
SYMBOL hours = b2         ' Clock hours setting.
SYMBOL minutes = b3       ' Clock minutes setting.
SYMBOL seconds = b4       ' Clock seconds setting.
SYMBOL temp = b5          ' Temporary variable used by display routine.
SYMBOL prnPos = b6        ' Printing position on LCD screen.
SYMBOL data = b7          ' 8-bit ADC result.
SYMBOL clocks = b8        ' Counter for ADC clock cycles.
SYMBOL I = 254            ' LCD Backpack instruction prefix.
SYMBOL clr = 1            ' LCD Backpack clear-screen instruction.
SYMBOL LCD = 7            ' Output pin for serial LCD.
SYMBOL CS = 0             ' ADC chip-select (activates ADC0831).
SYMBOL AD = pin1          ' Data output of ADC.
SYMBOL CLK = 2            ' Clock input to ADC.
SYMBOL clkIn = pin3       ' Input from time-base chip (2-Hz squarewave).
SYMBOL setRun= pin4       ' Switch: 0=set the time; 1=run clock.
SYMBOL set = 0            ' Set the time (0).
```

Figure 11-6 Program for time/temperature display.

```
SYMBOL run = 1          ' Run the clock (1).
SYMBOL hold = 1         ' Freeze time setting in set mode.
SYMBOL incMin= pin5     ' Button: 0=increment minutes; 1=hold minutes.
SYMBOL incHrs= pin6     ' Button: 0=increment hours; 1=hold hours.
SYMBOL L2 = 192         ' Location code for start of line 2 on LCD.
SYMBOL timePos = 135    ' Location code for displaying time on LCD.
SYMBOL tempPos = 202    ' "        "      "    "         temperature.


'===================================
'Display Setup/Initialization
'===================================
'The main program only displays the changing time/temperature
'information, so this setup code prints the unchanging labels
'"Time" and "Temp" on the display. The LCD will retain these
'labels until power is removed, so there's no need to print
'them again.
Begin:
 pause 1000                      ' Wait a sec for LCD initialization.
 serout LCD,n2400,(I,clr)        ' Clear the LCD screen
 serout LCD,n2400,("Time:")      ' Label top line of display.
 serout LCD,n2400,(I,L2,"Temp (F):")    ' Label bottom line of display.


'===================================
'MAIN PROGRAM LOOP
'===================================
'If the set/run switch is in the set mode, the program goes to
'a separate "set" routine. Otherwise it takes a temperature measurement
'and displays it, then checks for a transition on the clock-pulse
'input. If a transition occurs, it increments the variable ticks.
'The fourth transition of the 2-Hz clock means a second has passed.
'When that happens, the program calls a subroutine that updates the
'current time, and then writes that new time to the display.
doTiming:
 if setRun=set then setClock   ' If switch is closed, set the clock.
 gosub convert                 ' Get a temperature reading.
 serout LCD,n2400,(I,tempPos,#data,"   ")   ' Display it.
 if clkIn = bit0 then doTiming ' No change? Loop.
 ticks = ticks+1 & %11         ' Changed: increment ticks.
 if ticks  3 then doTiming     ' Loop if not 3 (4th count, 0,1,2,3).
 gosub incTime                 ' Fourth count: increment the time.
 gosub showTime                ' Display the time.
goto DoTiming                  ' Loop continuously.


'===================================
'Clock-setting routine
'===================================
'The program comes here if the setRun switch is in the set mode.
'If the incMin (increment-minutes) button is pushed, the program
'adds 1 to the count of minutes and redisplays the time. If the
'incHrs (increment-hours) button is pushed, then it adds 1 to
'the hours and redisplays the time. If both buttons are pushed,
'only the minutes respond. If neither button is pushed, the
'time freezes at its present count. Putting setRun into run mode
'restarts the clock.
setClock:
 if setRun = run then doTiming ' If run mode, go to main clock loop.
 seconds = 0                   ' Clear seconds.
 if incMin=hold then ckHours   ' If minutes button=hold, try hours.
  gosub incMinutes             ' If minutes button=set, increment minutes.
  goto setDone                 ' Finished setting minutes
ckHours:
 if incHrs=hold then setClock  ' If hours button=hold, continue.
  gosub incHours               ' If hours button=set, increment hours.
```

Figure 11-6 (*Continued*)

```
setDone:                     ' Finish setting—
 gosub showTime              ' Redisplay the time.
 pause 200                   ' Give user a chance to release..
goto setClock                ' ..the set button, and continue.


'================================
'Subroutines
'================================
'==showTime: Display the time (hours:minutes:seconds) on the serial
'LCD module. Uses subroutine showDigs to display the individual
'pairs of digits.
showTime:                    ' Display the time on serial LCD.
 serout LCD,n2400,(I,timePos) ' Start at hours position.
 let count = hours           ' Show hours digits.
 gosub showDigs
 serout LCD,n2400,(":")      ' Colon.
 let count = minutes         ' Now minutes.
 gosub showDigs
 serout LCD,n2400,(":")      ' Colon.
 let count = seconds         ' Now seconds.
 gosub showDigs
return


'==showDigs: Display the two-digit value stored in count on the LCD.
showDigs:
 let temp = count/10         ' Get the tens-place digit.
 serout LCD,n2400,(#temp)    ' Put it on the display.
 let temp = count//10        ' Get the ones-place digit.
 serout LCD,n2400,(#temp)    ' Put it on the display.
return                       ' Return to main program.

'==incTime: Increment the seconds. If seconds overflow (>59), increment
'minutes. If minutes overflow (>59), increment hours. And if hours
'exceed 12, reset to 1. Note that there are additional labels that
'allow the program to enter the subroutine at points to increment the
'minutes or hours. These entry points are used by the clock-setting
'part of the program.
incTime:
 seconds = seconds + 1 ' Increment seconds.
 if seconds  let seconds = 0 ' Wrap seconds around to 0.
incMinutes:                  ' —> Enter here to increment minutes.
 let minutes = minutes +1    ' Seconds overflowed, increment mins.
 if minutes  let minutes = 0 ' Overflow: wrap around to 0.
incHours:                    ' —> Enter here to increment hours.
 let hours = hours +1        ' Minutes overflowed: increment hours.
 if hours  let hours = 1     ' Overflow: wrap around to 1.
done: return                 ' Done: return to main program.

'==convert: Get an 8-bit analog reading from the ADC0831.
'With Vin- connected to ground and Vref to 2.55V, the 10mV/degree
'output of the LM34/35 directly translates to degrees.
convert:
 low CS                      ' Select ADC.
 pulsout CLK, 1              ' 10 us clock pulse.
 let data = 0                ' Clear data.
 for clocks = 1 to 8         ' Eight data bits.
  let data = data * 2        ' Perform shift left.
  pulsout CLK, 1             ' 10 us clock pulse.
  let data = data + AD       ' Put bit in LSB of data.
 next                        ' Do it again.
 high CS                     ' Deselect ADC.
return
```

Figure 11-6 (*Continued*)

If you decide to construct this project for permanent use, you will probably want to use an AC adapter to power it, since a standard 9V battery will only run the device for a couple of days. The parts list suggests a suitable adapter. You can either use a mating connector to the one attached to the adapter, or you can cut off the connector and solder the power wires directly to the Stamp. If you do this, check and double-check the polarity of the adapter with a meter. If you connect the adapter backwards—poof—the Stamp circuit will be damaged.

Don't be afraid to experiment with this project and modify it for your own purposes. You could incorporate the timekeeping techniques into a project of your own, or add temperature sensing to a completely different application. Or modify the clock to be a countdown timer. The point is that once you understand the principles involved, you've added power to your Stamp toolbox.

PARTS LIST

U1—LM34CZ Fahrenheit temperature sensor (Jameco PN 107094)

U2—ADC0831CCN 8-bit analog-to-digital converter (Jameco PN 116100)

U3—74HCT4060 or 74HC4060 oscillator/divider (Jameco PN 45962)

C1, C2—0.1µF monolithic ceramic capacitor (Jameco PN 25523)

C3, C4—15pF to 22pF ceramic disk capacitor (Jameco PN 15405)

XTAL—32,768-Hz tuning-fork-case crystal (Jameco PN 14584)

R1—10k potentiometer (Jameco PN 43001 or 94713)

R2—10 megohm resistor

R3—470k resistor

R4–R6—10k resistor

LCD—2x16 serial LCD module (Parallax PN 27910; Jameco PN 133129; or Scott Edwards Electronics PN BPI-216N)

S1, S2—normally open SPST pushbutton (Jameco PN 26622)

S3—SPST switch

AC adapter—9VDC output power supply (Jameco PN 100845)

11

TIME/TEMPERATURE DISPLAY

DATA-LOGGING THERMOMETER

BS2

This project uses the BS2 to create a thermometer with an LED display, and a data-logging thermometer that records the temperature at 10-minute intervals for downloading and analysis on your PC.

In the process of building these applications, you'll get acquainted with one of the BS2's most important features—its Shift instructions for interfacing synchronous-serial peripherals.

Getting Peripherals on the Bus

Although they can do a lot on their own, microcontrollers frequently need help from specialized peripherals, like displays, memory devices, and circuits that convert between analog and digital signals. The oldest and simplest way for a micro to exchange data with peripherals is a parallel-bus arrangement.

The idea is easy to understand; connect one signal wire from the micro to the peripheral for each bit of data you want to exchange. In Figure 12-1, we show eight wires, enough to exchange a byte at a time.

One additional wire, often called a strobe, tells the peripheral when to pick up data from the bus. At first the strobe line may not seem necessary, but consider one of the most common parallel interfaces, your PC's printer port. To print the word "see" the PC first places the eight-bit code for "s" on the bus and pulses the strobe. Next, it puts "e" on the bus and pulses the strobe. Then "e" again, and another pulse of the strobe.

If it weren't for the strobe line, the printer couldn't tell the difference between "se" and "see" or "seeeeee" for that matter. So the strobe line performs an important function in allowing transmission of multiple bytes.

The strobe can be used to give the bus other capabilities. Suppose you needed to send data to several peripherals, but didn't want to connect nine wires to each device. Connecting all devices to a common data bus, but giving each a separate strobe line, lets you selectively talk to individual devices, as shown in Figure 12-2.

DIFFERENT BUS WIDTHS

If you need to send 16 bits of data over an 8-bit bus, you simply send two bytes in a row. If you reduced the bus width to four bits, you could still send 16 bits, but it would take four transfers (and four strobe pulses) to do it. Taken to the logical extreme, you could reduce the bus width to one bit, and transfer 16 bits with 16 transfers, each accompanied by its own pulse of the strobe line.

That is the thinking behind synchronous-serial peripherals; that a one-bit bus can accomplish anything that a wider bus can. This significantly reduces the size of components and circuit boards by cutting the number of pins and wires needed to connect devices together. The downside is that data transfers over a single wire take longer than parallel transfers. But for many types of peripherals, blazing speed is less important than size and cost considerations.

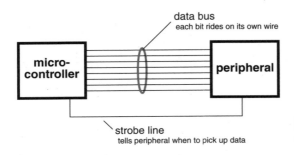

Figure 12-1 Each bit rides its own wire on a parallel bus.

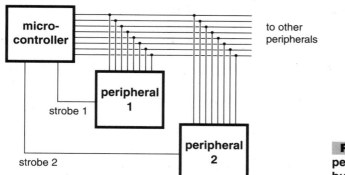

Figure 12-2 Multiple peripherals can share a bus.

Figure 12-3 shows a typical synchronous-serial[1] bus implementation. Note that the strobe gets a new name—clock—and that the job of selecting individual peripherals is moved to another line, typically called chip enable (CE) or chip select (CS). CE/CS serves a couple of purposes: (1) it allows the micro to selectively talk to individual devices on the bus, and (2) it gives each peripheral a point of reference in the incoming stream of bits. When we talk about a particular bit of a transfer—say bit 4—we mean relative to the number of clock pulses since the CE/CS line was activated for that peripheral.

PROGRAMMING FOR THE SERIAL BUS

So far, we've only looked at the physical hookup of the synchronous-serial bus. Getting a device connected is only part of the story. We also have to ensure that the microcontroller sends data in the format that each synchronous-serial device expects. Not all devices use the same format, but PBASIC's ingenious Shiftin and Shiftout instructions[2] can be configured for most any requirement.

As an introduction, let's look at a useful but relatively simple synchronous-serial device, the MAX7219 LED display driver. This IC accepts data from a microcontroller and displays it on common seven-segment LEDs, like the Panasonic LN543GKN8 shown in Figure 12-4. Internally, the display is wired in a multiplexed arrangement as shown previously in Figure 8-1 for the BS1 Magic Message project.

Driving such a display directly would be a headache. It would require 12 I/O lines to drive the 8-segment anodes (LED + connections) plus 4-digit cathodes (LED—connections wired together as in Figure 8-1). On the programming side, the Stamp would be required to continuously scan the display digits in order to produce the illusion of being continuously lit. This would leave little time for the Stamp to acquire interesting data to put on the display.

1. Synchronous means "occurring at the same time." In a synchronous-serial hookup, a pulse on the clock line coincides with valid data on the bus.

2. The Shiftin/out instructions take their name from a device called a shift register, which is quite often used as a synchronous-serial receiver. You can picture the operation of a shift register as being like a line of people moving sandbags. Each person hands a sandbag to the next person in line, and receives another sandbag from the previous person. A shift register does this with bits; a pulse of the clock causes each stage of the register to transfer its current bit to the next stage, and accept a new bit from the previous stage. In this way, bits are shifted into one end of the register and out the other, just like the sandbags.

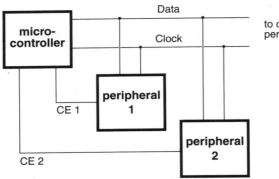

Figure 12-3 Reduce data bus to one wire, and you get synchronous serial.

PIN LOCATIONS

Pins: 24 23 22 21 20 19 18 17 16 15 14 13

Pins: 1   2   3   4   5   6   7   8   9   10 11 12

PIN FUNCTIONS   SEGMENTS

1.	PM	24.	–
2.	K, digit 4	23.	–
3.	Segment d	22.	Segment g
4.	DP 4	21.	K, DP 4
5.	K, digit 3	20.	Segment c
6.	DP 3	19.	Segment b
7.	Upper Colon	18.	Segment f
8.	K, digit 2	17.	K, colon/DP 3
9.	DP 2	16.	K, DP 2
10.	K, digit 1	15.	Segment a
11.	Segmente	14.	–
12.	Alarm	13.	–

Figure 12-4 Pinout of LED display.

The MAX7219 relieves the controller of most of these responsibilities. As you can see from Figures 12-5 and 12-6, it reduces the interface requirement for our four-digit LED display to just three I/O lines: Clock, Data, and Load.[3]

---

3. Load serves the same basic purpose as the strobe or chip-enable pins discussed at the beginning of the chapter. Why call it "Load?" Probably because the MAX7219 is designed to allow a daisy-chain method of connecting multiple chips to a microcontroller. Multiple MAX7219s can be connected with common Clock and Load lines. The data line of the first MAX7219 is connected to the microcontroller; the data line of the 2d MAX goes to the Data Out pin of the 1st MAX. Additional MAX7219s are connected in this same fashion. The microcontroller has to transfer (shift out) enough data for all of the connected MAX7219s, then pulse the Load pin to indicate "transfer complete." It's a clever, pin-saving scheme compared to using multiple chip enables, but not all synchronous devices support it.

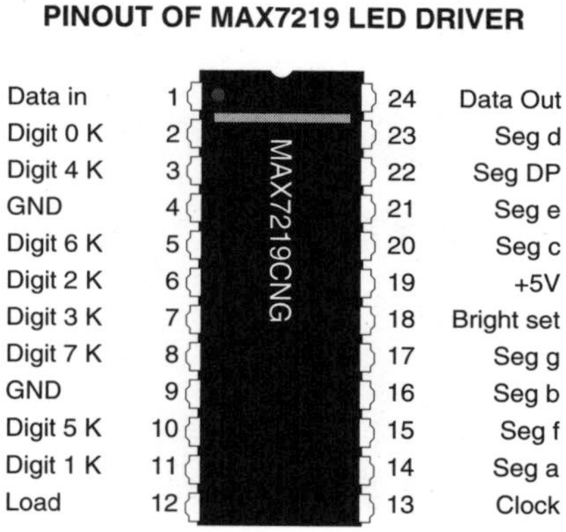

PINOUT OF MAX7219 LED DRIVER

Data in	1	24	Data Out
Digit 0 K	2	23	Seg d
Digit 4 K	3	22	Seg DP
GND	4	21	Seg e
Digit 6 K	5	20	Seg c
Digit 2 K	6	19	+5V
Digit 3 K	7	18	Bright set
Digit 7 K	8	17	Seg g
GND	9	16	Seg b
Digit 5 K	10	15	Seg f
Digit 1 K	11	14	Seg a
Load	12	13	Clock

(chip labeled MAX7219CNG)

Figure 12-5 Pinout of the MAX7219.

In light of our discussion of synchronous serial, the MAX7219 is a pretty typical receive-only device. It expects a microcontroller to send it two bytes of data for each transfer. The first byte represents a register address[4] and the second a value. Effectively, each transfer is a sentence that says, "Set this register to this value."

To write a PBASIC Shiftout instruction that talks to the MAX7219, we only need a single piece of information: When the BS2 sends synchronous data, which bit (the highest or lowest-valued) should come first? Turns out it's the most-significant bit (msb). So here's a skeletal Shiftout instruction for the MAX7219:

```
Shiftout DATA_n,CLK_n,msbfirst,[register,value]
```

...where DATA_n and CLK_n are constants set to the pin numbers of the data and clock lines, and msbfirst is a setting that tells shiftout to send the msb first. (Sometimes it's just impossible to make this stuff sound hard!)

After figuring out how to talk to the MAX7219, we need to think about what to say. Basically, our vocabulary consists of the MAX's register addresses, their range of possible values, and the effects of those values. Table 12-1 is a list. In the table, there's a small discrepancy in the way the MAX7219 refers to its digits. In some places they're numbered 0–7, but when it comes to register addresses, they're numbered 1–8. The MAX7219 reserves address 0 as a "no-op," a nonoperational placeholder to help in daisy-chain applications. When several MAX7219s are daisy chained, you have to send data to all of them at once. If you don't wish to disturb the data or settings of a particular MAX7219, send some dummy data to its register 0.

A few final details, then we'll look at the MAX7219 in action. In Figure 12-6, the schematic, I suggest that you use a husky voltage regulator to supply the MAX7219 circuit rather than the BS2's built-in voltage regulator. The reason is that the MAX7219 and LED display can draw more than 100mA, the max output of the BS2 regulator.

---

4. Don't be thrown by the term "register." It's just another name for memory, like a variable in PBASIC. The main reason for using the term register is to stress that this particular memory is part of the processor or device that uses it, and that it may have special properties that ordinary memory doesn't have (like faster access or the ability to control settings). To you as a PBASIC programmer, this is old hat. All of the variable memory, I/O pins, and I/O direction settings are technically registers.

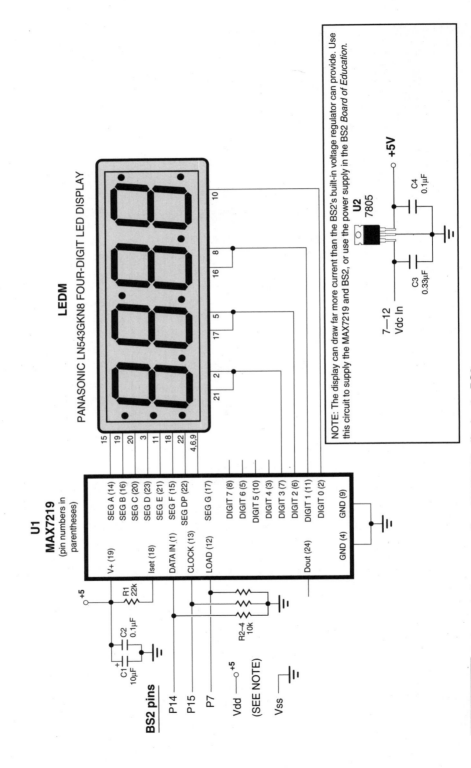

Figure 12-6 Schematic for connecting MAX7219, display to BS2.

TABLE 12-1 MAX7219 REGISTERS

REGISTER ADDRESS(ES)	VALUES	FUNCTION/MEANING
0	0–255	No-op. Any value written to register 0 is ignored. See text.
1–8	0–255	Digit addresses. Value is written to the specified digit (1–8). If the digit is configured for BCD decoding (register 9 below), the numeric value (0–9) is displayed on the 7-segment LED. Adding 128 to a number turns on the decimal point for that digit. If decoding is off, the bit pattern of the value is written to the LEDs, where a 1 bit5 LED on. LEDs are arranged in order: DP,A,B,C,D,E,F,G (see Fig. 12-4), so a value of 123 (binary 01111011) would light LEDs A,B,C,D, F and G.
9	0–255	Decode mode. Turns BCD decoding on for selected digits (0–7) of the LED display. Decoding interprets numbers 0–9 as their corresponding pattern on LED segments on the display. A 1 in a selected bit position turns decoding on for that digit. For example, a value of binary 00001111 would turn decoding on for digits 0–3.
10	0–15	Brightness. Sets the relative brightness of the display from 0 (off) to 15 (full brightness).
11	0–7	Scan-limit. The MAX7219 can drive as many as 8 LED digits. If fewer are actually connected, the scan limit should be reduced accordingly (otherwise, the MAX wastes time driving the unused outputs, which reduces brightness of the connected digits). For example, if 4 digits are connected, scan limit should be 3 (since digits 0,1, 2, and 3 are driven).
12	0,1	Shutdown. A value of 1 turns the display on; 0 turns it off.
15	0,1	Display test. A value of 1 puts the display into test mode, with all connected LED digits/segments turned fully on. A 0 sets the display for normal operation.

The current through each LED segment is set by R1 in Figure 12-6. Its value is selected from a table in the MAX7219 documentation, which says that for an LED forward voltage of 2V, a 24.6k resistor sets a current of 20mA per LED; a 15k resistor = 30mA. I selected 22k as a close standard value. Brightness can be fine-tuned by setting the brightness register (Table 12-1) to less than the maximum setting of 15.

Resistors R2-R4 in Figure 12-6 aren't strictly necessary, but they hold the I/O lines low when the BS2 resets. This prevents garbage data from appearing on the display when you reset or reprogram the BS2. Without these resistors, noise on the I/O lines could look like data to the MAX7219, which would obediently display random stuff on the LEDs.

When you build the circuit, don't leave out capacitors C1 and C2. They are important. The MAX7219 rapidly switches large amounts of current through the display LEDs, and it needs both a small (0.1μF) and large (10μF) capacitor to take care of its small/fast and large/slow current demands. Place these caps as close to the MAX7219 as feasible.

Figure 12-7 presents an example using the MAX7219 to display a counter on the LEDs.

```
' Program: MAXDEMO.BS2
' This program demonstrates the MAX7219 LED display driver using
' the BS2's Shiftout instruction to transfer data and instructions
' synchronously. The demo implements a simple counter on the 4-digit
' display.
' MODIFICATIONS: Check your understanding of the program by trying
' some modifications:
' -in the lookup table below, change the value after BRITE and notice
'  the change in display brightness
' -also in the lookup table, change the value after DECODE. 0s will
'  prevent the corresponding digits from being decoded into proper
'  LED patterns. Instead, the bit pattern of dispVal will directly
'  control the LEDs' on/off states (with bit0 controlling segment g
'  bit 1, f... and bit 7 controlling the decimal point.
' -in the main program loop, change the values of dPoint and zBlank
'  and see how these affect the decimal point and leading-zero
'  blanking functions
' -instead of a counter, use some other means of assigning a number
'  to dispVal. Set it to a value (dispVal=1234) or the result of a
'  math operation like Random or Sin(e).

' Hardware interface with the 7219:
DATA_n con   14     ' Bits are shifted out this pin # to 7219.
CLK_n  con   15     ' Data valid on rising edge of this clock pin.
LOAD_n con   7      ' Tells 7219 to transfer data to LEDs.

' Register addresses for the MAX7219. To control a given attribute
' of the display you write the register address followed
' by the data. For example, to set the brightness, you'd send
' BRITE followed by a number from 0 (off) to 15 (100% bright).
DECODE con   9      ' Control register; a 1 turns on BCD decoding.
BRITE  con   10     ' "       "   " intensity register.
SCAN   con   11     ' "       "   " highest # (0-7) of LED digit.
ON_OFF con   12     ' "       "   " on/off status of display (0=off).
TEST   con   15     ' Test mode (all digits/segment on, 100% bright)

' Variables used in the program.
max_dat    var   word   ' Word to be sent to MAX7219.
index var   nib    ' Index into setup table.
dPoint var  nib    ' Decimal-point position (4,3,or 2).
temp    var  byte   ' Temporary variable used in outputting digits.
zBlank var  bit    ' 0=blank leading zeros; 1=allow leading zeros.
nonZ   var  bit    ' Flag used in blanking leading zeros.
dispVal    var   word   ' Value to be displayed on the LEDs.
odd    var   index.bit0  ' Low bit of index (0=index is even; 1=odd)

' ====================== MAX7219 SETUP ========================
' Setting up the MAX7219 requires sending a series of register addresses
' paired with their desired values. After each address/value pair, the
' LOAD line must be pulsed. A tidy way to organize this job is to put all
' of the address/value pairs in a table, retrieve them one at a time, and
' shift them out to the MAX7219. After every odd-numbered entry (i.e.,
' value), pulse the LOAD line. The table below sets the following:
' SCAN = 3 (scan the lowest 4 digit outputs 0-3)
' BRITE = 10 (set brightness to 10 on a scale of 0-15; medium bright)
' DECODE = %1111 (decode 4-bit values into LED segments on all 4 digits)
' ON_OFF = 1 (turn the display ON)
```

Figure 12-7 MAX7219 demo program.

```
for index = 0 to 7              ' Retrieve 8 items from table.
  lookup index,[SCAN,3,BRITE,10,DECODE,%1111,ON_OFF,1],max_dat
  shiftout DATA_n,CLK_n,msbfirst,[max_dat]
  if odd = 0  then noLoad        ' If odd is 1,
  pulsout LOAD_n,1               ' pulse LOAD line.
NoLoad:                          ' Else, don't pulse.
next                             ' Get next item from table.


' ===================== MAIN PROGRAM LOOP =========================
' Now that the MAX7219 is properly initialized, we're ready to send it
' data. The loop below increments a 16-bit variable and displays it on
' the LEDs connected to the MAX. Subroutines below handle the details
' of converting binary values to binary-coded decimal (BCD) digits and
' shifting them out.

Loop:
    dPoint = 0         ' Turn off the decimal point (0).
    zBlank = 1         ' Turn off leading-zero blanking (1).
    gosub MaxDisplay ' Show dispVal on the LEDs.
    dispVal = dispVal+1      ' Increment dispVal
    if dispVal < 10000 then Loop
    dispVal = 0        ' Reset dispVal when it exceeds 4 digits.
goto loop


' ===================== DISPLAY SUBROUTINE ========================
' The MAX7219, as we have configured it, expects to be sent digit/value
' pairs to tell it what to display. For example, to display 0567, you
' would send the following pairs: 4/0 3/5 2/6 1/7 meaning "set digit
' 4 (lefthand digit) to 0, digit 3 to 5...and digit 1 to 7." The MAX7219
' decodes numeric values into the appropriate pattern of LEDs on the
' display. This capability is turned on by the setting of the DECODE
' register during initialization above.
' In addition to numeric values, the MAX's decode mode recognizes some
' additional wrinkles. If you add 128 (i.e., set bit 7) of a number,
' the MAX turns on that digit's decimal point. A value of 15 blanks
' a digit; 10 produces a minus (-) sign; and values 11-14 generate the
' characters E, H, L, and P respectively.
' The MaxDisplay subroutines rolls these capabilities together to
' display the four-digit value of dispVal on the LEDs. It can blank out
' leading zeros (i.e., display numbers like 99 as 99 rather than 0099)
' based on the value of zBlank. And it will display a decimal point
' at the digit specified by dPoint. (Only digits 4,3, and 2 are valid—
' we don't need a decimal point at position 1, so it's not wired up.)

MaxDisplay:
nonZ = zBlank                       ' If zBlank=1, display leading 0s.
for index = 4 to 1                  ' Work from digit 4 down to digit 1.
  shiftout DATA_n,CLK_n,msbfirst,[index]       ' Send digit position.
  temp = dispVal dig (index-1)      ' Get decimal digit (0-3) of dispVal.
  if temp = 0 then skip1            ' If digit = 1, set nonZ flag.
  nonZ = 1
skip1:
  if nonZ = 1 OR temp <> 0 OR index = 1 then skip2    ' If leading 0..
  temp = 15                        '..write a blank to that digit.
skip2:
  if index <> dPoint then skip3  ' Turn on decimal point in dPoint digit.
  temp = temp + 128
skip3:
  shiftout DATA_n,CLK_n,msbfirst,[temp] ' Send the digit.
  pulsout LOAD_n,1                       ' Load the displby.
next                                     ' Repeat for all 4 digits.
return                                   ' Done? Return.
```

Figure 12-7 (*Continued*)

A Synchronous-Serial Thermometer

I used the MAX7219 to show the basics of synchronous-serial interfacing because it's so simple and visible. The Stamp sends data, the MAX7219 receives it, and it shows up on the LEDs. Easy.

But synchronous serial can also be used to communicate with more complex peripherals, such as the DS1620 thermometer chip that we'll look at next.

The DS1620 is a complete thermometer on a chip. It measures the temperature in units of 0.5°C (0.9°F) over a range of −55° to +125°C (−67° to +257°F). Accuracy is quite good at ±0.5°C (±0.9°F) over much of the range [0° to 70°C (32° to 158°F)]. Compared to the usual methods of temperature measurement used with microcontrollers (involving a temperature sensor, analog-to-digital converter, and voltage reference as in the BS1 time-and-temperature project), the DS1620 is a bargain and a marvel of simplicity.

In addition to serving as a microcontroller-readable thermometer, the 1620 can also be programmed to act as a stand-alone thermostat, with separate outputs that turn on to signal when the temperature is above, below, or within a specified range.

To accomplish all of these wonders, the 1620 must be able to have long, meaningful conversations with a microcontroller. OK, so the conversations aren't all that long or profound, but the micro and the 1620 must each be able to talk and listen over the same data line.

A typical conversation between the Stamp and a 1620 goes like this:

1. Stamp sends DS1620 instructions to configure for thermometer operation and perform continuous temperature readings.
2. Stamp deactivates DS1620 and waits to allow it to reconfigure.
3. Stamp sends DS1620 instruction to begin operation, then deactivates it.
4. Stamp sends DS1620 instruction requesting the current temperature.
5. DS1620 sends Stamp a temperature reading.
6. Stamp deactivates DS1620.

Steps 1 through 3 are performed only once in a program, while steps 4 through 6 are performed every time the Stamp wants a new temperature reading. In order to get the reading, the Stamp must listen to data output by the 1620. It does that using the Shiftin instruction. Shiftin puts the data pin into input mode and acquires one bit of data at a time with each clock pulse—the reverse of Shiftout.

Let's see the DS1620 in action. Leave the MAX7219 connected to the BS2, and add the DS1620 as shown in Figure 12-8. Note that the two devices will now share data and clock lines as shown in Figure 12-3. Run the program of Figure 12-9. The program will get temperature measurements from the DS1620 and display them on the MAX7219, alternating between °C and °F. The program contains some interesting code that demonstrates two's complement signed numbers and the BS2 */ operator that can multiply a number by an integer fraction. For more information on signed numbers, see Appendix B.

BS2-Based Temperature Logger

Since the DS1620 is already hooked up, I thought I'd present one more temperature-oriented BS2 application, a data logger. This project takes temperature measurements at

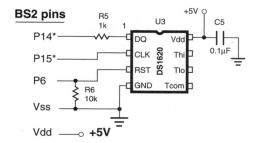

BS2 pins

NOTES:
• The data (DQ) and clock (CLK) lines of the DS1620 go to the same BS2 pins (P14 and 15) as the data/clock lines of the MAX7219.
• C5 should be located as close as possible to the DS1620.
• The reset (RST) pin of the DS1620 performs the same function as a chip-select/chip-enable pin.

Figure 12-8 Connecting the DS1620 digital thermometer.

```
' Program: LED_TEMP.BS2 (Display temperature on LED display)
' This program communicates with a DS1620 temperature sensor
' and a MAX7219 LED display chip. It displays the temperature,
' alternating between degrees C and F, on a 4-digit LED display.
' Program illustrates some useful concepts:
' -how asynchronous serial devices can share Data and Clock
' lines to save I/O pins
' -one way to deal with negative numbers in the Stamp's
' positive-integer math (see convertTemp routine)
' -tricks for simplifying your code (see displayTemp routine)
' ===Constants (pin assignments)
LOAD_n con     7          ' Load pin for MAX7219.
DSRST_n    con  6           ' Reset (enable) pin for DS1620.
CLK_n      con  15          ' Clock line (shared).
DATA_n con     14          ' Data line (shared).
' ===Constants (DS1620 instructions)
WCFG       con  $0C     ' Write-configuration instruction.
CPUCON con     $02     ' Computer-interface/continuous operation.
STARTC con     $EE     ' Start temperature measurements.
RTEMP      con  $AA     ' Read the current temperature.
' ===Constants (MAX7219 registers, symbols)
DECODE     con  9       ' Control register; a 1 turns on BCD decoding.
BRITE con    10     ' "        " intensity register.
SCAN con     11     ' "        "  " highest # (0-7) of LED digit.
ON_OFF con   12     ' "        "  " on/off status of display (0=off).
TEST con     15     ' Test mode (all digits/segment on, 100% bright)
BLANK      con  15     ' Digit value of 15 turns off all segments (decoded).
MINUS      con  10     ' Digit value of 10 makes a minus (-) sign (decoded).
C_SYM      con  %01001110     ' Pattern of LED segments, C (non-decoded).
F_SYM      con  %01000111     ' Pattern of LED segments, F (non-decoded).
' In non-decoded digits, the bit pattern determines which LED segments
' are on or off (1=on). Segments are arranged from bit 7 to bit 0
' left-to-right as follows: decimal point, segments g,f,e,d,c,b,a. So
' %01110111 would display capital A. See the figure for LED segments.)
```

Figure 12-9 Program listing for digital thermometer project.

```
' ===Variables
theTemp        var     word             ' DS1620 reading in 1/2 degr. C units.
tempByt        var     theTemp.lowbyte   ' Value portion of DS1620 temperature.
tempSgn        var     theTemp.bit8 ' Sign bit of DS1620 temperature.
mySign var     bit              ' Separate sign bit for calculations.
C_or_F var     bit              ' Selector bit: 0= readout in C; 1= readout in F
index var      nib              ' Index into setup table.
digit var      byte             ' Used in outputting digits.
odd var        index.bit0       ' Low bit of index (0=index is even; 1=odd)

' ===Begin Program: Initialize DS1620
low DSRST_n   ' Deactivate the DS1620
low LOAD_n    ' Deactivate the MAX7219.
high CLK_n    ' Start with clock high for DS1620.
pause 100     ' Give things a moment to settle.
high DSRST_n  ' Now, activate 1620 and write configuration to it:
shiftout DATA_n,CLK_n,lsbfirst,[WCFG,CPUCON]   ' "computer/continuous"
low DSRST_n   ' Deactivate the DS1620.
pause 50      ' Wait for it to digest configuration command.
high DSRST_n  ' Tell the 1620 to begin continuous conversions.
shiftout DATA_n,CLK_n,lsbfirst,[STARTC]
low DSRST_n   ' Deactivate.
pause 1000    ' Wait a second.

' ===Initialize the MAX7219
for index = 0 to 7           ' Retrieve 8 items from table.
   lookup index, [SCAN,3,BRITE,10,DECODE,%1110,ON_OFF,1],digit
   shiftout DATA_n,CLK_n,msbfirst,[digit]
   if odd = 0 then noLoad    ' If odd is 1,
   pulsout LOAD_n,1          ' pulse LOAD line.
NoLoad:                      ' Else, don't pulse.
next                         ' Get next item from table.

' ===Main Program Loop
' The program is very simple: it gets a temperature measurement from the
' DS1620, converts it to units (C or F) and sends it to the MAX7219-
' equipped display. With each loop it switches the units between
' degrees C and degrees F and pauses a few seconds.
again:
 high CLK_n               ' Raise clock (DS1620 likes high clock to start).
 high DSRST_n             ' Talk to DS1620.
 shiftout DATA_n,CLK_n,lsbfirst,[RTEMP]        ' Say "what's the temp?"
 shiftin DATA_n,CLK_n,lsbpre,[theTemp\9]       ' Store 9-bit temp in variable.
 low DSRST_n ' Deactivate 1620.
 gosub convertTemp
 gosub displayTemp
 C_or_F = ~ C_or_F ' Switch to opposite units (C->F or F->C)
pause 4000          ' Wait four seconds.
goto again          ' Do it again.

' ===SUBROUTINE: convert temperature
' I've separated this code as a subroutine only to make it easier to
' read; it's only used one place in the program. Most of the code
' is needed to deal with conversion from C to F, and with negative
' temperature values. The C-to-F conversion is fairly simple:
' F = C * 1.8 + 32. Since the DS1620 works in 1/2 degr C units,
' the formula is changed to 2F = 2C * 1.8 + 64. The one kink is
' the possibility of negative numbers, which are not directly
' compatible with the Stamp's integer math. To solve the problem,
```

Figure 12-9 *(Continued)*

```
' we use the absolute value of the temperature, and maintain the
' sign in a bit variable. After any math, we apply the sign to the
' result.
convertTemp:
  mySign = tempSgn          ' Copy the sign bit.
' The next line performs "sign extension" to convert the DS1620's 9-bit
' signed temperature into a 16-bit signed number. Then it takes the
' absolute value of the result. Now we have the sign (0= + and 1= -) in
' mySign and the value in theTemp.
  theTemp = ABS (theTemp ¦ (tempSgn * $FF00))
  if C_or_F <> 1 then noConvert   ' Convert to F if C_or_F = 1.
  theTemp = theTemp */ $1CC        ' Conversion step 1: theTemp=1.8*theTemp.
  if mySign <> 1 then noNeg         ' Restore the sign to theTemp.
    theTemp = -theTemp
noNeg:
  theTemp = theTemp + 64   ' Conversion step 2: Add 64.
  mySign = theTemp.bit15    ' Save sign in mySign.
  theTemp = ABS theTemp            ' And take absolute value again.
noConvert:
  theTemp = theTemp/2              ' Convert half degrees to whole degrees.
' At this point, theTemp contains the absolute value of the temperature,
' in C or F as dictated by the bit C_or_F, and mySign contains the sign bit.
return                       ' Return to main program.


' ===SUBROUTINE: Display temperature
' This routine outputs the temperature, contained in theTemp and mySign
' to the MAX7219-driven LED display. It works from left-to right
' across the display. First it gets digit 4, whose contents depend
' on whether the temperature is negative (mySign=1) and whether the
' temperature is 100 degrees or more. The logic goes like this:
' If mySign= 1 then digit= MINUS        ' Minus sign (-)
' If mySign= 0 then digit= theTemp dig 2    ' Hundreds digit of temperature
' If digit=0 then digit= BLANK           ' Show "099" as "99" (no leading 0)
' BS2 IF/THENs can only jump to a labeled location in the program, not
' execute other instructions, so the instructions above wouldn't work.
' Lookup to the rescue! The two lookup tables below encode the IF/THEN
' logic in a straightforward way. Note that you can put variables, even
' arithmetic/logic expressions, inside a lookup table. Elegant. After
' shifting out that digit, followed by the tens and ones digits of the
' temperature, the routine uses another lookup table to get the bit
' patterns for the C or F symbols according to which units are
' currently set. (Note that when we initialized the MAX7219 above
' we set numeric decoding for only the leftmost three digits. The
' last digit is not decoded and will display whatever pattern of bits
' we send, with 1=lit/0=dark.
displayTemp:
    lookup mySign,[(theTemp dig 2),MINUS],digit
    lookup digit,[BLANK],digit
    shiftout DATA_n,CLK_n,msbfirst,[4,digit]
    pulsout LOAD_n,1
    shiftout DATA_n,CLK_n,msbfirst,[3,(theTemp dig 1)]
    pulsout LOAD_n,1
    shiftout DATA_n,CLK_n,msbfirst,[2,(theTemp dig 0)]
    pulsout LOAD_n,1
    lookup C_or_F,[C_SYM,F_SYM],digit
    shiftout DATA_n,CLK_n,msbfirst,[1,digit]
    pulsout LOAD_n,1
return
```

Figure 12-9 *(Continued)*

programmable intervals and stores those measurements in the unused portion of the BS2's program memory.

For the hardware portion of the project, disconnect the MAX7219 and the external voltage regulator. Leave the DS1620 intact. Use a 9V battery to power the Stamp through its Vin pin, and make sure that the DS1620's +5 pin (pin 8) is tied to BS2 Vdd. Figure 12-10 shows the circuit; Figure 12-11 is the program that implements our data logger.

This project will use the Stamp programming cable as a convenient connection for PC-based terminal software. This will allow you to set the interval between measurements, and download the temperature log with your PC.

A small modification to the programming cable or carrier board may be required. Insert a switch in the BS2 ATN line as shown in Figure 12-10. ATN is the connection that allows the STAMP2 host program to reset the Stamp in preparation for programming. Unfortunately, most terminal software uses this same pin for handshaking with modems. Even if you turn off the handshaking, this pin is left in a state that holds the Stamp in its nonoperating reset mode. So it's necessary to make the ATN connection for programming, and to break it for communication.

The Parallax Windows Software does not require this modification. You can interact with the logger via its debug window.

Once you have constructed and programmed the project, you may connect it to your PC's serial port and communicate with it using terminal software—the same sort of program used to communicate with a modem. Terminal programs are included with Windows (Hyperterminal accessory). Make sure to configure your terminal software to match the settings outlined in the program. If you use the Parallax host program, no set-up is needed. Just pull up the debug window.

With the logger connected to your PC and the terminal software configured, you will get the following menu when you apply power to the BS2:

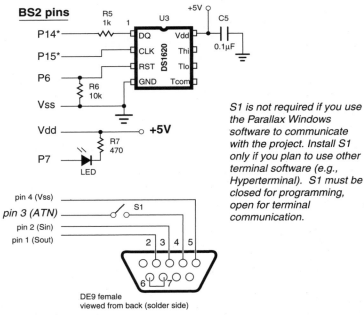

S1 is not required if you use the Parallax Windows software to communicate with the project. Install S1 only if you plan to use other terminal software (e.g., Hyperterminal). S1 must be closed for programming, open for terminal communication.

Figure 12-10 Circuit for temperature logger.

```
'Program: TEMP_LOG.BS2 (Log the temperature with BS2 and DS1620)
'This program creates a simple data logger that takes temperature
'measurements with the DS1620 digital thermometer chip and stores
'them in the BS2's EEPROM program memory. As written, the program
'sets aside 1kB of EEPROM for this purpose. Each measurement is
'stored as a 16-bit value, allowing a total of 512 measurements
'to be stored. The program uses the BS2's programming port to
'communicate with a PC (or other computer) serial port for
'initial setup and download of logged data. When the project is
'connected for communication, the connection to BS2 pin 3 (ATN)
'must be temporarily broken. (It must be restored to program
'the BS2.) Terminal communications software on the PC must be set
'as follows:
'          9600 bps, no parity, 8 data bits, 1 stop bit
'          Handshaking or flow control: "NONE"
'          Advanced UART features (buffers): OFF
'          ASCII setup, append line feeds to incoming data: ON
'When the BS2 is reset, it will place a text menu on the terminal
'screen, prompting you to check the current temperature, set the
'timing interval, view stored data, erase stored data, or begin
'logging. Old data must be erased before a new log is begun.
'===NOTE on timing accuracy:
'The measurement routine uses 'nap 5' (a delay of approximately
'576 ms with power reduced) instead of 'pause 500' to wait for the
'measurement. This reduces power consumption and may extend the
'life of the battery. However, you must take this half-second delay
'into account when you set logging intervals and review temperature
'logs—it's a good idea to record accurate start and end times to
'determine the _actual_ logging interval (if it's critical to
'your application). If you want to tighten up the timing accuracy
'of the measurement routine (takeTemp), substitute pause 500 for
'nap 5.

'===Constants (pin assignments, etc.)
DS_RST   con     6         ' Enable pin for DS1620.
CLK      con     15        ' Clock line.
DATA_n   con     14        ' Data line.
LED      con     7         ' LED, 0=on, blinks w/each measurement.
blink    con     2000      ' Duration (in 2us units) of LED blink.
'===Constants (DS1620 instructions)
WCFG     con     $0C       ' Write-configuration instruction.
CPUSHOT  con     $03       ' Computer-interface/1-shot operation.
STARTC   con     $EE       ' Start temperature measurements.
RTEMP    con     $AA       ' Read the current temperature.
'===Constants (Serial communication)
B96      con     $54       ' Baudmode for 9600-bps comms, pgm port.
pgm      con     16        ' Pin number of programming port.
timeout  con     30000     ' Serial timeout before automatic logging.
'===EEPROM Storage
samples  con     512           ' Maximum number of samples.
maxAddr  con     2*samples-1   ' Highest sample address.
theLog   data    (maxAddr+1)   ' Reserve space for samples.
time     data    word 60       ' Default time of 60.
timeH    con     time+1        '
'===Variables
theTemp  var     word              ' DS1620 reading in 1/2 degr. C units.
tempByt  var     theTemp.lowbyte   ' Value portion of DS1620 temperature.
tempSgn  var     theTemp.bit8      ' Sign bit of DS1620 temperature.
point5   var     theTemp.bit0      ' Last half-degree C of measurement.
sign     var     byte      ' Plus or minus sign to precede temp.
tenths   var     byte      ' Zero or five to follow decimal point.
cmd      var     byte      ' Command from user.
intval   var     word      ' Interval in seconds between measurements.
EEaddr   var     word      ' Address in EEPROM of a given measurement.
```

Figure 12-11 **Program listing for temperature logger.**

```
'===Begin Program: Initialize DS1620
high LED          ' Start with LED off.
low DS_RST        ' Deactivate the DS1620
high CLK          ' Start with clock high for DS1620.
pause 100         ' Give things a moment to settle.
high DS_RST       ' Now, activate 1620 and write configuration to it:
shiftout DATA_n,CLK,lsbfirst,[WCFG,CPUSHOT]      ' "computer/one-shot"
low DS_RST        ' Deactivate the DS1620.
pause 1000        ' Wait a second.

'===Show menu and get commands
'Send text to the terminal at 9600 bps, N81. Wait for a one-key
'command from the terminal. If none arrives within timeout, start
'logging automatically.
menu:
  serout pgm,B96,[CR,CR,"BS2 Temperature Logger"]
  serout pgm,B96,[CR,"(T)emp (I)nterval (R)ead (E)rase (L)og",CR,">"]
  serin pgm,B96,timeout,startLogging,[cmd]
  cmd = cmd min " " & $0DF        ' Convert to uppercase alphabet.
  lookdown cmd,["TIREL"],cmd      ' Convert cmd to # from 0-4.
  branch cmd,[tempNow,setInterval,readLog,eraseLog,startLogging]
goto menu

'===tempNow
'Take a temperature, report it to the terminal, and return to menu.
tempNow:
  gosub takeTemp
  gosub showTemp
goto menu

'===setInterval
'Prompt the user for logging time interval from 1 to 65535 seconds
'and return to menu. If the user doesn't respond within timeout,
'return to the menu. (If user still doesn't respond, the program
'will begin logging.)
setInterval:
  serout pgm,B96,[CR,"Interval in seconds (1 to 65535): "]
  serin pgm,B96,timeout,menu,[DEC intval]
  intval = intval min 1          ' Smallest interval is 1 second.
  write time,intval.lowbyte      ' Store interval in EEPROM.
  write timeH,intval.highbyte
goto menu

'===readLog
'Retrieve up to samples # of readings (two bytes each) from EEPROM.
'If the high byte of a reading is $FF (all ones) then we've
'passed the end of valid data, since the upper byte of a DS1620
'reading can only be 0 or 1. The erase routine loads the log
'with $FFs to make this work. When the routine finds invalid data
'or hits the end of the allocated storage space, it returns to menu.
readLog:
  for EEaddr = theLog to maxAddr step 2
    read EEaddr,theTemp.highbyte  ' Get upper byte of reading.
    if theTemp.highbyte = $FF then endRead        ' End-of-data marker?
    read (EEaddr+1),theTemp.lowbyte      ' Get lower byte.
    gosub showTemp                       ' Display the reading.
  next
endRead:
  serout pgm,B96,[cr,"-end of data",cr]
goto menu

'===eraseLog
'Write $FF to each location of the EEPROM data-log area. This provides
```

Figure 12-11 (*Continued*)

```
'an end-of-data marker for the readLog routine, since $FF is invalid
'as the high byte of a DS1620 measurement.
eraseLog:
 for EEaddr = theLog to maxAddr          ' For each byte of memory.
   write EEaddr, $FF                      ' Write $FF.
 next
 serout pgm,B96,[cr,"-log erased",cr]
goto menu

'===startLogging
'At the number of seconds set by the variable intval,
'take a temperature reading and record it to the EEPROM data log.
'Sleep for intval seconds and repeat until the log is filled.
'When log is full, flash LED to indicate this fact.
startLogging:
read time,intval.lowbyte          ' Get interval from EEPROM.
read timeH,intval.highbyte
 serout pgm,B96,[CR,"Logging at intervals of ",DEC intval," seconds."]
 for EEaddr = theLog to maxAddr step 2
   gosub takeTemp                         ' Take a reading.
   write EEaddr,theTemp.highbyte          ' Store the high byte.
   write (EEaddr+1),theTemp.lowbyte       ' Then the low byte.
   sleep intval                           ' Hit the hay.
 next                                     ' Next reading.
signalEnd:                     ' Done: signal user to download.
 nap 4                         ' Brief pause with power-down.
 pulsout LED,blink             ' Brief flash of LED.
goto signalEnd                 ' Keep signalling.

'===Subroutine: takeTemp
'Activate the DS1620 and tell it to perform a temperature measurement.
'Wait for the measurement to complete, read the result into variable
'theTemp and return.
takeTemp:
 high DS_RST     ' Tell the 1620 to perform one conversion.
 shiftout DATA_n,CLK,lsbfirst,[STARTC]
 low DS_RST      ' Deactivate.
 nap 5           ' Wait for DS1620 to perform measurement (500ms)
 high DS_RST     ' Activate the DS1620.
 shiftout DATA_n,CLK,lsbfirst,[RTEMP]     ' Ask, "what's the temp?"
 shiftin DATA_n,CLK,lsbpre,[theTemp\9]    ' Store 9-bit temp in variable.
 low DS_RST      ' Deactivate 1620.
 pulsout LED,blink        ' Brief flash of LED.
return

'===Subroutine: showTemp
'The DS1620 temperature is in 1/2-degree units, in a signed, 9-bit
'format. To display the temperature correctly, this routine stores
'the sign symbol (+ or -) in a variable according to the sign of
'the temperature, and stores the appropriate last half-degree
'character (to display ".0" or ".5") in another variable. To
'display the signed temperature, it sends the sign, the absolute
'value of the whole-degrees portion of the temperature, a decimal
'point, and the final half-degree, .0 or .5. Note that the
'DS1620 temperature reading in theTemp is not altered.
showTemp:
 lookup tempSgn,["+-"],sign     ' Put the sign symbol into sign.
 lookup point5,["05"],tenths    ' Put "0" or "5" into tenths.
'Now send the whole number serially in the format +78.5 or -13.0.
 serout pgm,B96,[cr,sign,DEC(ABS(theTemp|(tempSgn*$FF00))/2),".",tenths]
return
```

Figure 12-11 *(Continued)*

```
BS2 Temperature Logger
(T)emp (I)nterval (R)ead (E)rase (L)og
>
```

At the arrow prompt, enter the first letter of your menu choice. For example, to see the current temperature, press "T." The program accepts either upper- or lowercase commands. If you don't press a key within a specified timeout period (30 seconds in the example), the program will give up waiting and begin logging the temperature.

You can set the time interval between measurements to any value from 1 to 65535 seconds. The program uses the BS2's Sleep instruction to produce the time delays. Sleep serves two purposes: It produces a time delay, and it reduces the BS2's power draw to almost nothing during that delay. This makes if possible for our temperature logger to operate for weeks without a battery change.

The actual time interval between measurements may vary significantly from the interval you set, especially at the lower end of the range. For one thing, the Sleep instruction rounds your selection up to the nearest multiple of 2.3 seconds. This is a byproduct of the way that the BS2's internal timer works—see the manual for more information.

Another factor affecting the interval is the time required to take the temperature measurement. The DS1620 is set for one-shot conversion, meaning that it takes a measurement only when told to. This reduces its power demand, further extending battery life. But every time the program wants to know the temperature, it must ask the DS1620 to take a reading, wait a half-second for the reading to finish, then get the reading. That half-second conversion delay also figures into the interval between readings.

A final kink is that the accuracy of Sleep timing is approximately ± 1 percent. That sounds good, but 1-percent error over the course of a week amounts to almost 2 hours!

Since the logger only records the temperature, your reckoning of the times at which readings were taken depends entirely on the logging start time, the interval between readings, and the stop time. You can iron out most of the timing inaccuracy by taking the difference between the start and stop times, dividing it by the number of readings the logger took, and using that as your actual interval. While it's true that the interval between individual readings may vary, this approach will be fine for most applications.

Summary

The BS2's low power draw and easy connection to serial peripherals makes it a great starting point for custom data-logging projects. The first project I built with the BS2 was an elaborate data-collection prototyping board with multiple serial peripherals sharing a common bus.

PARTS LIST

LEDM—Panasonic 4-digit, common-cathode LED (Digi-Key part no. P456-ND)
U1—MAX7219 LED display driver (Digi-Key part no. MAX7219CNG-ND or Parallax 603-00001)

U2—7805T 5-volt positive voltage regulator (Digi-Key part no. NJM7805FA-ND, or construct project on Parallax Board of Education w/ built-in high-current regulator)

U3—DS1620 digital thermometer (Jameco part no. 114382 or Parallax 604-00002)

R1—22k resistor

R2, R3, R4, R6—10k resistor

R5—1k resistor

R7—470-ohm resistor

LED—any LED/any color

C1–10μF, 16V aluminum or tantalum electrolytic capacitor

C2, C4, C5—0.1μF ceramic disk or monolithic capacitor

C3—0.33μF ceramic disk or monolithic capacitor

S1—Any switch

AC adapter—9Vdc, 400mA (or greater) adapter (Jameco part no. 137331 or similar)

13

WORLDWIDE REMOTE CONTROL WITH C2TERM

BS2

CONTENTS AT A GLANCE

Want to control or monitor electronic equipment in a remote location? A BS2 and an inexpensive modem team up to create a dial-up communication and control terminal (C2TERM). Using your PC and communication software you can dial up C2TERM from any phone in the world, enter a password, and control lights and appliances throughout your home or business.

This project is a good opportunity to get familiar with asynchronous serial communication and the BS2 instructions Serin and Serout. It also demonstrates the powerful Xout instruction that transmits command signals through the AC wiring to control lights and appliances.

Asynchronous Serial Communication

Serial communication is the process of transmitting data one bit at a time. Asynchronous serial sends the data without a separate synchronizing signal to help the receiver distinguish one bit from the next. To make up for the lack of synchronization, asynchronous serial imposes strict rules for the timing and organization of bits. The reward for obeying these rules is an efficient, reliable way to send data over a single wire.

For the sake of brevity, let's agree that throughout the rest of this chapter "serial" means "asynchronous serial."

SERIAL TIMING AND FRAMING

The basic principle of serial communication is simple—to send multiple bits over a single wire, just place each bit on that wire for a fixed amount of time. For example, suppose you wanted to transmit the byte %10010101 to me in the next room.[1] In your room, you have a switch and a battery that are wired to control a light in mine. We agree beforehand that you will send one bit per second, starting with the lefthand bit, and that light on means 1 and off means 0.

With the rules established, you flick the switch on and off to match the pattern of bits:

Bits	1	0	0	1	0	1	0	1
Seconds:	0	1	2	3	4	5	6	7
Light:	ON	OFF	OFF	ON	OFF	ON	OFF	ON

After our experiment, we meet to see whether the message I received matches the one you sent. I show you my notepad:

0000000000000000000000000000000000010010101000000000000000000000000000

Your message was received faithfully, but it's buried in the middle of all those zeros I wrote before and after the actual message. We need another couple of rules: I start copying one second after an initial 1 (light on), and stop copying after receiving eight bits. We'll call that initial 1 the start bit. It's not part of the data, just a signal from you to me that precedes the data. We try again:

1. I'm using the % sign to indicate that the number is in binary notation. This is the same way that you represent binary in a PBASIC program.

Bits:	1	1	0	0	1	0	1	0	1
Seconds:	0	1	2	3	4	5	6	7	8
Light:	ON	ON	OFF	OFF	ON	OFF	ON	OFF	ON

Now you check my copy and it reads 10010101—a faithful replica of the data you sent. So you send me a series of bytes via our room-to-room communication system: %01010011 %01100101 %01110010 %01101001 %01100001 %01101100. When we compare notes, the first three bytes are OK, but the last three are strange:

11010011 11000011 11011000

These bytes are shifted to the left one place and have a 1 tacked onto the right-hand end. After some head scratching, we realize that my watch is a little slow. I copy the first few bytes correctly, but gradually my watch falls a whole second behind yours, making my copy a whole bit off from your transmission. The longer the message, the worse the problem gets.

We have to add one more rule: The sender will pause at least one bit time with the light in the 0 state (off) between bytes. That way, even if our watches are keeping slightly different time, the effect never accumulates for more than 10 seconds—the time required to send one byte.

In a nutshell, that's how asynchronous serial communication works. The whole process relies on the sender and receiver using the same timing (known as the bit rate or baud rate) and the same set of rules for organizing or framing the bits.

SERIAL PARAMETERS

Since it's vital that senders and receivers agree on the serial timing and parameters, a shorthand has developed to express this information. For instance, the serial setting for our project will be "2400 N81." That means 2400 bits per second (bps or baud), no parity bit, eight data bits, one stop bit. Let's look at each of these settings more closely.

BIT RATE (BAUD)

The bit rate of a serial link is the number of bits that can be sent in one second. It defines the length of time that any one bit will be placed on the wire that connects sender and receiver.[2] The term baud rate is often used interchangeably with bit rate, but there's a subtle difference when the terms are applied to modems. Don't worry about it. For our purposes, bits per second (bps) and baud mean the same thing.

In our project, we'll be using a bit rate of 2400 bps. That means that each bit is on the wire for 1/2400th of a second, or 416.6 microseconds (μs). Each byte (eight bits) we send will be framed by a start bit and a stop bit. That means that it takes 10 bit times to send a byte, so we can send up to 240 bytes per second.

PARITY

In our room-to-room communication example, we discovered errors in the data only after comparing what was sent to what was received. We could have made arrangements to

2. Of course, the serial connection doesn't have to be a wire at all, it can be a radio signal, light beam, etc.

check each incoming byte according to some agreed-upon rule. For example, we could add one more bit to the serial frame and call it the parity bit. You would count the 1s in the data to be transmitted to determine whether the total number was even or odd. For an odd number, you'd make the parity bit a 1; for an even number, 0.

When I received the data, I'd perform the same tally and compare my reckoning of odd/even to the parity bit. If they matched, I'd be more confident that my copy of that byte was correct. If they didn't match, I would figure that the data was possibly incorrect. If I had a way to signal you, I would do so and tell you to send that byte over. If not, I'd have to decide what to do with the doubtful byte.

The parity setup described above is called even parity, since the parity bit is set or cleared according to rules that make the total number of 1s (data bits plus the parity bit) an even number. The opposite arrangement is called odd parity. A third common option is no parity at all.

Why do without parity? Parity isn't a sure-fire check for errors. If one bit is incorrect, parity will detect it. If two bits are wrong, it won't. There's also no guarantee that the parity bit itself won't be received incorrectly. And then there's the dilemma of what to do when parity casts doubt on a particular byte.

For all of these reasons, many systems operate without parity. Instead they may send other kinds of double-check information as regular data.

DATA BITS

Since the byte is a common unit of storage for computers, it's natural to assume that serial communications would transfer data in groups of eight bits—one byte. But it's not always so. For example, if you only need to send ordinary text, six or seven bits can be perfectly adequate. In text-oriented applications, it is common to sacrifice one data bit in order to make room for the parity bit. This is one of the serial modes that the BS2 supports: 7E1 or seven data bits, even parity, one stop bit.

Unlike our room-to-room example, the data bits in most serial communication are sent least-significant-bit (lsb; the bit on the righthand end of a number like %11010010) first.

STOP BITS

The stop bit is a pause between the last data bit and the next start bit. In our room-to-room example, we saw that the stop bit allows the receiver to reset timing with each new start bit in order to prevent small timing errors from accumulating over multiple frames of data and eventually causing data errors.

The stop bit must be at least one bit-time long in order for this to work. However, some slow devices might need more than one stop bit between frames in order to do other processing. The specs for such slowpokes would call for one-and-a-half, two, or more stop bits. You may never encounter this specification, but you have to know that it's possible in order to understand why anyone would bother specifying the number of stop bits when it always seems to be 1. After all, the number of start bits is not specified, since it is always 1.

Although the BS2's Serout instruction does not directly support multiple stop bits, it does allow you to specify pacing in milliseconds. Pacing is a delay between frames of data, so it amounts to pretty much the same thing as multiple stop bits. For example, 2400-bps serial data sent with 1-ms pacing amounts to about 3.4 stop bits. There's one stop bit built into the data frame, plus a 1-ms delay, which amounts to about 2.4 additional bit times. When a specification calls for multiple stop bits, it's really saying "this is the minimum delay between frames that this device can handle," so it's OK to exceed that amount.

A SERIAL FRAME

Now that we're acquainted with all the elements, let's diagram a couple of typical serial frames. First, 2400 bps 7E1:

Time (μs):	0	416.6	833.3	1250	1667	2083	2500	2917	3333	3750
Bits:	start	bit0	bit1	bit2	bit3	bit4	bit5	bit6	parity	stop

Now 2400 bps 8N1, the format our project will use:

Time (μs):	0	416.6	833.3	1250	1667	2083	2500	2917	3333	3750
Bits:	start	bit0	bit1	bit2	bit3	bit4	bit5	bit6	bit7	stop

RS-232 Serial Signals

In addition to the timing and framing of data, serial senders and receivers must also agree on the electrical details of the connection. There are various standards for serial signaling, but the most common is RS-232. It uses signal voltages that are outside the range normally used by the Stamp, but can be readily interfaced by taking a crafty look at the specs.

The Stamp uses 5-volt logic. It outputs 0 volts for a 0 and 5 volts for a 1. When it accepts inputs from other circuits, it regards a voltage less than 1.5 volts as a 0 and greater than 1.5 volts as a 1.

This relationship of voltages to 1s and 0s is common to many digital-electronic devices, but it's not the only one. The serial port(s) on your PC conform to the RS-232 standard, which specifies wider signaling voltages. It makes intuitive sense—the 5-volt logic used to communicate between components separated by a few inches might not be ideal for communication between computers and peripherals separated by 50 feet of cable.

Under the RS-232 spec, a 0 is represented by a higher positive voltage, typically +10V, and a 1 by a negative voltage, typically −10V. The negative voltage is also the stop-bit state; positive is the start bit. The idea is that the larger the voltage difference between a 0 and a 1, the less likely that electrical noise picked up over a long cable would make an RS-232 device mistake one state for the other.[3]

To convert a 5-volt logic level to RS-232 and back normally requires components called RS-232 line drivers and line receivers. However, it's common practice to cheat on the RS-232 standard and do without these parts. Cheating requires a closer examination of the RS-232 rules.

3. Other serial standards, such as RS-422 and RS-485, use the voltage difference between two signaling wires to distinguish a 1 from a 0. This offers much better noise immunity over long wire runs without the need for separate positive and negative power-supply voltages.

An RS-232 sender is supposed to output +5 to +15 volts for a 0. An RS-232 receiver is required to recognize +3 to +15 volts as a 0. The reason for the reduced lower limit (+3V) is to allow for the voltage drop over a long wire run. Looking at this with cheating in mind, a 5-volt logic 1 is equivalent to an RS-232 logic 0.

For a 1, an RS-232 sender is supposed to output −5 to −15 volts. A receiver must recognize −3 to −15 volts as 1, again allowing some voltage drop through the wiring. There's no way to get −3 volts from our 5-volt logic without additional components, but most RS-232 devices also regard a signal that's close to 0 volts (ground) as a logic 1. So, provided that our 5-volt logic's 0 is close to 0 volts, it will suffice as an RS-232 logic 1.

The BS2 has built-in support for this kind of thinking, as you can program its serial-output (Serout) instruction to send inverted serial data. Connect the I/O pin directly to the input of an RS-232 receiver via a short cable and you're ready to go. I emphasize short cable (less than 10 feet), because even a small voltage drop or minor noise on the connection can cause communication errors.

What happens when the Stamp is on the receiving end of a ±15-volt RS-232 signal? Its I/O pins can be damaged by voltages outside the supply range of 0 to 5 volts. However, there's an easy fix. A 22k resistor in series with the ±15-volt line protects the I/O pin. See, each I/O pin is internally protected by a pair of diodes arranged to short out excessive voltages, typically zaps of static electricity. The series resistor limits the amount of current that the diodes have to handle, preventing them from overheating.

The BS2 serial-input instruction Serin has an option that allows reception of inverted serial data. To establish two-way RS-232 communication between the BS2 and a serial port requires a short cable and a cheap resistor. Who says cheaters never prosper?

An even easier route to get serial data into and out of the BS2 is to borrow the programming port. BS2 carrier boards have a DB9 connector that mates with a cable to your PC's serial port.

Figure 13-1 shows how a typical RS-232 frame would look on an oscilloscope screen. (An oscilloscope is a versatile instrument that graphs input voltages versus time.)

Typical RS-232 ports have 9 or 25 pins. Although we have limited our discussion so far to just the data lines, a serial port often includes a number of control lines used for various purposes. The most common use of these lines is handshaking, in which one serial device may indicate that it has data to send and the other signals whether or not it's ready to receive. In our application, we won't be using handshaking. It can be difficult to implement with a relatively slow computer like the Stamp, and is often unnecessary. We will use one of the other control lines—the ring—indicator output from the modem—to determine when to answer the phone.

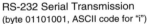

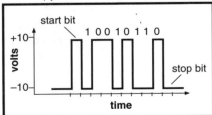

Figure 13-1 One RS-232 serial frame.

TABLE 13-1 RS-232 CONNECTOR PINOUTS

NAME/FUNCTION	DB25 PIN	DB9 PIN
Protective ground	1	—
Transmit Data (TD)	2	3
Receive Data (RD)	3	2
Ready to send (RTS)	4	7
Clear to Send (CTS)	5	8
Data Set Ready (DSR)	6	6
Signal Ground (SG)	7	5
Data Carrier Detect (DCD)	8	1
Data Terminal Ready (DTR)	20	4
Ring Indicator (RI)	22	9

Table 13-1 shows the pinouts of the two common styles of RS-232 connectors.

The names and functions of the RS-232 pins as listed in the table are slightly deceptive. They are valid only for what is known as data-terminal equipment (DTE). A PC is considered DTE, because it can serve as a terminal for sending data. The other flavor of RS-232 devices are called data-communications equipment (DCE). A modem is pretty much the definition of DCE, since its whole purpose is to facilitate communication.

The distinction between DTE and DCE becomes important when you are trying to figure out which pin does what on a particular device. Suppose you need to know which pin transmits serial data. On a PC with a DB25 connector, it's pin 2, the TD pin. On a modem with a DB25 connector, it's pin 3, which the table says is the receive-data pin. It makes a weird sort of sense—the pin through which the PC transmits data to the modem has to be the pin through which the modem receives data! This complementary relationship of inputs and outputs simplifies wiring up cables, since many times you just wire like pins together (1 to 1, 2 to 2, 3 to 3...), but it really messes up the names.

For more on RS-232 interfacing, see the reading list in Appendix E.

Serial Communication by Modem

With the exploding popularity of the Internet and online communication, it hardly seems necessary to define the word modem, but here goes: Modem is a contraction of the words modulator-demodulator, and it refers to a device for sending serial signals via a carrier, like audio tones transmitted by phone or radio. A phone modem typically consists of some analog circuitry for generating and detecting audio tones, and a small, fast computer for interpreting them.

The modem's computer is programmed with routines to communicate with a host computer serially, dial the phone, detect ringing and busy signals, establish a connection with another modem, exchange data, and hang up.

Most modems use the "Hayes" command set, named for the company that set the standard for modems in the early days of personal computing. This language is also sometimes called the AT command set, because most commands begin with AT for "attention."

A modem has two modes of operation—command mode and data mode. When the modem is not linked to another modem, it's in command mode, awaiting instructions from the local computer. The computer can tell the modem to dial or answer a call from another modem. When two modems make contact, they investigate each other through handshaking—a ritual exchange of tones and preliminary data that sets the ground rules for communication.[4]

Once handshaking is complete, the modem goes into data mode. In data mode it acts like a direct connection between local and remote computers. Once in data mode, a modem ignores commands unless it is first returned to command mode by a sort of secret knock (usually "+ + +" followed by a delay).When the exchange of data is complete, one of the computers issues the secret knock to return its modem to command mode, then instructs it to hang up. The other modem senses this loss of carrier, and hangs up too.

The modem you'll need for this project is an external modem—one designed to be connected through a serial port, not installed inside a desktop computer (internal). Feel free to use an older, "obsolete" modem, since this project uses no advanced features and communicates at only 2400 bps.

A benny of using an older modem is that it usually includes a good manual on the AT command set. In the old days users generally dealt directly with the modem, manually typing commands through a terminal program.

To really understand how a modem works, you should try using it manually with your PC and terminal software. If you have Windows installed, there's a simple terminal "accessory" program. Procomm is a popular commercial package, available in both DOS and Windows flavors.

Although the modem manual lists dozens of commands, you can get by with few important ones. Table 13-2 lists the ones I found helpful in getting the BS2 on line. Once your terminal program is set up and talking to the modem, you may send commands by typing them at the keyboard. For example, if you type "ATDT 5551234" followed by the Enter key, the modem will go off-hook (connect to the phone line) and send the touch-tone digits 555-1234.

To sum up, a modem is a small computer that transports your serial data across the phone lines. Next we'll look at a system for moving data through the AC power lines of your home.

X-10 Appliance Control

Some people dream about a Jetsons-style home in which every appliance and feature is under pushbutton control from the comfort of an easy chair. The technology has been around for years. The only thing missing is some sort of household data network to control all those appliances. Our homes come equipped with wiring for electricity, phones, and lately even cable TV, but no network.

---

4. The handshaking that modems do to establish communication is similar in concept and purpose to the hardware handshaking done through the control lines of the RS-232 port. But don't confuse the two. For example, when a terminal program offers you the option to turn off handshaking, it means the extra lines on the serial port, not the modem-to-modem handshaking process.

TABLE 13-2 USEFUL AT MODEM COMMANDS

INSTRUCTION	OPERATION
ATDT n	Dial n using touch tones. ATDT 4594802 dials 459-4802. The number to dial can include other symbols that change the way it is dialed, including:
	, Pause dialing
	W Wait for second dial tone
	; Stay in command mode after dialing
A/	Repeat last command entered.
ATH0	Hang up.
ATA	Answer the phone now (manual answer).
ATV0	Set modem responses to numbers (0–10).
ATV1	Set modem responses to words (e.g., CONNECT 2400).
ATZ	Return modem to default settings.
AT&W	Store configuration settings to nonvolatile memory.
ATS0 = 0	Turn off auto-answer function.
ATS0 = n	Turn on auto-answer and set for n rings.
ATM0	Silence speaker.
ATM1	Turn speaker on while dialing; off during communication.
ATE0	Do not echo commands.
ATE1	Echo commands.
+++	Escape command: shifts modem from data to command mode after a preset time delay (set by register S12).
ATS12 = n	Set escape (+++) delay to n number of 20-millisecond units. For example, ATS12=50 sets the escape delay to 1 second.

In 1978, Sears and Radio Shack introduced household remote-control systems that didn't need separate control wiring. Under this system, called X-10, the control signals ride on the existing AC power wiring. Since an appliance needs power anyway, control signals are available everywhere there is an appliance.

The X-10 code is an oddball kind of serial data format. It is sent as bursts of 120-kHz tones timed to coincide with the zero crossings of the AC power line. It's not necessary to know all of the gory details of this code, only how it is used. The BS2 instruction Xout generates the X-10 code automatically.[5]

5. I'm not providing a lot of detail on the X-10 code format because there's not much practical use for the information. You can't use it from within PBASIC except via the Xout instruction. If you're curious, you can contact X-10 Powerhouse for a detailed tech note; call 201-784-9700 or fax 201-784-9464.

To control an appliance with X-10, you must plug it into an X-10 control module. There are basically two categories of modules: on/off appliance/light controllers and on/off/dim light controllers. Within these groups, you'll find units with different power ratings and other features.

You assign each module an address by setting a pair of dials for a house code (A through P) and a unit code (also called a key, 1-16). Normally, all of the appliances in a given house would be assigned the same house code, but households with more than 16 appliances under X-10 control might use two or more house codes. If a nearby neighbor is using X-10, you want to make sure that your house codes or sets of house codes are different, since X-10 signals sometimes travel house-to-house.

To control X-10 modules from a manual controller, you'd set the house code, press one of 16 keys to send the unit code, then press a button for the desired action, such as ON, OFF, DIM, or BRIGHT.

The BS2 instruction Xout mimics this operation. Just tell it what house, unit, and action codes you want to send, and it takes care of the rest. Of course, before Xout can do anything, the signals from the BS2 must be connected to the power line. The only way to do this is through a safe, optically isolated interface like the X-10 Powerhouse PL-513 or TW-523. Figure 13-2, part one of the schematic for C2TERM, shows the hookup.

The two X-10 signal hookups are called zPin and mPin. The zPin is an output from the X-10 interface that sends a pulse to the BS2 at the instant the AC power waveform crosses

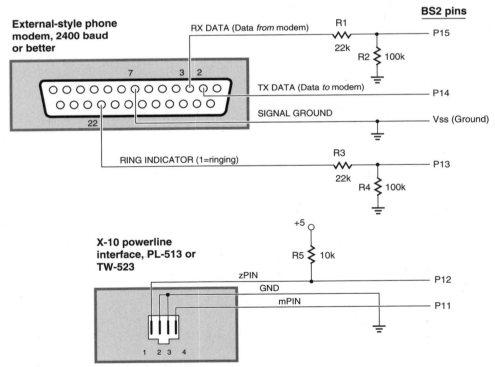

Figure 13-2 Connecting the modem and X-10 controller to a BS2.

zero volts. This cues the BS2 to take control of mPin, the pin that modulates (controls) the 120-kHz X-10 signal. Xout automatically generates the right signals with the right timing to broadcast your X-10 commands.

C2TERM

This is a fairly involved project, so we're going to take it a step at a time. We'll start with the X-10 interface.

INITIAL X-10 CHECKOUT

Connect the X-10 interface (PL-513 or TW-523) to your BS2 as shown in Figure 13-2. Do not connect the modem for now. You may plug the PL-513 or TW-523 into a wall outlet at any time; it's designed to provide a safe, isolated connection to the power line. But heed this warning:

You can be hurt or killed by AC line voltages! Do not open the case of the PL-513 or TW-523 for any reason. These units are designed to be safe, but only in their original, unmodified form.

Set an X-10 lamp or appliance control module to House A, Unit 1 and plug it into a wall outlet near the PL-513 or TW-523. Plug a lamp into the control module and turn it on so that when the module supplies power the lamp will light. Run the following short program:

```
zPin            con   12          ' zPin on P12.
mPin            con   11          ' mPin on P11.
houseA          con   0           ' 0=A, 1=B, 2=C...
Unit1           con   0           ' 0=Unit1, 1=Unit2...
xout mPin,zPin,[houseA\Unit1]     ' Talk to Unit 1.
xout mPin,zPin,[houseA\uniton]    ' Tell it to turn ON.
pause 1000                        ' Wait a second.
xout mPin,zPin,[houseA\unitoff]   ' Tell it to turn OFF.
stop                              ' End the program.
```

If all is well, when you run the program, the lamp will come on for 1 second, then turn off. If it doesn't, double-check your setup. In troubleshooting X-10, it can be very useful to have a manual X-10 control box on hand. If you can operate the modules manually, then any problem has to be with the BS2 hookup or programming. Less likely, but still possible, is a problem with the PL-513 or TW-523 powerline interface. But exhaust all other possibilities before letting yourself suspect this.

Once you have the setup working, try modifying the test routine above to address other house or unit numbers. Use additional control modules to check your modifications. Remember that once you have sent a given module's house and unit number, you have its attention. To send subsequent commands, you only need to precede them with the house number, as the test listing demonstrates. This is just like the manual X-10 controllers, so you may find it useful to rehearse X-10 communication by physically pressing the buttons.

MODEM CHECKOUT

Before you can use a modem with the BS2, you have to configure it properly and save those configurations in nonvolatile memory. This requires a temporary hookup to your PC.

You will need a modem cable and simple terminal communication software, such as the free Windows terminal accessory. Set the terminal program for 2400 bps, N81, no hand-shaking.

For the purposes of configuring the modem, you don't need to connect it to the phone line yet.

Once you have the modem connected and powered up and the terminal software configured, try typing AT <return>. The modem should respond with "OK" or "0" depending on whether it's currently set for text or numeric responses. If you don't get any sort of answer back from the modem, check your cabling and terminal software settings and try again. Some modems have lights that can help with troubleshooting; an "RD" light should flash briefly whenever you type a character in the terminal program.

Once the PC and modem are talking, you can type in the configuration commands shown in Table 13-3. In the table, <Enter> means press the enter key.

The last command, AT&W, causes the modem to commit this new configuration to non-volatile memory. Even with the power turned off, your modem will remember its new settings.

The next step is to download a test program to the BS2, connect the modem, and run a test. Unless you have two phone lines or a phone-line simulator (see parts list for source), you will need some help with this test. You may have to borrow a friend's computer and modem. If you have a really helpful (and computer literate) friend, you can have him or her dial into your BS2/modem test setup. Make sure that they set up for 2400 baud, N81. Here's the program that will enable your BS2 to answer the phone:

```
tLink        con    20000              ' Wait 20 seconds for linkup.
N2400        con    16780              ' Baudmode for 2400 bps inverted.
TxD          con    14                 ' Pin to output serial data.
RxD          con    15                 ' Pin to input serial data.
RI           var    IN13               ' Ring-indication output of modem.
name         var    byte(10)           ' String to hold user name.

waitForRing:                           ' When phone rings, RI goes high.
 if RI = 0 then waitForRing            ' Wait here while RI is low.

pickUpPhone:
 serout TxD,N2400,["ATA",cr]           ' Tell modem to pick up.
 pause tLink
 serout TxD,N2400,["Please enter your name: ",cr,lf]
 serin RxD,N2400,[str name\10\cr]             ' Get user name.
 serout TxD,N2400,["Thanks for calling, ", str name\10,cr,lf]
```

TABLE 13-3 CONFIGURING A MODEM FOR BS2 COMMUNICATION

TYPE THIS COMMAND	MODEM RESPONDS	PURPOSE
ATS0 = 0 <Enter>	"OK" or "0"	Disable auto-answer
ATS12 = 50 <Enter>	"OK" or "0"	Set "+++" response to 1s
ATV0 <Enter>	"0"	Set numeric responses
ATE0 <Enter>	"0"	Disable command echo
AT&W <Enter>	"0"	Memorize configuration

```
 pause 1000
 serout TxD,N2400,100,["Hanging up now.",cr,lf]

Disconnect:
 pause 2000
 serout TxD,N2400,["+++"]                 ' Switch to command mode.
 pause 2000
 serout TxD,N2400,["ATH0",cr]             ' Send hang-up command.
goto waitForRing                          ' Ready for another call.
```

This program will answer the phone, wait about 20 seconds for the modems to finish linking up, send a test message, get the user's name, send a customized response, and hang up. If the test message is not received completely, or the BS2 hangs up without the other computer receiving the test message, try increasing the value of the constant tLink. This will make the BS2 wait longer for the modems to finish linking up.

Modem-savvy readers may wonder why I used a time delay to wait for modem linkup when I could have employed the Serin instruction's WAIT option to look for the modem connect message. When modems establish a connection, they send the message "CONNECT" followed by the baud rate and other information to their host computer. When a modem is set for numeric responses, it sends a code number corresponding to the connection details.

In my experiments with BS2s and modems, I found that the BS2 often had trouble catching the connect message, which (with many modems) is accompanied by a bunch of random data. It's inelegant, but more reliable, to simply have the BS2 wait for the modems to do their thing before attempting to send any data.[6] (If it sends data too early, while the modem is still in command mode, the modem will hang up.)

DISPLAY AND BUTTON CHECKOUT

As a final check, we'll add the 4x20 serial display module and the four pushbutton switches that will serve as our user interface. The display works like a simplified, receive-only version of the PC terminal programs you have used to configure your modem. It understands many of the standard control characters, like carriage return, linefeed, tab and backspace, in addition to some specially suited to the LCD such as backlight control, fast cursor positioning, and automatic generation of four-line-tall numeric characters. Table 13-4 lists the control codes.

Connect the display and the buttons as shown in Figure 13-3. Set the LCD for 9600 baud, and Plus command set (configuration switches 1 up, 2 down; see the LCD instruction book for more details). Run the following short program:

```
clrLCD      con    12      ' Clear entire LCD screen.
posCmd      con    16      ' Position cursor.
colTen      con    74      ' Position 10 on the 4x20 screen.
lf          con    10      ' Linefeed control character.
bigNums     con    2       ' Begin big numbers.
N9600       con    $4054   ' Baudmode for inverted, 9600-bps output.
LCD         con    10      ' Serial LCD on P10.
i           var    nib     ' Temporary counter, 0-15.
state       var    bit     ' State of button pin.
```

6. Another explanation for unreliable results with Serin/WAIT during modem linkup stems from the fact that the BS2 cannot simultaneously send and receive serial data. Given that limitation, what happens if the BS2 initiates a Serin in the middle of an incoming byte? That's right; the first 0 in that byte will be mistaken for a start bit, and the byte will be garbled. Subsequent bytes may also be messed up, depending on the rate at which the data is being sent and the distribution of 0s that might be mistaken for start bits.

```
pause 1000                                  ' Wait for LCD startup.
serout LCD,N9600,[clrLCD]                   ' Clear the LCD screen.
for i = 1 to 4                              ' Print four labels: "Button #:"
 serout LCD,N9600,["Button ",dec i,":",cr]
next
  serout LCD,N9600,[posCmd, colTen]         ' Move to position 10.

checkBtns:                                  ' Check each button.
for i = 9 to 6                              ' For each pin, 9 to 6..
 state = 1                                  ' If pin=1 then state=1 else state=0.
 if ins & (DCD i) then isOne               ' —See text for explanation—
 state = 0
isOne:
  serout LCD,N9600,[dec state,lf,bksp]      ' Print state, goto next line
next                                        ' ..and backspace over old state.
goto checkBtns                              ' Do continuously.
```

ASCII VALUE	CONTROL CODE	FUNCTION
0	cntl-@	ignored before buffer; used for time delay
1	cntl-A	cursor to position 0 (home)
2	cntl-B	begin big-number display
3	cntl-C	ignored
4	cntl-D	blank cursor
5	cntl-E	underline cursor
6	cntl-F	blinking-block cursor
7	cntl-G	pulse buzzer output (ring bell)
8	cntl-H	backspace; back 1 space and erase character
9	cntl-I	tab (cursor to next multiple-of-4 position)
10	cntl-J	linefeed; cursor to line below
11	cntl-K	vertical tab; cursor to line above
12	cntl-L	formfeed; clear the screen
13	cntl-M	carriage return; cursor to start of line below
14	cntl-N	turn backlight on
15	cntl-O	turn backlight off
16	cntl-P	accept cursor-position data
17	cntl-Q	clear vertical column
18–31	—	ignored
32+	—	ASCII alphanumeric character set

TABLE 13-4 CONTROL CODES FOR 4X20 SERIAL LCD

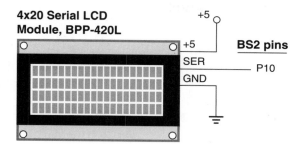

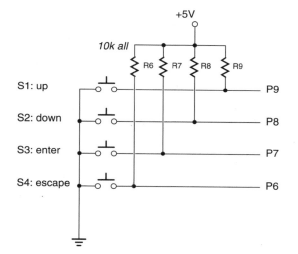

Figure 13-3 User interface consists of a 4x20 display and buttons.

When you run the program, the display will clear, then display the states of the four buttons in the format "Button 1: 1." When you press a button, the corresponding state shown on the display should change to 0. If it doesn't, or if the wrong state changes when you press a particular button, check and correct your wiring.

One part of the checkout program is worth extra attention. The program uses a For...Next loop to check the pins connected to the buttons. For each pin, we want to know whether it's 1 or 0 and print the appropriate state on the display. Unfortunately, there's no direct way to do this. You can write IF IN6 = 1 THEN... but you cannot use a variable to set the pin number to 6. No variable means no For...Next loop. Of course, it wouldn't kill you to use a separate group of instructions for each of the four pins, but that would bloat the program.

There is an indirect way to determine the state of individual bits of a variable using the Stamp's DCD (decode) function. DCD takes a number from 0 to 15 and returns a 16-bit value with a 1 in that position, and 0s in all other positions. Table 13-5 shows how this works.

As the table shows, we can get a number with a 1 in bit 6 by writing DCD 6. Next we can use that number to determine whether there's a 1 or a 0 in the same position of another number—such as the INS variable that holds the states of the BS2's pins. The logical AND operator (&) combines two values to give a result that contains a 1 in only those posi-

TABLE 13-5 THE DCD OPERATOR	
FUNCTION	**RESULT**
DCD 0	%0000000000000001
DCD 1	%0000000000000010
DCD 2	%0000000000000100
DCD 3	%0000000000001000
DCD 4	%0000000000010000
DCD 5	%0000000000100000
DCD 6	%0000000001000000
DCD 7	%0000000010000000
DCD 8	%0000000100000000
DCD 9	%0000001000000000
DCD 10	%0000010000000000
DCD 11	%0000100000000000
DCD 12	%0001000000000000
DCD 13	%0010000000000000
DCD 14	%0100000000000000
DCD 15	%1000000000000000

tions in which both input values contain 1. If you write INS & (DCD 6), the result will be %0000000000000000 if IN6 is 0 and %0000000001000000 if IN6 is 1. The states of the rest of the pins doesn't matter.

The final piece of the puzzle lies with the If...Then instruction. We normally use If...Then on comparisons, in the form IF x < > 0 THEN notZero. But PBASIC will also let you write IF x THEN notZero, which has the same effect.[7] If...Then regards 0 to mean false and any value other than 0 to mean true. That means that the line If ins & (DCD i) Then... means "If there's a 1 at the pin whose number is stored in variable i, then..."

This is a very valuable technique, and just a sample of the kinds of programming miracles that can be wrought with Boolean logic. For more on Boolean logic, see Appendix B.

COMPLETE C2TERM APPLICATION

Now that you have checked out each subsystem of C2TERM separately, it's time to load the complete program, shown in Figure 13-4, and give it a whirl. C2TERM operates in two modes, local and remote. Under local mode, the program lets you control X-10 devices by pressing the buttons to select options on the display. In remote mode, the program presents similar choices via modem to a remote computer.

7. Be careful when employing this form; the word-logic functions NOT, AND, OR, and XOR normally used with If...Then can produce unexpected results. See the BS2 manual on the CD-ROM for a complete explanation.

The program uses one house code, so it can control 16 X-10 devices. I have assigned example names to the devices which you are welcome to change. One of the program's strong points is its flexible storage and processing of strings, sequences of bytes that make up a text message or label. The BS2 lets us stash strings in unused portions of the program memory and assign names (constants) to their starting addresses. The program takes care of the rest.

Efficient handling of strings is vitally important to this program. In order to be user-friendly, the program is very chatty—substituting descriptive names for X-10 unit codes.

Embedding lots of text in a PBASIC program is a fast way to run out of program space. For one thing, each character of a text string takes a full byte of storage space. For another, many programmers embed each string in a separate Serout instruction. Think about that

```
'PROGRAM: X10CTL.BS2 (X-10 local and remote control)
'This program interfaces a BS2 to a modem, X-10 powerline
'device, serial LCD module, and switches to provide user-
'friendly remote and local control of 16 X-10 appliances.
'For local control, a user can view the name and ON/OFF
'status of the X-10 device on the LCD screen. By pressing
'UP/DOWN/ON/OFF buttons, the user can pick an appliance and
'command it on or off.
'For remote control, the program monitors the ring-indicator
'output of a modem. When the phone rings, the BS2 answers
'it and requests a password. If the password matches, it
'allows the logged-on user to view and change the states of
'the X-10 appliances.


'==============================================================
'               CONSTANTS FOR SERIAL LCD MODULE
'==============================================================
'These constants define characters that help format the 4x20
'serial LCD screen. Some formatting characters like CR
'(carriage return) aren't on this list, because they are already
'defined by the BS2 for Debug and Serout.
clrLCD   con     12      ' Clear entire LCD screen.
posCmd   con     16      ' Position cursor.
colTen   con     74      ' Position 10 on the 4x20 screen.
lf       con     10      ' Linefeed control character.
N9600    con     $4054   ' Baudmode for inverted, 9600-bps output.
LCD      con     10      ' Serial output pin for LCD (P10).
arrow    con     ">"     ' Pointer to highlight selected item.
statCol  con     17      ' Column in which to show ON/OFF status.
pntrCol  con     statCol-1    ' Column in which to show arrow.
'==============================================================
'            CONSTANTS FOR MODEM COMMUNICATION
'==============================================================
tLink    con     20000   ' # of milliseconds to wait for link up.
N2400    con     $418D   ' Baudmode for 2400 bps inverted.
TxD      con     14      ' Pin to output serial data to modem.
RxD      con     15      ' Pin to input serial data from modem.
FF       con     12      ' Form-feed code—clears terminal screen.
'==============================================================
'               X10 CONSTANTS
'==============================================================
myHouse  con     0       ' House code—0=A, 1=B, 2=C...
zPin     con     12      ' zPin on P12.
mPin     con     11      ' mPin on P11.
```

Figure 13-4 Program listing for C2Term.

```
'================================================================
'                    X10 DEVICE NAMES
'================================================================
'The Data directives below define the names of the X-10 devices.
'You can change these names—just make sure they are 16 characters
'or less, and end in the ASCII null character (0). Subroutines
'that use these strings start at the address constant (d0, d1, etc.)
'and continue reading data from EEPROM until they find a null.
'This is a common and efficient way to store and retrieve
'text strings of varying length.
' ADDRESS                    STRING DATA
'CONSTANT                    (CHARACTERS/CONTROLS)              NULL
'--------------------------------|--------------------
    d0          DATA      "Address",                         0
    d1          DATA      "Path",                            0
    d2          DATA      "Porch",                           0
    d3          DATA      "Pool/Spa",                        0
    d4          DATA      "Alleyway",                        0
    d5          DATA      "Garden",                          0
    d6          DATA      "Garage",                          0
    d7          DATA      "Kitchen",                         0
    d8          DATA      "Living Rm",                       0
    d9          DATA      "Dining Rm",                       0
    d10         DATA      "Master Bdrm",                     0
    d11         DATA      "Pool Pump",                       0
    d12         DATA      "Spa Heat",                        0
    d13         DATA      "Workshop Pwr",                    0
    d14         DATA      "Attic Fan",                       0
    d15         DATA      "Holiday lights",                  0
'================================================================
'                    OTHER TEXT STRINGS
'================================================================
'Note that longer strings can be broken into two or more
'lines. The only thing that matters is the end-of-string
'marker, ASCII null.
    ON          DATA      "ON ",                             0
    OFF         DATA      "OFF",                             0
    _ON         DATA      "  ON",cr,lf,                      0
    _OFF        DATA      "  OFF",cr,lf,                     0
    answPhone   DATA      "ATA",cr,                          0
    Logon  DATA      "X-10 WORLDWIDE CONTROL",cr,lf
    Prompt DATA      "Please enter your "
    Prompt2     DATA      "password: ",                      0
    Standby     DATA      clrLCD,"***Remote Access***"
    SB2         DATA      cr,cr, "  Please stand by",        0
    hangUpNow   DATA      cr,lf,"Hanging up now.",cr,lf,     0
    pwdOK  DATA      cr,lf,"Logged on.",cr,lf,       0
    offerChoices    DATA      cr,lf,"Enter a device # (1-16)"
    choice2     DATA      cr,lf,"to toggle its state",cr,lf
    choice3     DATA      "or 0 to log off",cr,lf,           0
    logOffNow   DATA      "Log off now (Y/N)? ",            0
    confirm     DATA      cr,lf,"Confirm (Y/N), device: ",        0
'================================================================
'                    STORAGE VARIABLES
'================================================================
'Some variables are used by both the LCD and modem routines.
'This works because the program does not attempt to service
'the LCD/buttons and modem at the same time.
    strAddr var     word      ' Address of string in EEPROM.
```

Figure 13-4 (*Continued*)

```
baud     var    word    ' Baud rate for stringOut.
serPin   var    nib     ' Serial output pin for stringOut
LCD_mdm var    bit     ' String to LCD (0) or modem (1).
char     var    byte    ' Character to send to LCD or modem.
reply    var    byte    ' User's reply to modem prompt.
item     var    nib     ' Selection from list of strings.
stats    var    bit(16) ' Status (ON/OFF) of the X10 devices.
slectn   var    nib     ' Currently selected item.
tempN1   var    nib     ' Nibble-sized temporary counter
tempN2   var    nib     '  "       "      "       "
'=============================================================
'                    I/O VARIABLES
'=============================================================
'Most I/O is done by instructions that use pin-number
'constants; see the constants for the LCD and modem. These
'variables are used in IF/THEN instructions for simple input.
RI       var    in13    ' Ring-indication output of modem.
upSw     var    in8     ' Button to move selection arrow up.
dnSw     var    in9     ' Button to move selection arrow down.
onSw     var    in7     ' Button to turn X10 device on.
offSw    var    in6     ' Button to turn X10 device off.


'=================================================================
'                     MAIN PROGRAM
'=================================================================
Initialization:
pause 2000              ' Wait for LCD startup.
gosub newScreen         ' Display first screen.
gosub showPntr          ' Show the selection pointer.
'Main program loop: The program continuously checks the states of
'the switches and the ring-indication input from the modem.
'When any of these become active, it jumps to appropriate routines
'(answer the modem, move the selection pointer, send an X-10
'command, etc.).
Main:
 if RI=1 then getModem
 if onSw = 0 then turnOn     ' Turn on an X-10 device.
 if offSw = 0 then turnOff   ' Turn off " " device.
 if (upSw & dnSw) = 1 then main' If neither up or down pushed, try again.
 if upSw = 1 then tryDown     ' If upSw isn't pushed, check dnSw.
  gosub hidePntr             ' Up switch: prepare to move pointer.
  tempN2 = tempN2+1          ' Increment temporary selection.
  if tempN2 & %11 <> 0 then posPntr    ' If user pressed "up" and new
  gosub newScreen            ' selection ends in %00, switch screens.
  goto posPntr               ' Reposition the pointer.
tryDown:                     ' Check the down switch.
 if dnSw = 1 then main       ' Not pressed? Back to main.
 gosub hidePntr              ' Pressed: prepare to move pointer.
 tempN2 = tempN2-1           ' Decrement temporary selection.
 if tempN2 & %11 <> %11 then posPntr   ' If user pressed "down" and new
 tempN2 = tempN2 & %1100     ' selection ends in %11, switch screens.
 gosub newScreen             ' Show new screen.
posPntr:                     ' Reposition the pointer.
 slectn = tempN2
 gosub showPntr
hold:
 pause 50                              ' Brief (50-ms) delay, then make
 if (upSw & dnSw) = 0 then hold        ' sure that switch is released
goto main                              ' before returning to main loop.
```

Figure 13-4 *(Continued)*

```
'===============================================================
'                        MODEM CONTROL
'===============================================================
getModem:
 strAddr = Standby              ' Put standby message on LCD.
 gosub stringOut
 LCD_mdm = 1                     ' Direct strings to modem.
 strAddr=answPhone              ' Tell modem to pick up.
 gosub stringOut
 pause tLink
 strAddr=Logon                  ' Send logon message.
 gosub stringOut
 serin RxD,N2400,5000,Disconnect,[WAIT ("X10_OK")]     ' Get PASSWORD.
 strAddr=pwdOK                  ' Confirm password OK.
 gosub stringOut
mdmStats:
 serout TxD,N2400,[FF]  ' Clear user's terminal screen.
 for item = 0 to 15            ' Display status of 16 devices.
   serout TxD,N2400,[DEC2 (item+1),". "]   ' Number device list 1-16.
   gosub pickStr              ' Get address of the device name.
   gosub stringOut            ' Print the name.
   lookup stats(item),[_OFF,_ON],strAddr ' Now print ON/OFF.
   gosub stringOut
 next                          ' Continue for all 16 devices.
choices:
 strAddr=offerChoices          ' Let the user pick an action:
 gosub stringOut               ' 1-16=toggle device; 0=log off.
 serin RxD,N2400,10000,Disconnect,[DEC reply]     ' Get reply.
 if reply = 0 then Done ' 0 is "quit"
 slectn = reply-1              ' Selection is 0-15; reply is 1-16.
 strAddr=confirm               ' Verify choice.
 gosub stringOut
 serout TxD,N2400,[DEC2 reply,"->"]     ' Show device number.
 lookup stats(slectn),[ON,OFF],strAddr  ' and new state.
 gosub stringOut
 serin RxD,N2400,10000,Disconnect,[reply]  ' Get reply.
 if reply = "Y" or reply = "y" then switchIt
 goto mdmStats
switchIt:
 if stats(slectn) = 1 then turnOFF
 goto turnON
Done:
 strAddr=logOffNow                      ' Confirm logoff.
 gosub stringOut
 serin RxD,N2400,10000,Disconnect,[reply]       ' Get reply.
 if reply = "Y" or reply="y" then Disconnect    ' If (Y)es, hang up.
 goto mdmStats                ' Else redisplay choices.
Disconnect:
 strAddr=hangUpNow            ' Send hang-up message.
 gosub stringOut
 pause 2000
 serout TxD,N2400,["+++"]     ' Put modem into command mode.
 pause 2000
 serout TxD,N2400,["ATH0",cr]  ' Tell it to hang up.
 LCD_mdm = 0                  ' Reroute serial to display.
goto Initialization          ' Re-initialize the display.
'===============================================================
'                         SUBROUTINES
'===============================================================
'==stringOut:  Output EEPROM strings to LCD or modem.
'The bit variable LCD_mdm picks the output device; 0=LCD
```

Figure 13-4 *(Continued)*

```
'1= modem. The address in variable strAddr is the starting point
'of the string in EEPROM. The routine outputs EEPROM bytes until
'it reaches the end-of string character (null).
stringOut:
  baud = N9600: serPin = LCD      ' Output to LCD if LCDmdm=0
  if LCD_mdm = 0 then getByte
  baud = N2400: serPin = TxD      ' Output to modem if LCDmdm=1
getByte:
  read strAddr,char               ' Get the character.
  if char <> 0 then continue      ' If char is 0, then return
return
continue:                         ' ..else continue
  serout serPin,baud,[char]       ' ..and send char to the LCD.
  strAddr = strAddr+1             ' Point to next character in string.
goto getByte                      ' Repeat until char = 0.
'================================================================
'==pickStr: Get starting address of names for X10 devices 0-15
'and place in variable strAddr.
pickStr:
lookup item,[d0,d1,d2,d3,d4,d5,d6,d7,d8,d9,d10,d11,d12,d13,d14,d15],strAddr
return
'================================================================
'==showStat: Display the status (ON or OFF) of the currently
'selected X10 device. That cluttered Serout instruction tells the
'LCD to expect a position value (posCmd), then calculates what the
'current position should be. It takes the last two bits of the
'selection (slectn & %11) to get the LCD line number. Multiplying
'that by 20 gets the beginning of one of the 4 lines, 0-3. Adding
'the constant statCol (17) gets the exact position—character 17
'of the selected line. Finally, adding 64 gives the single-byte
'value that the serial LCD expects. Once the cursor is in position,
'the routine looks up the address for the text that says "ON" or
'"OFF" depending on the state of the selected X10 device, and goes
'to stringOut to print that text on the LCD. Since stringOut is
'the last instruction in the routine, a Goto is used instead of
'a Gosub. This saves an unnecessary Return instruction.
showStat:
  serout LCD,N9600,[posCmd,(((slectn & %11)*20)+statCol)+64]
  lookup stats(slectn),[OFF,ON],strAddr
  goto stringOut
'================================================================
'==show/hidePntr: Show or hide the arrow that points to the currently
'selected X10 device status. Basically the same kind of cursor-
'positioning job as showStat above; see those comments for explanation.
showPntr:
  serout LCD,N9600,[posCmd,(((slectn&%11)*20)+pntrCol)+64,arrow]
return
'==
hidePntr:
  serout LCD,N9600,[posCmd,(((slectn&%11)*20)+pntrCol)+64," "]
return
'================================================================
'==newScreen: Display a screenful of X10 status information on LCD.
'This routine clears the LCD and then writes the names and states
'of four X10 devices to it. Which four devices is determined by bits
'2 and 3 of the variable tempN2, a nibble variable that keeps count
'of the arrow position on the screen.
newScreen:
  serout LCD,N9600,[clrLCD]         ' Clear the LCD screen.
    for tempN1 = 0 to %11 ' Display four devices/states.
      slectn = tempN2 | tempN1      ' Combine bits 2,3 of tempN2 with
```

Figure 13-4 *(Continued)*

```
      item = slectn                    ' ..bits 0,1 of tempN1.
      gosub pickStr                    ' Get the string.
      gosub stringOut                  ' Display it.
      gosub showStat                   ' Show device status
  next                                 ' ..for four devices
return                                 ' ..and return.
'===========================================================
'==turnON/OFF: Turn the selected X10 device on or off in response
'to the buttons. After the X10 code is sent, this subroutine
'makes sure that the user has released the on or off button
'in order to avoid sending redundant codes.
turnON:
 stats(slectn) = 1
 xout mPin,zPin,[myHouse\slectn]       ' Talk to unit
 xout mPin,zPin,[myHouse\uniton]       ' Tell it to turn ON.
 if LCD_mdm = 1 then mdmStats          ' If modem selected, skip LCD.
 gosub showStat                        ' Update the LCD.
hold1:
 if onSw=0 then hold1                  ' Wait til button up.
goto main                              ' Back to main loop.

turnOFF:
 stats(slectn) = 0
 xout mPin,zPin,[myHouse\slectn]       ' Talk to unit
 xout mPin,zPin,[myHouse\unitoff]      ' Tell it to turn OFF.
 if LCD_mdm = 1 then mdmStats          ' If modem selected, skip LCD.
 gosub showStat                        ' Update the LCD.
hold2:
 if offSw=0 then hold2                 ' Wait til button up.
goto main                              ' Back to main loop.
```

Figure 13-4 (*Continued*)

for a minute: each Serout instruction has to be stored as a tag that identifies the instruction as "Serout;" an I/O pin number from 0 to 16 (I/O pins 0 to 15, plus 16, representing the programming port); and a baudmode representing the serial data rate and format.

Making some educated guesses, let's assume that the name Serout is represented by 7 bits, the pin number by 5 bits and the baudmode by 16 bits—that's 28 bits (nearly 4 bytes) of program memory used—before Serout has said anything! This program has 22 occasions to send data serially, so it might have had more than 80 bytes of overhead associated with using Serout.

Instead, the program stores text as strings and uses a routine called stringOut to reduce 22 Serouts to just 10. Now there's also overhead involved with Gosub and Return, but chances are good that the technique saved perhaps 50 bytes of program memory.

The program also manages to be frugal with program memory in another way. It uses a bit variable called LCD_mdm to switch stringOut and other subroutines between the LCD and the modem. When LCD_mdm is 0, the subs talk to the LCD; when it's 1, they talk to the modem. This saves us from writing near-identical, but separate, code for the LCD and modem.

Going Further

Why the mania about conserving code space in this program? Well, I just know that after the initial novelty wears off, you're going to want to add more features to this program, and I wanted to leave you plenty of room. The same philosophy applies to the hardware design; I could have interfaced the LCD directly (as in the RS-485 terminal project), but that would have deprived you of both code space and I/O pins.

An exciting use for those extra I/Os might be to add some monitoring capability to the program. Add a temperature sensor here, an intrusion detector there, and pretty soon you have a state-of-the-art home-control system—all based on a computer that fits in less than one square inch.

PARTS LIST

Resistors (all 1/4W, 10% or better)
R1, R3—22,000 ohms
R2, R4—100,000 ohms
R5–R9—10,000 ohms
Other Components
S1–S4—any normally-open pushbutton switches
External-type phone modem, AT command set, 2400 baud or better
X–10 powerline interface, model PL-513 or TW-523 (Parallax or home-automation suppliers)
4x20 Serial LCD—model BPP-420L (Parallax, Jameco, JDR Microdevices, or Scott Edwards Electronics, Inc.)
Phone-line simulator kit—Party Line by Digital Products Company.

SHORT-RANGE SONAR

BS2

Although "sonar" conjures up images of submarine warfare (or at least fancy fishing-boat accessories), our sonic detection and ranging project works on dry land. It grafts the BS2's digital intelligence to some analog electronics to precisely measure short distances.

This BS2 sonar can be put to work in a small robot's collision-avoidance system, or as a close-up presence detector for displays, alarms, toys, etc. A bonus version of the project creates a sonar-based musical instrument—wave your hand in front of the sonar detectors to control eerie science-fiction sounds.

The parts list is fairly long, but the overall cost of the project is quite modest—about $10 (excluding the BS2). The project requires no adjustments or calibration.

Teachers and students, take note: this sonar project is an ideal opportunity to tie together elements of physics, electronics, programming, and math.

Sonar Principles

If you've ever heard an echo, you already have a pretty good idea of how sonar works. Sound travels relatively slowly, about 1130 feet per second. When it strikes a hard surface, like a rocky cliff face, much of it is reflected. When conditions are just right, you can holler out "HELLO" and hear your own voice reflected back to you across a canyon or city street.

Of course, your voice and other sounds are always being reflected by hard surfaces around you, but you don't notice it because the delays are so short. For example, if you're 5 feet from a wall, it takes only 5/1130 = 4.4 milliseconds (thousandths of a second, abbreviated ms) for your voice to reach the wall, and 4.4 ms for it to return.

An 8.8-ms round-trip time is far too short for you to consciously notice an echo. But to a computer like the BS2, 8.8 ms is an eternity. Consider the BS2's pulse-timing instructions, Pulsin, Pulsout, and RCtime. Each of these works with units of 2 microseconds (millionths of a second, µs). With that kind of timing, the BS2 can split 8.8 ms into 4400 parts!

Let's look at it another way. Sonar determines the distance to an object by measuring the time taken for a sound to travel to the object and back. The total distance the sound travels is twice the distance to the object. Sound travels 1130 feet per second, which we can look at as 565 round-trip feet per second, or 6780 round-trip inches per second. The round-trip time for a sonar echo is therefore 1/6780 = 147 µs per round-trip inch. This is approximate, since the speed of sound varies with temperature, but it's a good starting point.

The significance of that value, 147µs, is that the BS2's fast timing functions with their 2µs units could provide sonar ranging resolution down to 1/73rd of an inch—about equal to the thickness of three pages of this book.

FREQUENCY AND WAVELENGTH

So far, we've talked about sound in terms of human hearing. Sounds audible to humans with keen hearing range from 20 to 20,000 cycles per second [expressed as units called hertz (Hz)]. Sound travels through the air as rapid variations in pressure (vibrations), which we picture as something like ripples on the surface of a pond.

Sound waves are moving through the air and vibrating at the same time. To understand this on a gut level, help me out with a simple demonstration. Get a scrap of paper and a pencil or pen. Scribble up and down in one place on the paper—you end up drawing a heavy line. Now, continue scribbling, but move your hand across the paper at the same time. You end up drawing a profile of those pond ripples, like Figure 14-1. If you move your hand at the same rate, but scribble faster, the peaks and valleys of the ripples are closer together. Scribble slower, and the ripples are farther apart.

You can turn the experiment around by scribbling at the same rate but moving your hand slower or faster. What you are demonstrating is the relationship between a wave's frequency (how fast you scribble), the speed at which the wave moves (how fast you move your hand across the paper), and the wavelength (the width of the ripples you drew).

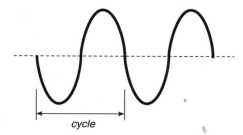

cycle

A sound consists of a series of peaks and valleys representing compression and expansion of the air through which it travels. The ripple pattern repeats, and the distance between two identical points on the ripples is called a cycle. Frequency is the number of cycles that occur in a second. Period is the number of seconds per cycle. Wavelength is the distance traveled by the sound during the time it takes to complete one cycle.

Figure 14-1 Frequency and wavelength.

14

SHORT-RANGE SONAR

Since sound travels at a more-or-less constant speed, its wavelength is determined almost entirely by frequency. The wavelength of sound can be calculated by dividing speed by frequency: 1130/frequency = wavelength in feet. For example, a 400-Hz sound like a telephone dial tone has a wavelength of 1130/400 = 2 feet, 10 inches. A 2000-Hz tone, like the beeps made by electronic buzzers in watches, has a wavelength of about 7 inches. A 40,000-Hz tone, like the ping generated by our sonar project, has a wavelength of 0.34 inches.

Wavelength is important in sonar applications for a couple of reasons. Long wavelengths tend to spread out and bend around obstacles instead of being reflected cleanly. Think about hearing music playing in the next room—all you can hear are the low (bass) frequencies. The high frequencies have been reflected and absorbed, but the lows have spread and bent and traveled on.

Since sonar involves bouncing a burst of sound off a target, high frequencies are better because they don't spread out as much as low frequencies do. Shooting a high-frequency sonar pulse at an object is like throwing a small rubber ball at a wall. Sending out a low frequency pulse would be like throwing a mattress!

The other desirable characteristic of high-frequency sonar has to do with detecting the echoes. The receiving electronics may require several cycles to recognize a sound. For the sake of argument, let's say it's 10 cycles. So 10 wavelengths of a given echo would have to reach the receiver for it to recognize the echo. At 40,000 Hz, 10 wavelengths is 3.4 inches. At 2000 Hz, it's almost 6 feet (70 inches). If the object to be detected were closer than 3 feet away, a 2000-Hz sonar wouldn't be finished sending a 10-cycle pulse before the beginning of the same pulse was already starting to return. No good.

Now that we've discussed some of the theory behind sonar, let's take a look at a practical implementation.

The Sonar Circuitry

The circuit for our sonar isn't terribly complicated, but it has a lot of parts. That sounds like a contradiction, but it's true. The circuit consists of nothing more than a transmitter for sending short bursts of 40,000-Hz (40-kHz) sound, and a receiver for converting incoming

40-kHz sound into digital signals. These functions are implemented with a pair of analog-electronic building blocks whose behavior is determined by voltage levels and resistance ratios, so there are quite a few resistors in the circuit. Fortunately, resistors are cheaper by the pound than peanuts, so it's still a very reasonable design.

Let's look first at the sonar receiver, Figure 14-2. A logical starting point is RCVR, the 40-kHz receiver. RCVR is a microphone that's tuned to pick up 40-kHz signals. It consists of a crystal attached to a foil diaphragm. When sound strikes the diaphragm, the crystal vibrates. The crystal's piezoelectric properties cause it to generate a voltage when it is mechanically stressed. The piezo crystal converts sound into a faint electrical signal.

The signal goes into the first stage of an LM358 dual operational amplifier (op amp) IC. This IC has two complete op amps in a single 8-pin package. As the name suggests, an op amp's primary function is to amplify—increase the strength of—an electronic signal. A ratio of two resistor values sets the degree of amplification, the gain.

In this case, it is the ratio of R1 (10k) to R4 (100k) that sets the gain of U1a: 1:10 or 10x amplification. That means that if the signal from RCVR is 0.01V (10 millivolts, abbreviated mV), it will leave U1a 10 times as strong—100mV. Since we're talking about an alternating current (ac) signal, which is a series of ripples like Figure 14-1, these signal voltages are expressed as volts peak-to-peak (Vp-p). That's the vertical measurement in volts from the top of a peak to the bottom of a valley. It's still called peak-to-peak, though.

Just as small ripples can ride on top of a deep pond, so the ac signal ripples ride on top of a direct-current (dc) voltage. Resistors R2 and R3 form a voltage divider that sets this voltage to approximately 1.6V. If this voltage were not set above ground (0V), the op amp would be required to output a negative voltage in order to reproduce the valleys of the ripples. Since the op amp has no negative voltage supplied to it, it would cut off the bottom portion of the signal. Adding a dc offset voltage prevents this cutoff.

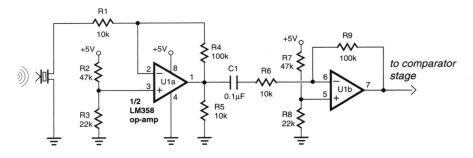

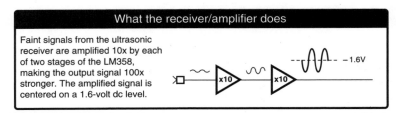

Figure 14-2 Sonar receiver/amplifier stage.

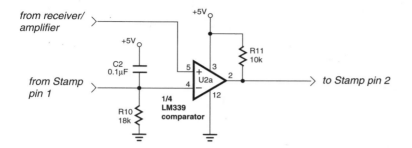

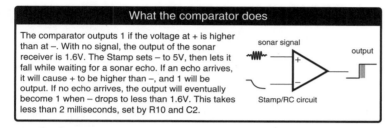

Figure 14-3 Sonar comparator stage.

U1b is essentially the same circuit—another 10x amplifier. However, before feeding the signal from U1a into it, we have to get rid of the 1.6V dc offset. Otherwise, U1b would be obliged to amplify it 10x, which would call for a 16V output. The power supply is only 5V, so the maximum that U1b can output is 5V—another potential cutoff problem. Capacitor C1 fixes this by allowing the small ac voltage to pass, leaving the dc offset behind. So U1b amplifies only the ac signal. U1b adds its own 1.6V dc offset, so now we have 100x the signal riding on top of 1.6 volts. So, assuming that the original signal from RCVR was 10mVp-p, it's now 1Vp-p.

You might be wondering why the circuit doesn't use one op amp set to a gain of 100x. After all, it's just a ratio of two resistors. There are several reasons, but here are the two most important ones:

■ Inexpensive op amps have limited high-frequency response. The LM358 has a maximum gain of 25x at 40kHz.

■ When amplifiers are set for high gain, delayed feedback from their outputs that gets into their inputs can cause them to oscillate. The squealing that public-address systems produce when sound from the speakers reaches the microphone is an example of oscillation. When an amp oscillates, the signal that it's supposed to amplify gets lost. Oscillation is less likely with two low-gain amplifiers than one high-gain amp.

The output of the second amplifier passes to a comparator stage, Figure 14-3. A comparator is a simple-minded version of an op amp. Its job is to compare two input voltages. If the voltage at + is higher than the voltage at −, then the output is high. Otherwise, the output is low.

The amplified sonar signal, plus the 1.6V dc offset, go into the + terminal of the comparator. The − terminal is connected to BS2 pin 1 (P1) and to C2 and R10.

What's happening here is pretty interesting. When the BS2 wants to take a sonar measurement, it has the pinger send out a pulse. It puts a high on P1 to set the bottom leg of C2 to +5V, then disconnects P1 by making it an input. While the Stamp waits for the echo, the voltage at U2a's – terminal is falling in accordance with the RC time constant of C2 and R10 (the little curve shown in Figure 14-3 and explained in more detail in Appendix C).

The amplified sonar signal is compared to the falling voltage at U2a–. The longer the circuit waits for an echo, the fainter that echo will be. But since the voltage that the echo will be compared to is also decreasing, the circuit has a better chance of detecting the echo. Finally, when the voltage at U2a– drops below 1.6V, the dc level of the signal at U2a+, the circuit outputs a 1. This ends the BS2's timing cycle whether or not an echo has been received.

The last stage we'll look at is the pinger, Figure 14-4. U2b is a crystal-controlled oscillator. Since we want a 40-kHz tone, we use a 40-kHz crystal, and that's exactly the frequency that emerges from U2b pin 1. The various components surrounding U2b set up the ideal conditions for oscillation and were taken from the manufacturer's data book.

The output from U2b is a 5Vp-p, 40-kHz signal. Resistor R16 and R17 form a voltage divider that cuts that signal in half, to 2.5Vp-p. That allows us to use another comparator section to turn the signal on and off. When BS2 pin P0 outputs 1, the U2c– sees 5V. U2c+ is getting a 2.5Vp-p signal, so it never exceeds U2c–, and the output stays 0.

When BS2 P0 outputs 0, the voltage at U2c– is approximately 0.5V. Now the 2.5Vp-p signal from the oscillator is allowed to pass through to the output, which drives XMTR, the 40-kHz transmitter. XMTR is similar to RCVR in that it contains a piezo crystal connected

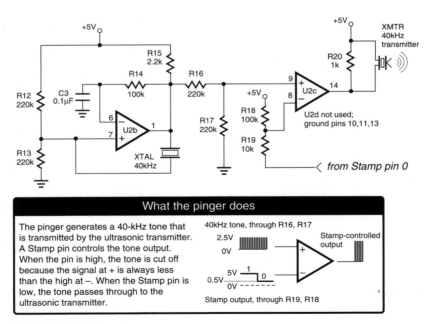

What the pinger does

The pinger generates a 40-kHz tone that is transmitted by the ultrasonic transmitter. A Stamp pin controls the tone output. When the pin is high, the tone is cut off because the signal at + is always less than the high at –. When the Stamp pin is low, the tone passes through to the ultrasonic transmitter.

Figure 14-4 Sonar pinger stage.

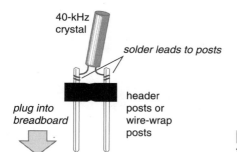

Figure 14-5 Mounting the 40-kHz crystal for breadboarding.

to a diaphragm. XMTR uses the flip-side of the piezo effect—voltage across the crystal causes it to flex. This makes the diaphragm vibrate and sends a 40-kHz tone into the air.

Construction Notes

You can build the sonar circuitry using any method you prefer—waffle board, solderable prototyping board, point-to-point wiring, etc. I set up the prototype on waffle board so that I could experiment with the op-amp gains and RC time-constant input to the comparator. There's nothing difficult or critical about the circuit, but it has lots of parts and therefore lots of opportunities to connect parts incorrectly. Double- and triple-check your work to avoid frustration.

If you use solderless waffle board, you will have to attach heavier wires to the tiny 40-kHz crystal. I snapped off a two-pin section of a header-post strip and soldered the tiny leads to it. See Figure 14-5. This gave me a sturdy set of posts to plug into the breadboard. Don't try to plug the crystal directly into the breadboard without some kind of mounting arrangement; the small diameter of the wire leads will prevent them from connecting securely with the contact strip.

Make sure to mount the 40-kHz transmitter and receiver about 1 inch apart and facing the same way. For testing, place the sonar on an uncluttered table top and use a large, flat object as a target. A book standing on end works fine.

The Sonar Program

With custom-designed sonar circuitry at its disposal, the BS2 doesn't need much of a program in order to measure short distances. Figure 14-6 is the program listing. The heart of the program lies in these instructions:

```
high ping              ' Start with 40kHz xmitter off.
high compRC            ' Raise C2 to +5 volts.
 pause 1               ' Allow time for C2 to reach +5V.
input compRC           ' Disconnect pin from C2.
pulsout ping,pingLen   ' Send a short 40kHz pulse.
rctime rcvr,0,echTime  ' Wait for echo; save time to echTime.
```

Now you can see how the circuit serves the needs of the program. The ping pin (P0) turns the pinger on and off, the compRC pin (P1) sets the voltage at C2 and R10, and the

```
'Program: SONAR.BS2 (short-range sonar using BS2)
'This program, in cooperation with the 40kHz send/receive circuitry
'shown in the text, creates an inexpensive short-range (10") sonar
'system suitable for collision avoidance in small robots. The program
'instructs the circuit to emit a short burst of 40kHz sound (ping)
'and waits to hear a return echo. The BS2 RCtime instruction times
'the ping-to-echo interval to the nearest 2us. Raw sonar data
'is quite noisy, so the sonar subroutine actually takes several
'samples (5 or more; set by the nSmp constant below). It scales
'the samples to byte size (0-255 units of 4us each), stores
'them in an array, and sorts them from high to low. The program
'then averages the five middle samples. The result is a reading
'that's quite stable (assuming a fixed distance from the sonar
'to the object). Best accuracy is in the range of 2 to 7 inches
'(about 58 to 225 units).
'========CONSTANTS
nSmp      con    10               ' Number of samples (NOT LESS THAN 5).
maxSmp    con    nSmp-1   ' Array index # of last sample.
maxSrt    con    nSmp-2   ' Maximum index # to sort.
s1        con    nSmp/2-2         ' 1st sorted sample to include in average.
s2        con    nSmp/2-1         ' 2nd "       "       "  "      "   " "
s3        con    nSmp/2           ' 3rd "       "       "  "      "   " "
s4        con    nSmp/2+1         ' 4th "       "       "  "      "   " "
s5        con    nSmp/2+2         ' 5th "       "       "  "      "   " "
ping      con    0                ' Output to activate pinger.
pingLen   con    200              ' Duration of ping in 2-us units.
compRC    con    1                ' Output to set comparator RC circuit.
rcvr      con    2                ' Input from 40kHz receiver/comparator.
'========VARIABLES
echTime   var    word             ' Time to echo return.
smp       var    byte(nSmp)       ' Storage for multiple readings.
index     var    byte             ' Counter for sampling.
swapTmp   var    byte             ' Temporary storage for swapping.
swap      var    bit              ' Flag to indicate whether sort is done.
'========PROGRAM
'The "again" loop takes sonar ranges continuously and displays
'them on the PC's debug screen.
high ping                         ' Turn pinger off initially.
again:                            ' Loop.
 gosub sonar                      ' Take the sonar reading
 debug "Echo time (0-255 units): ", dec echTime,cr     ' Display it.
goto again                        ' Repeat endlessly.
'========SONAR SUBROUTINE
'It takes only five instructions to get a quick sonar snapshot of the
'distance to the closest sonar-reflective object. However, you can get
'better, more consistent results by taking several sonar readings,
'discarding the highest and lowest ones and averaging the middle.
'This routine takes the number of samples specified by the constant
'nSmp, sorts them, and averages the middle. Each reading takes
'only a few milliseconds (owing to the sonar's short range).
sonar:
for index = 0 to maxSmp ' Take nSmp samples.
 high compRC                      ' Raise C2 to +5 volts.
 pause 1                          ' Allow time for C2 to reach +5V.
 input compRC                     ' Disconnect pin from C2.
 pulsout ping,pingLen             ' Send a short 40kHz pulse.
 rctime rcvr,0,echTime            ' Wait for echo; save time to echTime.
 smp(index) = echTime/2 max 255   ' Save to array smp() as byte (0-255).
next                              ' Get another sample.
```

Figure 14-6 Program listing for short-range sonar.

```
'At this point, there are nSmp sonar samples stored in the bytes of
'the smp() array. One way to discard the lowest and highest samples
'is to sort the array so that the lowest index values contain the
'largest numbers. The code starting with "sort" does this using
'a technique called "bubble sort." The idea is simple—compare
'adjacent bytes in the array, for instance smp(0) and smp(1).
'If the value stored in smp(0) is greater than or equal to that in
'smp(1), do nothing. Otherwise, swap the values so that smp(0)
'gets the contents of smp(1), and vice versa. Keep doing this
'with each pair of values in the array. The larger values in the
'array will migrate toward the lower index values—they rise
'like soda bubbles. Repeated passes through the array will
'completely sort it. The routine is done when it makes a loop
'through the array without swapping any pairs.
sort:
 swap = 0                          ' Clear flag that indicates swap.
for index = 0 to maxSrt ' For each cell of the array...
 if smp(index) >= smp((index+1)) then noSwap  ' Move larger values up.
  swapTmp = smp(index)  ' ..by swapping them.
  smp(index) = smp(index+1)
  smp(index+1) = swapTmp
  swap = 1                         ' Set bit if swap occurred.
noSwap:
next                              ' Check out next cell of the array.
if swap = 1 then sort             ' Keep sorting until no more swaps.
'The line below just averages particular cells of the array. If you
'use my values of the constants s1 through s5, it averages readings
'from the middle of the range. By assigning other values to s1-
's5, you can alter this.
echTime = smp(s1)+smp(s2)+smp(s3)+smp(s4)+smp(s5)/5 max 255
return                            ' Done: return to program.
```

Figure 14-6 *(Continued)*

rcvr pin (P2) reports the state of the comparator output. To take a sonar reading, the program sets compRC high to establish a +5V level on C2. It pulses ping briefly to send out a short 40-kHz tone. Then the RCtime instruction waits for an echo to be detected, as indicated by a change in the state of the rcvr pin. It records the waiting time in 2µs units in the variable echTime.

The program listing goes a step further. Sonar readings can vary due to multiple echoes, detection delays, etc. The sonar program cleans up these variations by taking several readings in quick succession and recording them in an array, then sorting and averaging the readings.

An array is a variable with multiple cells that can be selected by an index number. Arrays are useful when you want to perform the same operations on several variables. For instance, if you wanted to add the number 23 to 10 different byte variables, you might write:

```
a = a + 23
b = b + 23
c = c + 23
'...and so on, through the 10th variable
```

A better way would be to create an array and use a loop to perform the operation on each of the cells.

```
alpha      var      byte(10)     ' Create a 10-byte array
index      var      nib          ' Use a nibble (0-15) as a counter
for index = 0 to 9
  alpha(index) = alpha(index) + 23
next
```

14

SHORT-RANGE SONAR

The sonar program stores echo times in an array, and then sorts the array so that the largest values are in the lowest-numbered cells of the array. It uses a classic algorithm called a bubble sort, described in the program comments.

The purpose of sorting the array is to allow the program to take the average of the readings that fall in the middle of the range, ignoring the highest and lowest readings. This dramatically improves the consistency of the sonar readings.

And just what are those readings? They're not inches.

First, let's define a reference point. The sonar is pretty much blind to objects closer than 2 inches from the fronts of the 40-kHz transducers. By the time the BS2 gets set to look for an echo, it has already come and gone. So our zero point is actually set at 2 inches, the sonar's minimum detection range. All measurements are relative to this point.

The RCtime instruction takes the initial measurement in 2μs units. The prototype produced raw measurements from close to 0 to about 520 units. In order to store such values into a byte, the program divides by 2 and limits the result to a maximum of 255.

The scaled units correspond to 4μs of echo time. Since there are 147 μs in a round-trip inch, each unit represents about 1/37th of an inch. However, various influences make it difficult and possibly futile to convert these units to inches (or centimeters) of absolute distance.

First of all, the speed of sound varies with temperature. The figure we've been using, 1130 feet/second, is valid only at 60°F (15.6°C). The actual speed of sound is calculated by the formula in Figure 14-7.

Over the operating temperature range of the BS2, 32° to 122°F (0° to 70°C), the sonar round-trip-inch time varies from 153.3 μs to 136.8 μs.

Detection time also influences sonar readings. We've been assuming that the sonar circuit detects a return echo at the instant of its arrival. Depending on the strength of the echo and other factors, that may not be true. Suppose that the detector misses just one cycle of the returning 40-kHz ping; that's a timing error of 1/40,000 = 25 μs, and a distance error of 0.34 inches.

Now consider the sonic environment. We've been assuming that a single, distinct echo returns with each ping we send out. In a quiet room, and with the sonar facing a hard, flat surface, that's a reasonable assumption. In a cluttered environment of irregular surfaces with varying reflective properties, we can expect the returning echo to be a tattered mess.

The BS2 contributes some inaccuracy, too. Its time base is a ceramic resonator, good to approximately ±0.5 percent. Its timing measurements are only as accurate as the resonator itself. This is not a particularly large error contributor, but it's there.

Most of the sonar's limitations could be handled with more hardware and more programming. With enough work, we could use 50 bucks worth of electronics to make a slightly less accurate version of a 25-cent plastic ruler! However, the sonar would make a great collision-avoidance instrument for a small robot just the way it is. Knowing whether the distance to a wall is 1 inch (unsafe) or 6 inches (safe) would be extremely valuable. And although the sonar readings don't scale directly to absolute distance, changes in the readings are a pretty

$$V = \frac{1087 \sqrt{(273 + t)}}{16.52}$$

V is velocity in feet/second (fps)
t is temperature in °C

Figure 14-7 Formula for the speed of sound in air.

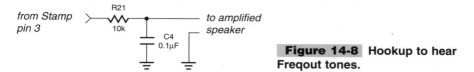

Figure 14-8 Hookup to hear Freqout tones.

good indication of a change in distance. If the sonar reads 136 units to an object now, and 100 units in the next reading, we can be pretty sure that it's about an inch closer.

Just for Fun—The "Sonamin"

An obscure musical instrument called a Theremin produces musical notes in response to the position of the musician's hands. To play it, you stand in front of a pair of antennas and wave your hands as though conducting an invisible orchestra.

The sounds that come from a Theremin are otherworldly; the oooohh-WEEEEE-ooh music from old science-fiction movies is the most familiar example of Theremin music. Theremins have always been popular electronics projects, though, because of the magical aspect of converting hand gestures into sound.

I thought it would be fun to use our sonar as a sonic Theremin—a Sonamin, if you will. This just-for-fun application adds an amplified speaker to the sonar hardware shown previously; see Figure 14-8. The Sonamin program appears in Figure 14-9. This program is quite a bit shorter than the one that takes distance readings. It eliminates all of the sorting and averaging of the previous program. Instead, it takes one sonar shot, and uses that value to set the frequency of a tone generated by the BS2 instruction Freqout.

The program takes a sonar reading, which generates a value from 0 to around 500. It multiplies this value by 2 and sets it to a minimum of 200. This value is used by Freqout to generate tones from 200 to 1000 Hz. Each tone is played for a fixed period of time (300 ms in the example), then another sonar shot is taken and another note played.

Freqout can generate two sine-wave tones simultaneously. A sine wave is a so-called "pure" tone with the ripple-shaped profile shown in Figure 14-1. When two sine waves mix, they produce a sound that includes the sum and difference of the two input frequencies. For example, if you mix a 1000-Hz tone with a 1500-Hz tone, the result contains frequencies of 1000 Hz, 1500 Hz, 2500 Hz (1000 + 1500), and 500 Hz (1500 − 1000).

The Sonamin program exploits this effect by making the BS2 generate two sine-wave sounds based on the sonar return, one shifted 8 Hz away from the other. Suppose the sonar-based frequency is 500 Hz. Freqout mixes 500 Hz with 500−8 = 492 Hz to produce additional frequencies of 992 and 8 Hz.

That 8-Hz sound produces an interesting quaver in the output sound, what musicians call vibrato. Experiment with this value and listen to the result—you can get some really interesting effects.

Going Further

I hope that this project captures the imagination of some science or electronics instructors and becomes the basis for many interesting lab exercises. Look at all of the subjects we've

```
'Program: SONAMIN.BS2 (sonar-theremin)
'This program is a variation on SONAR.BS2. Instead of displaying
'numerical data on the screen, it converts sonar readings to
'tones played through an amplified speaker. See the text for
'hookup details. When you run this program, you should get a
'soft ticking sound from the speaker initially. When you pass
'your hand within approximately 10" of the sonar transducers,
'you will hear a series of tones whose frequency is proportional
'to the distance between your hand and the transducers.
'========CONSTANTS
ping     con    0       ' Output to activate pinger.
pingLen  con    200     ' Duration of ping in 2-us units.
compRC   con    1       ' Output to set comparator RC circuit.
rcvr     con    2       ' Input from 40kHz receiver/comparator.
spkr     con    3       ' Output to amplified speaker.
len      con    300     ' Length of notes in milliseconds.
vib      con    8       ' Vibrato offset (Hz).
'========VARIABLES
echTime  var    word    ' Time to echo return.
freq     var    word    ' Frequency output to speaker.


'========PROGRAM
'The "again" loop takes sonar ranges and uses them to control the
'frequency of tones generated with the Freqout instruction. Freqout
'plays the tone for the # of milliseconds set by the constant len,
'adding a second tone whose frequency differs by the # of Hz set
'by the constant vib. Assuming vib is a low value (< 20), this
'causes a 'vibrato' effect, an eerie quavering in the tone, in
'keeping with the sci-fi nature of this project.
high ping                   ' Turn pinger off initially.
again:                      ' Loop.
 gosub sonar                ' Take the sonar reading
 echTime = echTime * 2 min 100 ' Convert to range of 150-1000.
 if echTime < 900 then makeTone      ' Silence the speaker..
  echTime = 0                        ' ..if hand is out of range.
makeTone:
 freqout spkr,len,echTime,(echTime-vib) ' Play tone 300ms w/8Hz vibrato.
goto again                  ' Repeat endlessly.
'========SONAR SUBROUTINE
'This code takes a single, quick sonar measurement. The range of
'echTime with the circuit shown in the text is from 0 to about 500.
sonar:
 high compRC                ' Raise C2 to +5 volts.
 pause 1                    ' Allow time for C2 to reach +5V.
 input compRC               ' Disconnect pin from C2.
 pulsout ping,pingLen       ' Send a short 40kHz pulse.
 rctime rcvr,0,echTime      ' Wait for echo; save time to echTime.
return                      ' Get another sample.
```

Figure 14-9 Program listing for sonic Theremin.

touched on: properties of sound and wave propagation, design of analog electronics, sorting and sampling of data, and issues of precision and accuracy.

PARTS LIST

Resistors (all 1/4W, 10% or better)
R1, R5, R6, R11, R19, R21—10k
R2, R7—47k

R3, R8—22k
R4, R9, R14, R18—100k
R10—18k
R12, R13, R16, R17—220k
R15—2.2k
R20—1k

OTHER COMPONENTS

C1–C4—0.1μF ceramic capacitor, 50WVDC
RCVR—40kHz ultrasonic receiver (Jameco 136653 contains RCVR/XMTR)
XMTR—40kHz ultrasonic transmitter (Jameco 136653 contains RCVR/XMTR)
U1—LM358AN dual operational amplifier (Jameco 120862)
U2—LM339N quad comparator (Jameco 23851)
XTAL—40-kHz quartz crystal (Digi-Key SE3316-ND)
SPKR—Amplified speaker, such as those sold for use with portable stereos, or Radio
   Shack model 277-1008c.

NETWORK TERMINAL WITH KEYPAD AND DISPLAY

BS2

When they debuted, personal computers were exciting because they allowed users to break free of sharing time on a remote and mysterious mainframe and have their very own computer. Now everyone wants to connect their PCs together in networks or via the Internet and share time on remote and mysterious servers. Just goes to show how fickle computer users are.

Microcontrollers can be networked, too, opening up all sorts of interesting applications. A classic example is building control and security. Microcontrollers installed throughout a building can gather data on entry/exit attempts, thermostat settings, fire-detector status, elevator position, electrical demand, water pressure, etc., and report back to a central computer. This computer, with or without human supervision, can issue instructions over the network to lock/unlock doors, adjust air-conditioning vents, trigger sprinklers and alarms, reroute elevators, turn off nonessential systems, notify maintenance workers of trouble, etc.

For our BS2 network project, we're going to simulate one of those functions, an ID-checking security-door system. In the process, we will see how to interface the BS2 to a liquid-crystal display (LCD), keypad, and network (RS-485) driver chip. You can adapt the resulting BS2 network terminal to any application that might require up to 32 points at which a central computer can communicate with users through an 80-character display and a 16-button keypad.

If you don't require networking capabilities, don't worry. A stand-alone version of the terminal demonstrates how to use the LCD and keypad without a network connection.

Alphanumeric LCD Modules

As the name suggests, alphanumeric LCD modules display both letters and numbers. LCDs are available in many sizes, ranging from 1 line of 8 characters to 4 lines of 40 characters. Each character consists of a 5-by-8 grid of liquid-crystal dots (pixels) that are turned on or off in a bitmap representation of the number or letter to be displayed.

Controlling all those pixels is a tall order. Consider the 4-line by 20-character display used in this project. It displays 80 characters, each consisting of 40 pixels—a total of 3200 pixels to control. If we assigned one BS2 I/O pin to each pixel, it would take 200 BS2s (almost $10,000 worth!) to drive the display. Obviously, these $50 display modules have some tricks up their sleeve.

The primary trick is a specialized LCD controller IC. Several manufacturers make them, but the most popular is the Hitachi HD44780. This IC contains 1536 bytes of ROM holding bitmap pictures of 192 alphanumeric characters; 80 bytes of display RAM to store the contents of the display; 64 bytes of character-generator RAM to hold up to 8 custom symbols; an interface that allows a microcontroller to transfer data or instructions in 4- or 8-bit chunks; multiplexed outputs for driving LCD pixels; and additional outputs for communicating with expansion chips to drive more LCD pixels.

If you set out to design an LCD module, you would have to understand all those aspects of the controller's operation. Using a ready-made LCD module is far simpler.

The LCD controller is designed to interface with a 4- or 8-bit data bus. A bus is a set of connections shared by the peripherals of a microcontroller or microprocessor. The idea is that all peripherals (memory devices, displays, I/O ports, etc.) receive the same data bits all the time. Separate control signals tell a particular peripheral whether to respond to data on the bus or ignore it. Additional control lines modify the peripheral's response. For example, sometimes the signals on the bus represent data, other times they are commands.

LCD modules have three control lines, enable (E), read/write (R/W), and register select (RS). Table 15-1 summarizes their functions.

To write data to the LCD—for example, to display a character on the screen—takes these steps:

TABLE 15-1 LCD CONTROL LINES AND FUNCTIONS

CONTROL LINE	STATE	MEANING
E (enable)	0	LCD disabled (ignores bus)
	1	LCD enabled (communicates with bus)
R/W (read/write)	0	Bus writes data to LCD
	1	LCD writes data to bus
RS (register select)	0	bus is instruction or status
	1	bus is data (e.g., character to display)

- Clear R/W to 0 (write)
- Set RS to 1 (data)
- Write the data (e.g., character code) to the bus
- Set E to 1 briefly, then clear to 0

That sequence assumes that the data bus is 8 bits wide, since the character codes are bytes corresponding mostly to the ASCII character set (see chart, Appendix B). LCDs can also accept bytes over a 4-bit bus the same way you get a piano through a doorway—by cutting it in half and moving one half at a time.[1] The procedure is the same as for bytes, but the final two steps are done twice—first with the upper four bits (high nibble) on the bus, and again with the lower four bits (low nibble). The LCD controller reassembles the nibbles into a byte.

After you write data to the LCD, the display controller points to the next location in memory. This normally means that data is written from the bottom of the LCD memory upward, making characters appear from left to right on the display. You can also instruct the LCD to print from right to left, or to scroll the screen instead of moving the cursor (printing position).

Writing instructions to the LCD follows the same sequence as writing data, except that RS is cleared to 0 in the second step. Instructions allow you to show or hide the cursor, move the cursor, clear the display, scroll the display horizontally, and set up (initialize) low-level details of the LCD's operation. That last item, initialization, must be done before anything else, since the LCD controller wakes up without a clue as to what bus width you're using (4- or 8-bit) or what kind of display it's driving.

The controller can read data and status information back from the LCD. For instance, if you needed to find out what character is written to a particular position on the display, you would set the display-RAM address to that position by writing the appropriate instruction, then read the data. The steps for reading data from the display are:

- Set R/W to 1 (read)
- Set RS to 1 (data)
- Set E to 1
- Read the data from the bus
- Clear E to 0

You can also read back the cursor location and display status (busy or idle). The procedure is the same as reading data, but RS is cleared to 0 in step 2.

1. That's not really the right way to move a piano.

Table 15-2 is the instruction set for the LCD controller. Some of the instructions have multiple options that are enabled by setting or clearing particular bits. This calls for constructing an 8-bit number with the appropriate bits set or cleared and writing that number to the LCD. Take Display Control as an example: The lowest three bits control display on/off, cursor on/off, and cursor blinking/underline. If you want the display on (1), cursor on (1), and underline cursor (0), make the three lowest bits of the Display Control instruction %110. So the complete instruction would be %00001110.

Not all of the possible instructions are particularly useful. Table 15-3 is a quick-reference guide to values for the most commonly used LCD instructions.

TABLE 15-2   LCD INSTRUCTION SET

INSTRUCTION	D7	D6	D5	D4	D3	D2	D1	D0	DESCRIPTION
Clear display*	0	0	0	0	0	0	0	1	Clear the display and move cursor to top-left character position.
Home cursor*	0	0	0	0	0	0	1	—	Move cursor to top-left character position. If display has been shifted, undo shift.
Entry mode	0	0	0	0	0	1	I/D	S	Direction of cursor movement or display shift I/D: 0=decrement (right-to-left) 1=increment (left-to-right) S: 0=cursor moves 1=display moves (scrolls)
Display Control	0	0	0	0	1	D	C	B	D: 0=display off 1=display on C: 0=hide cursor 1=show cursor B: 0=underline cursor (if C=1) 1=blinking cursor (if C=1)
Cursor/ Display Shift	0	0	0	1	S/C	R/L	—	—	Shift the cursor or the screen right or left. S/C: 0=move cursor 1=move display R/L 0=right one position 1=left one position
Function Set	0	0	1	BW	N	F	—	—	Set bus width (BW), number of screen lines (N), and font (F). BW: 0=4-bit bus width 1=8-bit bus width N: 0=1-line display 1=2- or 4-line display F: 0=5x7 pixel characters 1=5x10-pixel characters

INSTRUCTION	D7	D6	D5	D4	D3	D2	D1	D0	DESCRIPTION
TABLE 15-2 Continued									
Set CG RAM Address	0	1	A	A	A	A	A	A	Set address in character-generator RAM. Set or clear A bits to make up a 6-bit number (0-63). Subsequent data writes/reads address CG RAM until a DD RAM address is set.
Set DD RAM Address	1	A	A	A	A	A	A	A	Set address in data-display RAM. Set or clear A bits to make up a 7-bit number (0-127). Setting the DD RAM address controls the printing position (cursor location) on the screen. Subsequent data writes/reads address DD RAM until a CG RAM address is set.
Read Status and Address	BF	A	A	A	A	A	A	A	Set R/W to 1, clear RS to 0 to read the busy flag (BF) and current address in either CG or DD RAM (whichever was last addressed) as a 7-bit number from 0 to 127. BF: 0=not busy 1=controller busy–wait

NOTES: All instructions are written with R/W=0 and RS=0. To write data, R/W=0, RS=1. To read status, R/W =1, RS=0. To read data, R/W=1, RS=1. In the table, bits marked with—are "don't care" states; they can be either 0 or 1 without affecting the function of the instruction. Most instructions execute in less than 40µs, but clear and home (marked *) take as much as 1.64ms. These instructions require a pause to allow time for completion.

CURSOR POSITIONING

In most practical applications, like our network terminal, 90 percent of the game is cursor positioning. When you initialize the LCD, you tell it how you want the screen to react to a newly written character via the Entry Mode instruction. The most common setting is %00000110, which means move the cursor from left to right one space for each character received—just like the cursor on a computer screen. With that setting, the LCD takes care of itself until the cursor reaches the end of a line. Then your program has to reposition the cursor to the beginning of the next line.

Figure 15-1 shows how the LCD controller's DD RAM addresses match up to the physical layout of common LCDs. It's probably not what you expect, but it's not hard to fix with a little programming. The network terminal program uses a counter to track the cursor. When the counter reaches a multiple of 20 (the width of the 4x20 LCD screen), the program positions the cursor to the beginning of the appropriate display line.

TABLE 15-3   MOST COMMONLY USED LCD INSTRUCTIONS

INSTRUCTION/ACTION	CODE
Clear screen	1
Home cursor (undo scrolling, movement)	2
Scroll screen one character left	24
Scroll screen one character right	28
Move cursor one character left	16
Move cursor one character right	20
Blank display (retaining contents)	8
Turn on underline cursor (and unblank display)	14
Turn on blinking-block cursor (and unblank display)	13
Turn off cursor (and unblank display)	12
Set display (DD) RAM address	128 + addr
Set character-generator (CG) RAM address	64 + addr

Keypad Interface

Both the BS1 and BS2 have a Button instruction that lets programs accept input from a pushbutton switch. A keypad is a collection of pushbuttons, so you might think that you should read a keypad with multiple Button instructions.

This idea falls flat after a couple of quick calculations. First, each switch read by Button must be on a separate I/O pin. Our network terminal interfaces to a 16-button keypad, so it would require all of the BS2's I/O lines. Secondly, a separate byte variable is required for each Button instruction, so we would need 16 bytes of RAM. That leaves just 10 bytes for the rest of the project's functions.

Button is not the way to read a keypad. We have to take matters into our own hands and write a PBASIC routine for the purpose. Our first job is to reduce the number of I/O pins required. Keypad manufacturers provide a big hint here; they wire their pads in a matrix arrangement, which automatically cuts the required number of I/O lines in half.

Matrix keypads are wired in rows and columns, as shown in the network terminal schematic, Figure 15-2. For example, the buttons 1, 4, 7, and * of our keypad all have one leg connected to a common column wire. The buttons 1, 2, 3, 4 are all connected to a common row. If we number the rows and columns from top-left to bottom-right starting with 0, we can assign each button a unique coordinate. For example, the 1 key is at column0/row0; the 6 is at column2/row1; and so on.

Since the 16-button keypad is broken into 4 rows and 4 columns, it only takes eight I/O lines to scan all of the possible row/column combinations. Here's how that works.

The keypad scanning routine starts with the row connections to the keypad set to input. The rows are set high (1) by pullup resistors built into the LCD data lines. The columns are initially all output high.

The keypad scanner's goal is to determine whether one of the rows of the keypad is shorted to one of the columns. The scanner outputs a 0 to one of the columns, then looks at the row inputs to see whether that 0 shows up. If not, it sets that column to 1 and tries the next column.

When the scanner finds a 0 on a row input, it can easily determine which key is pressed from the row/column coordinate. A Lookup table matches the coordinates to the actual number or symbol printed on the key.

Stand-alone LCD/keypad demo

Before we discuss networking, let's use our LCD and keypad knowledge to create a stand-alone demonstration. If you are thinking of building a project requiring a display and keypad,

LCD controllers have a total of 80 bytes of RAM to store the characters that appear on the screen. If the screen has less than 80 character positions, some of this data is not visible. Further, the 80 bytes aren't arranged sequentially; the first 40 bytes are 0–39 and the remaining 40 bytes are 64–103. Your program must reposition the cursor (set the DD RAM address) whenever it passes a break in the memory map.

1-line LCD

Initialize with Function Set N=0
(one-line operation)

To move to a character position, clear RS to 0 and send 128+position. When the cursor reaches the end of the line, program should reposition it to 0.

**pseudo
1-line LCD**

1x16 models

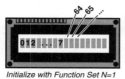

Initialize with Function Set N=1
(two-line operation)

Many 1x16 LCDs are organized as 2x8s. The next screen location after 7 is 64. To print there, clear RS and send 192 (128+64).

2-line LCD

Initialize with Function Set N=1
(two-line operation)

To print to the 2d line of a 2-line display, clear RS and send 192 (64+128).

4-line LCD

*up to 4x20;
4x40 displays
have two
controllers*

Initialize with Function Set N=1
(two-line operation)

To print to a given line of a 4-line display, use the addresses shown at left. For example, to print to the 3d line, clear RS and send 148 (128+20).

Figure 15-1 **Memory and screen layouts of LCDs.**

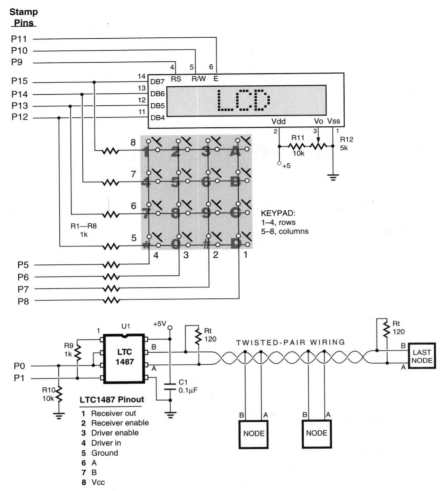

Figure 15-2 Schematic for the network terminal project.

take a close look at this program. Since all of the LCD/keypad features are well documented and corralled into subroutines, you can easily modify them for your own applications.

Use the circuit shown in Figure 15-2, minus the networking components (those associated with the LTC1487). Run the program of Figure 15-3 for an interactive demo.

The program demonstrates the principles of LCD and keypad operation we discussed
earlier. It combines often-needed display or input services into reusable subroutines. For
instance, instead of forcing the program to laboriously write one character of text at a time,
the program uses a subroutine called printEEstring to retrieve data from the BS2's EEP-
ROM program memory and write it to the screen.

That subroutine, in turn, depends on another, called printChar, that reads individual
bytes and interprets them as text, control characters such as carriage return and clear-
screen, or cursor-positioning instructions, depending on their value.

```
'Program: LCD_KEY.BS2 (Using an LCD and keypad with the BS2)
'This program demonstrates how to initialize and use a 4x20
'LCD module and scan a 16-key matrix keypad with the BS2.
'The demo shows how to use these two devices together to
'create a user-friendly interface for your programs.
'For the demo, it prompts you to enter a resistor value,
'then responds with the three-band color code for that
'value. Use the * key to backspace and the # to enter.
'=========Constants=========
E        con     11      ' LCD Enable pin (1=active;0=inactive).
RS       con     9       ' Register-select pin (1=data; 0=instr).
RW       con     10      ' Read/write pin (1=read; 0=write).
clr      con     12      ' Clear the LCD screen (CTL-L).
crsF     con     4       ' Turn off LCD cursor (default, CTL-D).
crsN     con     5       ' Turn on LCD cursor (CTL-E).
EOS      con     3       ' End-of-string marker (CTL-C).
line0    con     134     ' Position code for line0, character 6.
line1    con     154     ' "         "    "  line1, "          "
line2    con     174     ' "         "    "  line2, "          "
line3    con     194     ' "         "    "  line3, "          "
kMx      con     8       ' Maximum # of keys to buffer
'=========Variables=========
busOut   var     OUTD    ' 4-bit data bus, output.
busIn    var     IND     ' 4-bit data bus, input.
busDir   var     dirD    ' Direction for 4-bit data bus.
bytOut   var     byte    ' Byte to be output to LCD.
indexN   var     nib     ' General-purpose counter, 0-15.
indexW   var     word    ' General-purpose counter, 0-65535.
adrsW    var     word    ' Address word for strings in EEPROM.
kTimeW   var     word    ' Timeout for keyed input in 10ms units.
buffer   var     nib(kMx) ' Storage for up to kMx keystrokes
bPntr    var     nib     ' Pointer into buffer.
scrPos   var     byte    ' Position on the LCD screen, 0-79.
char     var     byte    ' Character to be printed.
key      var     nib     ' Button on keypad (0-15)
db       var     bit     ' Debounce flag for key scanner.
press    var     bit     ' Key-press flag, 1= key pressed.
'=========EEPROM Strings=========
'These strings (sequences of bytes in EEPROM, ending with the
'end-of-string character, EOS), contain data and control characters
'to be sent to the LCD. The idea is to allow the host to call up
'whole screens of text without having to send it over the serial
'connection. String addresses are stored in a lookup table later
'in the program. Note that strings can include control characters
'such as clr, and cursor-positioning values 128-207 (for screen
'locations 0-79). Note that strings can continue on another line;
'it's the EOS character that determines where the string ends.
intro   data    clr,130,"LCD/Keypad Demo",171,"press any key...",EOS
enter   data    clr,"RESISTOR COLOR CODES"
enter2  data    150,"enter a resistor",169," value in ohms:",193,EOS
sorry   data    clr,148,"Sorry, value must be 10 ohms or more.",EOS
result  data    clr, 131,"—COLOR CODE—",EOS
bk      data    "black",EOS          ' RESISTOR COLOR CODES, 0-9.
bn      data    "brown",EOS
rd      data    "red",EOS
og      data    "orange",EOS
yl      data    "yellow",EOS
gn      data    "green",EOS
bl      data    "blue",EOS
```

Figure 15-3 Program listing demonstrating LCD and keypad.

15

NETWORK TERMINAL

```
vi      data    "violet",EOS
gy      data    "gray",EOS
wh      data    "white",EOS
'====================================
'=========LCD Initialization=========
'Before it will function properly, the LCD must be initialized.
'This consists of sending some wakeup signals and settings that
'tell the LCD how to behave (4- or 8-bit bus; 1- or multi-line
'display; cursor; screen on/off; font (if applicable), etc.
'There are two phases to initialization—in the first phase,
'the LCD bus may be set up for 4 or 8 bits, so a 4-bit 'attention'
'signal is written three times, followed by the 4-bit bus-set
'instruction. Once this is done, the rest of the setup can be
'sent via the normal 4-bit write method that will be used
'throughout the rest of the program.
'==Phase 1: Get LCD's attention.
pause 100                   ' Wait for power to settle.
DIRH = %11111110            ' Set LCD data, control pins to output.
for indexN = 0 to 3         ' Send initialization bits from list.
  lookup indexN,[%0011,%0011,%0011,%0010],busOut
  pulsout E,10:pause 10     ' Pulse the enable line to write to LCD,
next                        ' Phase 1 init done.
'==Phase 2: Configure the LCD for operation.
'In phase 2 of the initialization, we send the LCD instructions via
'the normal write-LCD subroutine. These instructions are Function
'Set (4-bit, 4-line), Display Control (display on, cursor off),
'Entry Mode (increment, cursor moves), and Clear. For more detailed
'info on the instructions, see the instruction set table in text.
for indexN = 0 to 3         ' Send each of 4 table entries (0-3) below.
  lookup indexN,[%00101000,%1100,%0110,1],bytOut
  gosub writeInstruction ' Write bytOut to LCD as instruction.
next
pause 1                               ' Allow time for screen to clear.
'At this point in the program, the LCD screen is ready to go, but
'blank. Let's display one of the EEPROM messages on the LCD until
'we receive other instructions from the master unit. This will
'confirm that the unit is working, but hasn't yet received any
'data/instruction.
start:
  adrsW = intro             ' Point to intro message.
  gosub printEEstring       ' Print intro to LCD screen.
'==============================
'=========MAIN PROGRAM=========
'This program leads the user through the process of entering a
'resistor value via the keypad, and displays the corresponding
'color code on the LCD.
again:
  press=0                         ' Reset key-press indicator.
  gosub inKey                     ' Check the keypad.
  if press = 0 then again         ' If no press, look again.
getEntry:                         ' Key pressed: get an entry.
  adrsW = enter                   ' Point to entry message.
  gosub printEEstring             ' Print it to the LCD.
  kTimeW = 400                    ' Set for up to 4 seconds per keypress
  gosub keyInput                  ' and get a string from keypad in buffer.
  if bPntr = 0 then start         ' No keys pressed? Start over.
  if bPntr = 1 then tooSmall      ' Handle < 10-ohm entries.
  adrsW = result                  ' Else print the result screen to LCD.
  gosub printEEstring
```

Figure 15-3 *(Continued)*

```
'Here's where the program prints the actual color code. The numbers
'in cells 0 and 1 of the buffer array are the first and second
'numbers entered, so they determine the first two colors.
 char = line1:gosub printChar    ' Set print location to line1.
 indexN = buffer(0)              ' Print color for 1st digit.
 gosub showColor
 char = line2:gosub printChar    ' Set print location to line2.
 indexN = buffer(1)              ' Print color for 2nd digit.
 gosub showColor
'The third color, the multiplier, can be taken as the number of
'digits entered minus 2. For example, a 4700-ohm resistor
'has a red multiplier band, because there are two zeros after the
'two leading digits (4 and 7). There are four digits in 4700;
'4 - 2 = 2; and 2 yields the red color code.
 char = line3:gosub printChar    ' Set print location to line3.
 indexN = bPntr-2                ' Color code is # digits, minus 2.
 gosub showColor                 ' Print the color for multiplier band.
 pause 4000                      ' Leave info on screen 4 seconds.
goto getEntry                    ' Prompt for another entry.

'==tooSmall
'If the user enters just 1 digit, the color code doesn't fit the
'simple pattern used in this demo. This code displays an error
'message and allows a new entry.
tooSmall:
 adrsW = sorry                   ' Print sorry message to LCD.
 gosub printEEstring
 pause 2000                      ' Wait 2 seconds.
 goto getEntry                   ' Let 'em try again.


'=============================
'========SUBROUTINES=========
'=============================
'========Color Lookup & Print======
'This code looks up the address in EEPROM of the strings that spell
'out the color codes. The number to look up must be in indexN,
'and must be in the range 0-9. The text (e.g., "orange") will
'print at the current screen location. Note that this subroutine
'uses all 4 levels of the BS2's Gosub capability; it should not
'be used from within another Gosub-ed routine.
showColor
 lookup indexN,[bk,bn,rd,og,yl,gn,bl,vi,gy,wh],adrsW    ' Look up address.
 gosub printEEstring             ' Print color code string.
 return                          ' Return to program.
'========Keypad Subroutines======
'The inKey routine scans the 4x4 keypad. If a key is pressed, it
'sets the press bit to 1. Code that responds to a keypress should
'clear press to 0 before using inKey again. The number of the key
'pressed (0-15 from upper left to lower right) is in the variable
'key. The routine has a strict debounce method—the last key
'pressed must be released before another key will register.
inKey:
 busDir=0                        ' Set the bus to input.
for indexN = 5 to 8             ' For each of the keypad columns:
 low indexN                      ' Make one column output low.
 key = ~ busIn                   ' Get inverse of row bits.
 key = NCD key                   ' Determine bit # of lowest bit = 1.
 if key <> 0 then push           ' Key is 0 unless a button is pressed.
 high indexN                     ' Make column high.
```

■ Figure 15-3 (*Continued*)

```
next                   ' And try the next column.
  db = 0               ' Key is up, so we can accept new key.
  busDir= %1111        ' Set bus back to output.
return                 ' And return.
push:                  ' A key has been pressed. If this key
  if db = 1 then done  ' has already been detected, ignore it.
  db = 1: press = 1    ' New key: set up debounce and tell program.
'The purpose of the formula below is to convert the row/column
'coordinates key and indexN to a number from 0 to 15.
'Key is the column, and ranges from 1-4, so we subtract 1
'to get 0-3. IndexN is the row/pin number and ranges from 5-8,
'starting with 8 at the top of the keypad. We subtract 5 to
'get 0-3 and invert (~) to put the 0 value at the top of the
'pad. ANDing with %11 strips off all but the two lowest bits,
'leaving us with 0-3, top to bottom. Finally, multiplying the
'row by 4 and adding the column yields the key number, 0-15.
  key = (key-1)+((~(indexN-5) & %11) * 4)
done:
  high indexN          ' Cleanup: make column pin high again,
  busDir= %1111        ' and set bus back to output
return                 ' before returning.


'This routine, keyInput, works like the INPUT instruction of
'most dialects of BASIC. The LCD cursor should be positioned
'on the screen where the input is to take place before
'entering this routine. A timeout value (in units of 10ms,
'from 1 to 65535) determines how long the routine will
'wait for input to be entered. The routine will turn on the
'cursor and accept input through the keypad. The * key works as
'a backspace; # key is enter. ABCD are ignored. When the enter
'key is pressed or the routine has timed out, it returns with
'a zero-terminated string stored in buffer.
keyInput:
  press = 0            ' Reset the press flag.
  char = crsN          ' Turn on the cursor.
  gosub printChar
  bPntr = 0            ' Clear buffer pointer to 0.
  indexW=kTimeW        ' Set indexW to timeout value.
keyLoop:
  if indexW<>0 then skip0      ' If time is up, end routine now.
  bPntr = 0:goto kiDone
skip0:
  indexW = indexW-1            ' One tick toward timeout.
  pause 10                     ' Wait 10ms.
  gosub inKey                  ' Scan the keypad.
  if press = 0 then keyLoop    ' No keypress? scan again.
  indexW=kTimeW        ' Key pressed: reset index to timeout value.
  press = 0            ' Reset keypress flag.
  lookup key,[1,2,3,$A,4,5,6,$B,7,8,9,$C,$F,0,$E,$D],key  ' Get keycode.
  if key = $E then kiDone      ' If $E (the # key) then done.
  if key <> $F then skip1      ' If $F (*), then backspace.
    if bPntr = 0 then keyLoop  ' If no entries to erase, exit.
    char = BKSP: gosub printChar ' Else send backspace to LCD.
    bPntr = bPntr-1            ' And reduce the character count by 1.
    goto keyLoop               ' Backspace finished—continue.
skip1:
  if key > 9 then keyLoop      ' Key > 9, but not * or #, so ignore it.
  char = key + "0"             ' Key is 0-9—convert to text.
  gosub printChar              ' Print text character on LCD.
```

Figure 15-3 (*Continued*)

```
 buffer(bPntr) = key            ' And store its code in buffer array.
 bPntr = bPntr + 1              ' Point to the next buffer location.
 if bPntr = kMx then kiDone     ' Limit buffer to max # of chars.
 goto keyLoop                   ' If buffer not full, get more keys.
kiDone:                         ' Done (timed out, buffer full, or "#" pressed).
 bPntr = bPntr max kMx ' Make sure pointer doesn't exceed max # of chars.
 char = crsF                    ' Turn off the cursor.
 gosub printChar
return                          ' Return to main program.
'=========LCD Subroutines=========
'Dealing with the LCD is a lot easier and more efficient if you
'break the tasks involved down into small pieces. For example,
'to write a string of text to the LCD screen, you must be able to
'print characters and position the cursor. To do those things,
'you must be able to write data and instructions to the LCD.
'In using these routines, you must be careful not to exceed the
'BS2's 4-level limit for Gosubs. For example, printEEString Gosubs
'to printChar which Gosubs to writeData. That's 3 levels of
'"nesting." There's 1 level of Gosub left, allowing printString
'to be used from within a Gosubed routine. But that's it.
'==Print a string stored in EEPROM starting at address adrsW to the
'LCD. String must end with the EOS character
printEEstring:
 read adrsW, char               ' Get a character from EEPROM.
 if char = EOS then gDone       ' If end-of-string char, we're done.
 gosub printChar                ' Else print it to the LCD.
 adrsW = adrsW+1                ' Point to the next EEPROM address.
goto printEEstring              ' And repeat the process.
gDone:                          ' Done: return to program.
return
'==Print a character (in variable char) to the LCD and update the
'cursor position.
printChar:
 if char < 32 OR char > 127 then ctlChar  ' Handle control characters.
 bytOut = char         ' Write character to LCD. Increment screen
 gosub writeData       ' position, making it wrap around, 0-79.
 scrPos = scrPos+1//80 ' If position is a multiple of 20, then
 if (scrPos//20)=0 then moveCrs  ' move cursor to start of next line.
return                ' Otherwise, return to program.
'==Handle control characters. Understands the same five screen
'controls as BS2 Debug, plus cursor-on, cursor-off, and a single-byte
'cursor-positioning command.
ctlChar:
 bytOut = 255
 lookdown char,[clr,HOME,crsF,crsN,BKSP,TAB,CR,128],bytOut
 branch bytOut,[clrLCD,homeCrs,offCrs,onCrs,backSp,tabCrs,cRet]
position:
 scrPos = char-128//80  ' Subtract 128, limit to 0-79.
 goto moveCrs                      ' Move the cursor.
clrLCD:
 scrPos =0                         ' Clear screen position.
 bytOut =1:goto writeInstruction   ' Send cls (1) to LCD
homeCrs:
 scrPos =0:goto moveCrs            ' Move cursor to 0.
offCrs:
 bytOut =12:goto writeInstruction  ' Send off-cursor instruction.
onCrs:
 bytOut =13:goto writeInstruction  ' Send on-cursor instruction.
backSp:
```

Figure 15-3 (*Continued*)

```
  scrPos=scrPos-1//80           ' Move back 1; limit 0-79.
  gosub moveCrs                 ' Move the cursor.
  bytOut=32: gosub writeData    ' Write a space to screen.
  goto moveCrs                  ' Move the cursor and return.
tabCrs:
  scrPos = ((scrPos+4) & $FC)//80 ' Jump to next multiple of 4 pos.
  goto moveCrs
cRet:
  scrPos = ((scrPos/20)+1)*20   ' Move to next multiple of 20 pos.
  goto moveCrs
'==Update the cursor position. Used when text wraps from one line
'to another, or control characters move the cursor.
moveCrs:
  bytOut = scrPos/20 & %11      ' Convert position to lines 0-3.
  lookup bytOut, [128,192,148,212],bytOut ' Lookup LCD DD-RAM address.
  bytOut = scrPos//20 + bytOut  ' Add offset 0-19 to line address.
  gosub writeInstruction ' Write it to the LCD.
return
'==Write instructions or data to the LCD.
writeInstruction:         ' If RS=0, the LCD interprets data
  low RS                  ' sent as an instruction. If RS=1,
  goto wContinue          ' data is interpreted as data and
writeData:                ' written into memory (i.e. the screen).
  high RS                 ' This routine sets or clears RS depending
wContinue:                ' on whether you gosub "writeInstruction"
  low RW                  ' or "writeData." In either case, RW=0
  busOut = bytOut.highnib ' (write), and data is transferred 4 bits
  pulsout E,10            ' at a time. The high nibble is first,
  busOut = bytOut.lownib  ' then the low nibble. The actual transfer
  pulsout E,10            ' happens when the enable pin (E) is pulsed
return                    ' high. When done, the routine returns.
```

Figure 15-3 (*Continued*)

The keypad routines have the same sort of structure. There's a subroutine called keyInput that gets an entry of up to eight digits from the user. The user presses * to backspace and # to enter. The routine places the digits in the cells of an array of nibbles, and stores the number of keys pressed in another variable. KeyInput depends on a simpler routine, inKey, to read individual key presses.

Although this kind of hierarchy makes for an efficient program, you have to be careful to avoid exceeding the BS2's four-level Gosub limit. For example, in the program of Figure 15-3, there's a subroutine called showColor. It contains a Gosub to printEEstring, which Gosubs printChar, which Gosubs writeData. That's four levels of Gosub. If another part of the program contained a subroutine called bigMistake with a Gosub to showColor, the BS2 would never find its way back to bigMistake.

To adapt the demo for your own purposes, spend some time playing with the program until you understand how it works. The user interface is often the most difficult part of an application, and this program contains all of the most important elements.

RS-485 Networking

Most computer users are at least passingly familiar with the term RS-232; it has become synonymous with communication via a desktop computer's serial communication ports.

RS-232 designates a family of electrical specifications for signals from such ports. The signals are appropriate for connecting a computer to an accessory, like a modem or bar-code reader, located nearby (up to 50 feet away). RS-232 assumes that there will be only one device at each end of the line, normally a computer and a peripheral.

RS-485 is another electrical spec for serial signals. Its purpose is different from that of RS-232; it's meant to allow many computers and/or accessories (up to 32) to communicate over a single set of wires, up to 4000 feet long. In order to meet these goals, RS-485 signaling is designed for greater immunity to the electrical noise that gets picked up on long wire runs, and to allow multiple devices to drive the signal wires (albeit one at a time). It even allows for mistakes. If two or more devices try to drive the wires at the same time, their RS-485 interfaces won't be damaged.

We're used to the idea of digital data, 1s and 0s, being expressed by voltage levels. The Stamps regard voltages below 1.5V as 0 and above as 1. RS-485 signals express 1s as 0s not as absolute voltages, but as the voltage difference between signals on a pair of wires. The two wires of the pair are designated A and B; when the voltage on A is greater than the voltage on B, that's a digital 1. When B is greater than A, that's a 0.

This differential signaling is nearly immune to noise pickup over long wire runs. The catch is that the local ground potentials of the RS-485 devices cannot differ by more than ±7 volts. If they do, the system won't work.

Since one wire pair can be used by many devices, RS-485 devices have to take turns using it. This is called half duplex. If two or more RS-485 units transmit at the same time, the error condition is called bus contention. It won't hurt them, but their transmission won't get through.

An RS-485 network is like a group of people using walkie-talkies on the same frequency. There are two major limitations:

- When a unit is listening, it cannot talk; when talking, it can't listen.
- If two units talk at the same time, neither will be heard correctly.

Those limitations make it vital to coordinate units' behavior. The easiest ways to do this is to appoint one unit master and all other units slaves. No slave unit transmits until told to do so by the master. This prevents units from transmitting at the same time. It requires the master to check in with all of the slave units periodically for updates. If a slave unit has urgent information, it must still wait until called upon by the master before transmitting.

RS-485 INTERFACE

A single chip, an LTC1487, serves as the RS-485 interface, as shown in the schematic, Figure 15-2. Pin P4 of the BS2 controls the transmit/receive function; P3 serves as both input and output for serial data. The program normally leaves the LTC1487 in receive mode (P4=0). It can monitor data sent over the network by executing a Serin instruction on P3. When it needs to talk, it puts a high on P4 to put the LTC1487 into transmit more, and executes a Serout instruction to P3. When it's done talking, it returns to receive mode.

In the schematic, only one LTC1487 is shown, but each box labeled "node" is short-hand for another identically equipped BS2. Note that the first and last nodes on the wiring pair should have a 120-ohm resistor connected across their A/B terminals. The resistor serves to terminate the network, taming reflected signals that might otherwise distort the serial signals.

15

NETWORK TERMINAL

Limitations of the Demo Application

Let me stress right at the start that our example application is just that—an example. It's meant to demonstrate three important subsystems: LCD, keypad, and RS-485 interface. By separating these functions into subroutines, I've made it possible for you to extract portions of the program for use in your own applications.

What this demo is not is a ready-to-go commercial terminal. As we discuss the program, we'll see some limitations arising from characteristics of the BS2 and the nature of RS-485 communication itself.

The biggest limitation is that the BS2 cannot execute other instructions while it is receiving serial data. This means that our network terminal slaves must wait for instructions from the master, carry out those instructions, and then respond. The network is tied up during this time.

Depending on the final application, and your ingenuity, this may be either a minor problem or a showstopper. In our example, the network is tied up while a user enters data at a keypad. Although there's a timeout that ends the transaction if no key is pressed for some period of time, mischievous (or inept) users could tie up the system for a long time by entering and deleting numbers.[2]

If you're a newcomer to electronics, you may be surprised at this discussion of an application's drawbacks. Aren't projects like this supposed to be recipes that can be followed blindly? Frankly, no. As you progress in electronics, you will encounter all kinds of application notes, example circuits, code snippets, etc. They are meant as broad sketches that describe a technique or supply a starting point. But it's still up to you to ensure that your final design meets your needs. That may require combing through data books, making calculations, or sitting in a quiet place and thinking through all of the situations your project may encounter.

That said, let's take a look at our network terminal and an example program that runs it through the hoops.

The Network Terminal

Figure 15-4 is the program listing for the network terminal. Most of the program consists of a series of subroutines that operate the LCD and keypad. The terminal is a general-purpose device. Although our demo will use it to request and process a fictitious user ID number, the terminal could be used in any application requiring remote displays and keypads.

The terminal has a very simple protocol. It listens to the RS-485 network for an attention character (tilde ~) followed by a two-letter ID, such as "AA." When a terminal recognizes its own ID (using Serin's Wait function), it prepares to receive a string of up to 8 bytes. Once it receives 8 bytes, or a shorter sequence of bytes ending with the end-of-string character EOS (ASCII 3/control-C), it begins processing the received bytes. Table 15-4 shows how it responds to the various byte values. For bytes containing 0–127, the terminal's responses track pretty closely with the standard ASCII character set (Appendix A). That is, the ASCII code for "A" prints an "A" to the LCD; a carriage return makes the cursor drop down to the beginning of the next line; and so on. This is the same kind of response you would expect from a PC running terminal-communication software.

---

2. Think of the people who are always ahead of you in the line at the ATM.

```
'Program: NETERM.BS2 (BS2 network terminal)
'This program implements a simple but useful network terminal with
'the BASIC Stamp II microcontroller. It communicates via RS-485,
'and allows a host computer to remotely display data on a 4x20
'LCD screen and acquire data from a keypad. The BS2 has three
'I/O pins left over, so additional functions could be added to
'this framework.
' Multiple terminals can share a single RS-485 party line. The
'host starts all conversations with a given terminal by sending
'a unique ID, followed by a 1- to 8-character string of data
'and/or instructions. Data consists of text to be displayed on
'the LCD, ASCII codes 32-127. Some of the ASCII codes below
'32 have special meanings, like clear-screen, carriage return
'backspace, etc. The tilde (~) is reserved as a prefix for
'the unit's ID. Codes from 128 to 207 position the cursor to
'positions 0-79 of the screen. Code 208 means scan the keypad and
'return a single character; 209, input a string from the keypad;
'210-215, display one of 6 strings stored in EEPROM. Codes 216-255
'are not used in this application.
'=========Constants=========
myID1    con     "A"    ' 1st character of this unit's 2-char ID.
myID2    con     "A"    ' 2nd character of this unit's 2-char ID.
tilde    con     "~"    ' Prefix for ID; DO NOT USE IN NORMAL TEXT.
ackChar  con     "`"    ' Acknowledgment; DO NOT USE IN NORMAL TEXT.
baud     con     $54    ' 9600-baud comms.
serIO    con     1      ' Pin P1 used for serial input/output.
serDir   con     0      ' Pin P0 sets serial direction 1=talk, 0=listen.
E        con     11     ' LCD Enable pin (1=active;0=inactive).
RS       con     9      ' Register-select pin (1=data; 0=instr).
RW       con     10     ' Read/write pin (1=read; 0=write).
clr      con     12     ' Clear the LCD screen (CTL-L).
crsF     con     4      ' Turn off LCD cursor (default, CTL-D).
crsN     con     5      ' Turn on LCD cursor (CTL-E).
EOS      con     3      ' End-of-string marker (CTL-C).
'=========Variables=========
busOut   var     OUTD   ' 4-bit data bus, output.
busIn    var     IND    ' 4-bit data bus, input.
busDir   var     dirD   ' Direction for 4-bit data bus.
bytOut   var     byte   ' Byte to be output to LCD.
indexN   var     nib    ' General-purpose counter, 0-15.
indexW   var     word   ' General-purpose counter, 0-65535.
adrsW    var     word   ' Address word for strings in EEPROM.
kTimeW   var     word   ' Timeout for keyed input in 10ms units.
buffer   var     nib(6) ' Storage for up to 6 keystrokes.
bPntr    var     nib    ' Pointer into buffer.
scrPos   var     byte   ' Position on the LCD screen, 0-79.
char     var     byte   ' Character to be printed.
key      var     nib    ' Button on keypad (0-15).
db       var     bit    ' Debounce flag for key scanner.
press    var     bit    ' Key-press flag, 1= key pressed.
hostBuf  var     byte(8) ' Buffer for input from host.
hostN    var     nib    ' Nibble counter for processing host string.
'=========EEPROM Strings=========
'These strings (sequences of bytes in EEPROM, ending with the
'end-of-string character, EOS), contain data and control characters
'to be sent to the LCD. The idea is to allow the host to call up
```

Figure 15-4 Program listing for network terminal.

15

```
'whole screens of text without having to send it over the serial
'connection. String addresses are stored in a lookup table later
'in the program. Note that strings can include control characters
'such as clr, and cursor-positioning values 128-207 (for screen
'locations 0-79). Note that strings can continue on another line;
'it's the EOS character that determines where the string ends.
intro    data    clr,134,"Stamp II",151,"RS485 Network",174,"Terminal",EOS
enter    data    clr,132,"-FOR ACCESS-",151,"press and hold",174,"any key",EOS
accept   data    clr,154,"CLEARED",169," Have a nice day!",EOS
reject   data    clr,151,"ACCESS DENIED",EOS
moment   data    clr,152,"One momement",174,"please...",EOS
prompt   data    clr,189,"* bksp   #enter",129,"ENTER YOUR"
prompt2 data    149,"ID NUMBER: ",EOS
'========================================
'=========LCD Initialization=========
'Before it will function properly, the LCD must be initialized.
'This consists of sending some wakeup signals and settings that
'tell the LCD how to behave (4- or 8-bit bus; 1- or multi-line
'display; cursor; screen on/off; font (if applicable), etc.
'There are two phases to initialization—in the first phase,
'the LCD bus may be set up for 4 or 8 bits, so a 4-bit 'attention'
'signal is written three times, followed by the 4-bit bus-set
'instruction. Once this is done, the rest of the setup can be
'sent via the normal 4-bit write method that will be used
'throughout the rest of the program.
'==Phase 1: Get LCD's attention.
pause 100                ' Wait for power to settle.
DIRH = %11111110         ' Set LCD data, control pins to output.
low serDir               ' Set serial direction to listen.
for indexN = 0 to 3      ' Send initialization bits from list.
 lookup indexN,[%0011,%0011,%0011,%0010],busOut
 pulsout E,10:pause 10   ' Pulse the enable line to write to LCD,
next                     ' Phase 1 init done.
'==Phase 2: Configure the LCD for operation.
'In phase 2 of the initialization, we send the LCD instructions via
'the normal write-LCD subroutine. These instructions are Function
'Set (4-bit, 4-line), Display Control (display on, cursor off),
'Entry Mode (increment, cursor moves), and Clear. For more detailed
'info on the instructions, see the instruction set table in text.
for indexN = 0 to 3      ' Send each of 4 table entries (0-3) below.
 lookup indexN,[%00101000,%1100,%0110,1],bytOut
 gosub writeInstruction ' Write bytOut to LCD as instruction.
next
pause 1                              ' Allow time for screen to clear.
'At this point in the program, the LCD screen is ready to go, but
'blank. Let's display one of the EEPROM messages on the LCD until
'we receive other instructions from the master unit. This will
'confirm that the unit is working, but hasn't yet received any
'data/instruction.
adrsW = intro
gosub printEEstring


'================================
'=========MAIN PROGRAM=========
'The program implements an ultra-simple protocol. It listens to the
'RS-485 bus for an ID, consisting of the tilde (˜7E) character followed
'by a two-character ID (myID1 and myID2). Each terminal on the net
'must have a different myID pair. After receiving its ID, the
```

Figure 15-4 (*Continued*)

```
'program accepts up to 8 bytes into the array hostBuf. If the
'host wants to send fewer characters, it can end transmission early
'with the EOS character. Once data is received, each character
'is processed appropriately (displayed on the LCD or interpreted
'as an instruction. When the terminal is done processing the
'string it sends the ackChar to the host, indicating that it's
'ready for more.
again:   ' Wait for tilde/ID, then store up to 8 bytes in hostBuf.
 low serDir                    ' Set to receive.
 Serin serIO,baud,[WAIT (tilde,myID1,myID2), str hostBuf\8\EOS]
   for hostN = 0 to 7          ' Process up to 8 bytes.
     char = hostBuf(hostN)      ' Get one byte.
     if char = 0 then ack       ' If 0, end of hostBuf.
     if char > 215 then nexChar ' 216-255: undefined, ignore.
     if char > 209 then EEstring' 210-215: Show an EE string.
     if char = 209 then inputString   ' 209: Get a numeric string.
     if char = 208 then singleKey     ' 208: Scan keypad once.
     gosub printChar                  ' 0-207: LCD text or control.
   nexChar:next                ' Process next byte.
 ack:   pause 10
        high serDir:pause 1        ' Set to transmit
        Serout serIO,baud,[ackChar]  ' Send acknowledgement.
        low serDir
goto again                      ' Repeat forever.
'========Terminal Services==========
'The main program recognizes three special instructions that scan
'for a single keypress, wait for an input string, or display a
'string stored in EEPROM on the LCD. These sections of code handle
'those service requests. They are not really subroutines in the sense
'that they don't use gosub/return. They're just chunks of code that
'have been set aside to avoid cluttering up the main program loop.
'==Scan for a key press.
'SingleKey uses the subroutine inKey to scan the keypad for a key
'press at that particular moment. If no key is pressed, it
'returns "-". If a key is pressed, it sends the appropriate
'character, such as "3" or "*". It does not use the debounce
'feature of inKey, since that requires repetitive scanning of the
'keypad. The relatively slow rate of serial communication makes
'debouncing unnecessary anyhow.
singleKey:
 db=0                       ' No need for debounce, so turn it off.
 press = 0                  ' Reset the press flag.
 gosub inKey                ' See if a key is down.
 if press <> 0 then sendKey    ' If key pressed, send it.
 high serDir:pause 1           ' Set to transmit
 Serout serIO,baud,["-"]       ' No key pressed: send "-"
 low serDir                    ' Set to receive
  goto again                   ' Await more instructions.
sendKey:
 lookup key,["123A456B789C*0#D"],char    ' Get keycode.
 high serDir:pause 1           ' Set to transmit
 Serout serIO,baud,[char]      ' Send to host.
 low serDir                    ' Set to receive
 goto again                    ' Await more instructions.
'==Get a numeric string from the keypad.
'InputString uses the routine keyInput to get a series of keys
'from the keypad. The user must press each key within 2 seconds,
'or the routine aborts and sends "0" to the master. It uses
```

Figure 15-4 *(Continued)*

15

NETWORK TERMINAL

```
'"0" instead of "-" for the abort response in order to allow the
'master to use SERIN's automatic number-reception instead of
'forcing it to receive and dissect a string. Naturally, you
'may change this if your priorities are different; see the 1st
'SEROUT instruction below.
inputString:
  kTimeW = 200              ' Allow 200 x 10ms = 2 seconds for input.
  gosub keyInput ' Get user input.
  if bPntr <> 0 then sendString 'If a string was entered, send it.
  high serDir:pause 1        ' Set to transmit
    Serout serIO,baud,["0",cr]  ' No string: send "0" + carriage return.
  low serDir                 ' Set to receive
    goto again               ' Await more instructions.
sendString:
  bPntr = bPntr-1            ' Adjust to start with 0th nib of array.
  for indexN = 0 to bPntr          ' Send each entry...
  high serDir:pause 1        ' Set to transmit
    Serout serIO,baud,[buffer(indexN)+"0"]      ' ...as a number, "0,1..."
  next                       ' Send all entries
    Serout serIO,baud,[cr]   ' Terminate # with carriage return.
  goto again                 ' Await more instructions.
'==Display a string from EEPROM.
'Codes 210 through 215 make the terminal display one of six strings
'stored in EEPROM on the LCD screen.
EEstring:
  char = (char max 215) - 210   ' Move to range 0-5.
  lookup char,[intro,enter,accept,reject,moment,prompt],adrsW
  gosub printEEstring
  goto nexChar                     ' Done, back to main program.


'=============================
'=========SUBROUTINES=========
'=============================
'=========Keypad Subroutines======
'The inKey routine scans the 4x4 keypad. If a key is pressed, it
'sets the press bit to 1. Code that responds to a keypress should
'clear press to 0 before using inKey again. The number of the key
'pressed (0-15 from upper left to lower right) is in the variable
'key. The routine has a strict debounce method—the last key
'pressed must be released before another key will register.
inKey:
  busDir=0                 ' Set the bus to input.
for indexN = 5 to 8        ' For each of the keypad columns:
  low indexN               ' Make one column output low.
  key = ~ busIn            ' Get inverse of row bits.
  key = NCD key            ' Determine bit # of lowest bit = 1.
  if key <> 0 then push    ' Key is 0 unless a button is pressed.
  high indexN              ' Make column high.
next                       ' And try the next column.
  db = 0          ' Key is up, so we can accept new key.
  busDir= %1111            ' Set bus back to output.
return                     ' And return.
push:                      ' A key has been pressed. If this key
  if db = 1 then done      ' has already been detected, ignore it.
  db = 1: press = 1        ' New key: set up debounce and tell program.
'The purpose of the formula below is to convert the row/column
'coordinates key and indexN to a number from 0 to 15.
'Key is the row, and ranges from 1-4, so we subtract 1
```

Figure 15-4 *(Continued)*

```
'to get 0-3. IndexN is the column/pin number and ranges from 5-8,
'starting with 8 at the top of the keypad. We subtract 5 to
'get 0-3 and invert (~) to put the 0 key at the top of the
'pad. ANDing with %11 strips off all but the two lowest bits,
'leaving us with 0-3, top to bottom. Finally, multiplying the
'column by 4 and adding the row yields the key number, 0-15.
 key = (key-1)+((~(indexN-5) & %11) * 4)
done:
 high indexN               ' Cleanup: make column pin high again,
 busDir= %1111             ' and set bus back to output
return                     ' before returning.


'This routine, keyInput, works like the INPUT instruction of
'most dialects of BASIC. The LCD cursor should be positioned
'on the screen where the input is to take place before
'entering this routine. A timeout value (in units of 10ms,
'from 1 to 65535) determines how long the routine will
'wait for input to be entered. The routine will turn on the
'cursor and accept input through the keypad. The * key works as
'a backspace; # key is enter. ABCD are ignored. When the enter
'key is pressed or the routine has timed out, it returns with
'a zero-terminated string stored in buffer.
keyInput:
 press = 0                 ' Reset the press flag.
 char = crsN               ' Turn on the cursor.
 gosub printChar
 bPntr = 0                 ' Clear buffer pointer to 0.
 indexW=kTimeW             ' Set indexW to timeout value.
keyLoop:
 if indexW<>0 then skip0       ' If time is up, end routine now.
 bPntr = 0:goto kiDone
skip0:
 indexW = indexW-1             ' One tick toward timeout.
 pause 10                      ' Wait 10ms.
 gosub inKey                   ' Scan the keypad.
 if press = 0 then keyLoop     ' No keypress? scan again.
 indexW=kTimeW             ' Key pressed: reset index to timeout value.
 press = 0                ' Reset keypress flag.
 lookup key,[1,2,3,$A,4,5,6,$B,7,8,9,$C,$F,0,$E,$D],key  ' Get keycode.
 if key = $E then kiDone       ' If $E (the # key) then done.
 if key <> $F then skip1       ' If $F (*), then backspace.
  if bPntr = 0 then keyLoop    ' If no entries to erase, exit.
  char = BKSP: gosub printChar ' Else send backspace to LCD.
  bPntr = bPntr-1              ' And reduce the character count by 1.
  goto keyLoop                 ' Backspace finished—continue.
skip1:
 if key > 9 then keyLoop       ' Key > 9, but not * or #, so ignore it.
 char = key + "0"              ' Key is 0-9—convert to text.
 gosub printChar              ' Print text character on LCD.
 buffer(bPntr) = key          ' And store its code in buffer array.
 bPntr = bPntr + 1            ' Point to the next buffer location.
 if bPntr > 6 then kiDone     ' Limit buffer to 5 chars max.
 goto keyLoop                 ' If buffer not full, get more keys.
kiDone:                       ' Done (timed out, buffer full, or "#" pressed).
 bPntr = bPntr max 6          ' Make sure pointer doesn't exceed max # of chars.
 char = crsF                  ' Turn off the cursor.
 gosub printChar
return                        ' Return to main program.
```

Figure 15-4 *(Continued)*

```
'=========LCD Subroutines=========
'Dealing with the LCD is a lot easier and more efficient if you
'break the tasks involved down into small pieces. For example,
'to write a string of text to the LCD screen, you must be able to
'print characters and position the cursor. To do those things,
'you must be able to write data and instructions to the LCD.
'In using these routines, you must be careful not to exceed the
'BS2's 4-level limit for Gosubs. For example, printEEString Gosubs
'to printChar which Gosubs to writeData. That's 3 levels of
'"nesting." There's 1 level of Gosub left, allowing printString
'to be used from within a Gosubed routine. But that's it.
'==Print a string stored in EEPROM starting at address adrsW to the
'LCD. String must end with the EOS character
printEEstring:
  read adrsW, char            ' Get a character from EEPROM.
  if char = EOS then gDone    ' If end-of-string char, we're done.
  gosub printChar             ' Else print it to the LCD.
  adrsW = adrsW+1             ' Point to the next EEPROM address.
goto printEEstring            ' And repeat the process.
gDone:                        ' Done: return to program.
return
'==Print a character (in variable char) to the LCD and update the
'cursor position.
printChar:
  if char < 32 OR char > 127 then ctlChar  ' Handle control characters.
  bytOut = char              ' Write character to LCD. Increment screen
  gosub writeData            ' position, making it wrap around, 0-79.
  scrPos = scrPos+1//80      ' If position is a multiple of 20, then
  if (scrPos//20)=0 then moveCrs  ' move cursor to start of next line.
return                       ' Otherwise, return to program.
'==Handle control characters. Understands the same five screen
'controls as BS2 Debug, plus cursor-on, cursor-off, and a single-byte
'cursor-positioning command.
ctlChar:
  bytOut = 255
  lookdown char,[clr,HOME,crsF,crsN,BKSP,TAB,CR,128],bytOut
  branch bytOut,[clrLCD,homeCrs,offCrs,onCrs,backSp,tabCrs,cRet]
position:
  scrPos = char-128//80      ' Subtract 128, limit to 0-79.
  goto moveCrs                        ' Move the cursor.
clrLCD:
  scrPos =0                              ' Clear screen position.
  bytOut =1:goto writeInstruction        ' Send cls (1) to LCD
homeCrs:
  scrPos =0:goto moveCrs       ' Move cursor to 0.
offCrs:
  bytOut =12:goto writeInstruction        ' Send off-cursor instruction.
onCrs:
  bytOut =13:goto writeInstruction        ' Send on-cursor instruction.
backSp:
  scrPos=scrPos-1//80          ' Move back 1; limit 0-79.
  gosub moveCrs                ' Move the cursor.
  bytOut=32: gosub writeData   ' Write a space to screen.
  goto moveCrs                 ' Move the cursor and return.
tabCrs:
  scrPos = ((scrPos+4) & $FC)//80 ' Jump to next multiple of 4 pos.
  goto moveCrs
cRet:
```

Figure 15-4 *(Continued)*

```
 scrPos = ((scrPos/20)+1)*20    ' Move to next multiple of 20 pos.
 goto moveCrs
'==Update the cursor position. Used when text wraps from one line
'to another, or control characters move the cursor.
moveCrs:
 bytOut = scrPos/20 & %11        ' Convert position to lines 0-3.
 lookup bytOut,[128,192,148,212],bytOut ' Lookup LCD DD-RAM address.
 bytOut = scrPos//20 + bytOut    ' Add offset 0-19 to line address.
 gosub writeInstruction ' Write it to the LCD.
return
'==Write instructions or data to the LCD.
writeInstruction:          ' If RS=0, the LCD interprets data
 low RS          ' sent as an instruction. If RS=1,
 goto wContinue  ' data is interpreted as data and
writeData:                 ' written into memory (i.e., the screen).
 high RS                   ' This routine sets or clears RS depending
wContinue:                 ' on whether you gosub "writeInstruction"
 low RW         ' or "writeData." In either case, RW=0
 busOut = bytOut.highnib ' (write), and data is transferred 4 bits
 pulsout E,10            ' at a time. The high nibble is first,
 busOut = bytOut.lownib  ' then the low nibble. The actual transfer
 pulsout E,10            ' happens when the enable pin (E) is pulsed
return                    ' high. When done, the routine returns.
```

Figure 15-4 (*Continued*)

In order to use the terminal, another computer (such as a BS2) must be connected to the same RS-485 net. It can send an ID sequence followed by up to eight characters. If the sender wants to transmit fewer than 8 characters, it must end the string early with the EOS (end-of-string) character. A single BS2 instruction handles all of this—recognizing the ID, acquiring a string of a given length, and optionally ending the string on the EOS character. Here's that versatile instruction:

```
Serin serIO,baud,[WAIT (tilde,myID1,myID2), str hostBuf\8\EOS]
```

Basically, the instruction works like this: Serin looks at pin serIO for serial data at the baud rate set by baud. It waits for the three-byte sequence of tilde, myID1, myID2, which in our case consists of ~AA. When ~AA is received, the characters that follow are stored in the cells of an array called hostBuf (for host buffer) until either eight characters have been received, or the EOS character is recognized. If EOS causes the instruction to end before eight characters are stored, the remaining cells of the array contain 0s.

Obviously, Serin was designed for just this kind of job.

Once the string is received, the terminal program goes to work on it. It examines each byte and determines what to do based on its meaning from Table 15-4. If a byte is an ordinary ASCII character like 65 (the code for "A"), the terminal prints it to the LCD. If it's a control code like 12 (ASCII form feed, interpreted by terminals as clear screen), the terminal takes the appropriate action. And if it's a terminal instruction, such as 208 for scan keypad, the terminal carries out that instruction and, if appropriate, transmits back a response.

When the program is finished processing a string, it sends an acknowledgment character (ackChar, the ` character) to the host. A well-behaved host won't send any more data or instructions until it sees this response. If it doesn't receive this response within a reasonable amount of time, it has to assume that the terminal it has been talking to is

TABLE 15-4 NETWORK TERMINAL INSTRUCTIONS

CODE	MEANING	REMARKS
0	ignored	
1	HOME*	move cursor to top/left of LCD
2	ignored	
3	EOS	end string short of 8 bytes
4	crsF	turn LCD cursor off
5	crsN	turn LCD cursor on
6,7	ignored	
8	BKSP*	back up 1 space and erase
9	TAB*	go to next multiple-of-4 screen location
10,11	ignored	
12	CLR	clear the LCD screen
13	CR*	carriage return; go to start of next line
14–31	ignored	
32–127	ASCII characters	standard text; may use characters in quotes like "A" to send these codes. Do not use the "~" or "`" characters; these are reserved as ID prefix and acknowledgment.
128–207	screen position	move to screen position 0 (128) thru 79 (207)
208	scan keypad	scan keypad for 1 character, return "0" thru "9", "*", "#", or "A" thru "D"; if no key pressed, return "-"
209	get input	let the user enter a series of keys and return corresponding string
210–215	display string	display one of several strings stored in the BS2's EEPROM
216–255	ignored	

NOTE: entries marked with * are predefined constants in PBASIC2 and may be used without a separate "con" directive defining them. They are normally used in Debug and Serout instructions.

turned off or not connected to the net. The serial-timeout feature of the Serin instruction makes this easy:

```
Serin serIO,baud,T_O,enRet,[WAIT (ackChar)]
```

Serin listens to pin serIO at the baud rate set by baud. It is waiting for ackChar to arrive. If data does not arrive within the time set by the constant T_O, the program will go to the label enRet.

There's one caution to observe with this kind of instruction—the timeout value does not dictate how long to wait for ackChar, just for any serial data. If any serial data arrives at intervals less than the timeout time, Serin will continue to wait for ackChar. However, the way our network terminals are programmed to behave, the only serial activity on the net should be an ackChar from the terminal that last received data. But when you start programming your own network applications, you should be aware of this subtlety involving timeouts.

The program listing of Figure 15-5 governs the behavior of an example network master. It sends data and instructions, and waits for ackChar to indicate that its wishes have been carried out. The program instructs a set of terminals (with IDs AA through AD) to scan their keypads, and, if a key is pressed, to get a bogus user-ID number from the person

```
'Program: NETMSTR.BS2 (BS2 network master)
'This program demonstrates how to use the network terminal NETERM.BS2.
'One of these network masters (wired as shown for the terminals,
'but without keypad and LCD) can manage multiple slaves. Slave IDs
'follow the pattern AA, AB, AC...AZ. This program polls each
'of its slaves, from AA through A? (?=maxID constant below) for a
'keypress, has it get data from a user, checks that data against
'a constant, and display accept or reject messages on the slave
'depending on whether the data matches the constant. Although the
'program is flawed as a password-access system for reasons described
'below, it is an excellent example of interfacing with the NETERMs.
'========Constants=========
tilde    con    "~"      ' Prefix for ID; DO NOT USE IN NORMAL TEXT.
ackChar  con    "`"      ' Acknowledgment; DO NOT USE IN NORMAL TEXT.
OKcode   con    123      ' Dummy passcode for demo.
baud     con    $54      ' 9600-baud comms.
T_O      con    300      ' Number of ms to wait for terminal response.
maxID    con    "D"      ' Highest ID of connected terminals (A-?)
serIO    con    1        ' Pin P1 used for serial input/output.
serDir   con    0        ' Pin P0 sets serial direction 1=talk, 0=listen.
getKey   con    208      ' Tell terminal to get one key (- = no press).
getNum   con    209      ' Tell terminal to get a number (- no press).
intro    con    210      ' Terminal: display intro screen.
enter    con    211      ' "            "        entry "
accept   con    212      ' "            "        accept "
reject   con    213      ' "            "        reject "
moment   con    214      ' "            "        one moment please "
prompt   con    215      ' "            "        prompt for input "
EOS      con    3        ' end-of-string marker (ctl-C).
'========Variables=========
reply    var    byte     ' Single-byte response from terminal.
cmd      var    byte     ' Go/no-go instruction for terminal.
slave    var    byte     ' ID of slave to check.

'========MAIN PROGRAM=========
'After setting up each of the slaves with the entry screen,
'the program goes into a loop, interrogating each slave for
'a keypress. When a keypress is detected, the slave is walked
'through the procedure of getting an ID number, which is checked
'against OKcode. If it matches, the slave displays the "accept"
'screen; if it doesn't, the slave displays the "reject" screen.
for slave = "A" to maxID        ' Display entry screen on each of
```

Figure 15-5 Program listing for network master.

```
      gosub entryScreen                   ' the slave terminals.
    next
    'Check each unit for a keypress. If a particular unit does not
    'respond to this request, after T_O milliseconds the program
    'skips to the next unit. This ensures that the system will
    'still work (albeit more slowly) even if one or more slave units
    'are absent or broken.
    poll:
      for slave = "A" to maxID                ' Poll each slave for keypress.
        high serDir:pause 1            ' Set to transmit.
        Serout serIO,baud,["~",slave,getKey,EOS]      ' Check for keypress.
        low serDir                             ' Set to receive.
        Serin serIO,baud,T_O,nextUser,[reply]        ' Get response.
        if reply <> "-" then handleRequest     ' Key pressed? Take care of it.
    nextUser: next
    goto poll
    'If a keypress is detected, the program continues here to handle it.
    'Accepting data from the user is the Achille's heel of this demo.
    'During data entry, the slave unit cannot listen to the master.
    'And the master cannot talk to other slaves because it must listen
    'for a response. So the system is at the mercy of users who keep
    'entering and deleting keystrokes within the keypad timeout period.
    'For the demo, it's acceptable to understand this limitation. In
    'a real application, some strategy would be required—perhaps the
    'slave terminal could count keystrokes and quit after a given
    'number.
    handleRequest:
      high serDir:pause 1                     ' Set to transmit.
      Serout serIO,baud,["~",slave,prompt,getNum,EOS]        ' Display prompt.
      low serDir                              ' Set to receive.
      Serin serIO,baud,[DEC reply]            ' Get the number.
      cmd = accept                            ' IF reply = OKcode then
      if reply = OKcode then good             ' ..cmd = accept ELSE
      cmd = reject                            ' ..cmd = reject.
    good:
      high serDir:pause 1                     ' Set to transmit.
      Serout serIO,baud,["~",slave,cmd,EOS]   ' Display result.
      pause 2000                              ' Wait for user to read it.
      gosub entryScreen
    goto nextUser                             ' Next user.
    'Display the entry screen on the currently selected slave unit.
    'In this demo, the program waits for an acknowledgment for
    'a time, and continues on, whether or not it receives one.
    'Something to consider in a real application might be how to
    'handle missing "acks." Call maintenance? Delete them from
    'the list of units to poll?
    entryScreen:
      high serDir:pause 1                     ' Set to transmit.
      Serout serIO,baud,["~",slave,enter,EOS]        ' Display main screen.
      low serDir                              ' Set to receive.
      Serin serIO,baud,T_O,enRet,[wait (ackChar)]    ' Receive ack.
    enRet: return                             ' Return to program.
```

Figure 15-5 *(Continued)*

pressing the keys. If the ID is "123" the host tells the terminal to display a welcome message. If it's anything else, the host tells the terminal to display a bug-off message. When the transaction is done, the host goes back to scanning the keypads.

Although this is hardly a realistic application, it is a suitable springboard for designing your own applications that communicate with the network terminals.

PARTS LIST

Resistors (all 1/4W, 10% or better)
R1–R9—1k
R10,R11—10k
R12—5k potentiometer
Rt—120 ohms
Other Components
C1—0.1µF, 0.1µF ceramic capacitor, 50WVDC
U1—LTC1487 (Digi-Key LTC1487CN8-ND)
Keypad—16-key keypad (Digi-Key GH5003-ND)
LCD—4x20 alphanumeric display (Digi-Key 73-1091-ND)

USING THE POWERFUL BS2-SX

The BASIC Stamp II SX (BS2-SX) is the hotrod of the Stamp family. Hopped up with extra speed, memory, and program space, it's designed to tackle problems too tough for the other Stamps.

Like an automotive hotrod, the BS2-SX is not the best vehicle for beginners. You should learn the fundamentals on a BS2, then step up to the BS2-SX when a particular project requires it.

This chapter assumes that you have some experience with the standard BS2. We will build on that knowledge and get you up to speed on the high-performance BS2-SX.

Most of what we will cover also applies to the BS2-E, which has all of the BS2-SX features except warp speed. The program examples presented in this chapter will run properly on both the BS2-SX and BS2-E.

SX-Powered Stamp

A BS2-SX is essentially a standard BS2 with a more powerful engine—a rip-snorting, 50-MHz SX microcontroller[1] replaces the tame, 20-MHz PIC micro used in the standard BS2. This engine swap, coupled with a larger program EEPROM, gives us performance boosts in three areas:

■ Speed. The BS2-SX runs 2.5 times faster. This allows finer timing resolution for delays and pulse-generation/measurement, faster serial communication, higher frequency sound synthesis, and just plain faster overall operation. (This does not apply to the BS2-E, which is clocked at 20MHz.)
■ Memory. The BS2-SX has an additional 64-byte block of RAM. This so-called scratchpad RAM, which is outside the memory used for variables, can be used like EEPROM data space, but without the slow access and write limitations.
■ Multiple Program Storage. The BS2-SX can store eight full (2kB) BS2 programs in EEPROM and switch between those programs on the fly. With thoughtful design, you can use this capability to create projects with multiple modes of operation. This is an important capability, which we'll demonstrate in examples programs here (Figures 16-1 through 16-4) and put to good use in the capacitance-test project (Chapter 17).

The SX enhancements are built on the standard BS2 chassis, so a BS2-SX will run standard BS2 programs with few (sometimes no) changes. However, to get the maximum performance boost, programs should be designed from the get-go for the BS2-SX. To that end, let's take a more detailed look at the SX advantages of Speed, Scratchpad Memory, and Multiple Programs.

SPEED: MORE THAN JUST FAST

A standard BS2 executes about 4000 BASIC instructions per second (ips); the BS2-SX is 2.5 times faster, at 10,000 ips. The benefit of additional speed goes beyond processing more instructions in less time, though. It also makes several PBASIC instructions work better or more precisely.

---

1. In a nutshell, the SX microcontroller is similar to the PIC micro used in the original Stamps but reengineered for superior speed and new capabilities. The BS2-E gets a 20-MHz SX processor, giving it the same speed as the original BS2, but additional program space and scratchpad memory. For more information on the remarkable SX, see http://www.sxtech.com/

Note that the speedup applies only to the BS2-SX. The other SX-based Stamp, the BS2-E, has been restricted to 20MHz for complete timing compatibility with the standard BS2. The idea is to give BS2 programmers a device with more memory (described later) without requiring them to modify existing programs' timing.

Timing: Pause, Pulsout, Pulsin, RCTime, and Count   All of the timing instructions have similar characteristics. They have a specified unit of time[2] (e.g., 1 millisecond for Pause) and a maximum time of operation of 65,535 units (the highest number that PBASIC-2's 16-bit arithmetic can handle). In the BS2-SX, all of the timing instructions except Pause and Count have had their timing units cut from 2 microseconds (μs) to 0.8 μs.

This lets you generate or measure timing events more precisely, but limits you to a maximum duration of 65,535 $\times$ 0.8 μs = 52.4 ms. We'll spell out the effect on each of the timing instructions:

Pause, the time-delay instruction, works in units of 1 millisecond (ms), just like the original BS2. So PAUSE 1000 still gives a 1-second delay.

Pulsout generates a pulse on a pin by inverting its state for a specified time, then changing it back. On the BS2-SX, Pulsout works in units of 0.8 μs for a maximum pulse length of 52.4 ms. For example, PULSOUT 3,7734 generates a pulse on pin 3 of 0.8 $\times$ 7734 = 6187 μs (6.19 ms) duration.

Pulsin measures the width of a pulse on a given pin in units of 0.8 microseconds, up to a max of 52.4 ms. To avoid getting stuck if a pulse doesn't occur or is too long, Pulsin gives up after 52.4 ms. For example, PULSIN 6,1,pWidth would watch pin 6 for a pulse that begins with a change in state from 0 to 1. If the pulse occurs within 52.4 ms and is less than 52.4 ms long, the length in units of 0.8 μs is stored in the variable pWidth.

RCTime is similar to Pulsin, but instead of a complete pulse it measures the time in 0.8 μs units for a pin to change out of a given state (1 or 0). If the pin doesn't change within 52.4 ms, RCTime quits. For example, RCTime 4,0,theTime would measure the time for pin 4 to change state from 0 to 1. If this change occurs within 52.4 ms, the time in units of 0.8 μs is stored in the variable theTime.

Note that RCTime requires a small amount of setup time before it starts its actual timing measurement. On the BS2, that time is approximately 200 μs; the speedier BS2-SX cuts it to 80 μs. So more than 80 μs must elapse between RCTime's execution and the change in state.

Count measures the number of complete cycles (change in state from 1 to 0 and back to 1 or 0 to 1 to 0) that occur during an interval of time in units of 0.4 ms. Maximum interval is 65,535 $\times$ 0.4 ms = 26.2 seconds. In the BS2-SX, Count can recognize transitions (changes in state 1-0 or 0-1) as close as 1.6 μs apart. Since a cycle consists of two

---

2. Units of time used in this chapter:
    1 millisecond (ms) is one-thousandth of a second;
    1 microsecond (ms) is one one-millionth of a second (one one-thousandth of a millisecond).
    1 megahertz (MHz) is one million cycles per second.
    1 kilohertz (kHz) is one thousand cycles per second.
    1 kilobit per second (kbps) is one thousand bits per second. (Bits per second is often called 'baud.')

transitions, cycle times of 3.2 μs are acceptable, making Count capable of handling square-wave signals[3] of up to 312,500 cycles per second (hertz, abbreviated Hz; or 312.5 kilohertz, kHz).

Serial communication: Shiftin, Shiftout, Serin, and Serout   BS2s support two forms of serial communication—synchronous (communication with peripheral chips; see Chapter 12), and asynchronous (e.g., RS-232; see Chapter 13).

The primary difference between sync and async has to do with timing. Synchronous communication uses a separate clock pin to supply a "beat" to which each transmitted data bit is synchronized. Asynchronous scraps the clock pin, requiring the sender and receiver to agree precisely on the bit timing.

Shiftin and Shiftout are synchronous-communication instructions and as such are not functionally affected by the SX speedup. Bits being sent or received are synchronized to the clock pin. As long as the clock pulses don't exceed some maximum frequency—usually greater than 1 megahertz (MHz; millions of cycles per second)—the actual speed is irrelevant. The original BS2 clock rate was 16,667 cycles/second (16.67 kilohertz; kHz), so the BS2-SX's is 2.5 × 16.67kHz = 41.67 kHz.

Compared to 1 MHz (or more), 42kHz is relatively slow, so most peripheral devices will have no trouble keeping up with the BS2-SX.[4]

Serin and Serout are asynchronous serial instructions. As such, they require you to set a baudmode value that controls how fast bits will be sent or received (bits per second; bps). While the original BS2 maxes out at about 50,000 bps (50 kbps), the BS2-SX can achieve 115.2 kbps. BS2-SX baudmode values for a given baud rate are different than those for the original BS2. The BS2-SX manual contains a chart, but it's missing some of the higher speeds; see Table 16-1.

I've added a couple of columns to Table 16-1 to indicate the actual timing of serial data with the BS2-SX. As you can see, timing accuracy is very good up to 57,600 baud. At 115,200 the closest baudmode value produces timing that is more than three percent off. A timing error of more than five percent almost inevitably produces errors in serial

---

3. A square wave is a signal that switches back and forth between two fixed voltage levels, spending equal time at each. For example, if you wrote the following Stamp program—

```
repeat:
  toggle 1   ' Flip pin 1 state 0-to-1 or 1-to-0
  pause 1
goto repeat
```

—you'd be cycling the state of pin 1 between 1 and 0 every 1 ms. This would produce a complete cycle (1-0-1 or 0-1-0) every 2 ms. Since an equal amount of time would be spent in each state (1 and 0), the output of pin 1 would be a square wave. Some techs loosely use the term square wave to mean any signal that switches between two fixed voltages regardless of timing, but that's not the true definition.

4. Some devices state both a maximum clock rate and a minimum pulse width. You must comply with both. For example, the maximum clock rate might be 1 MHz, but only if the clock is a true square wave (see note 3). Each cycle of a 1 MHz square wave takes 1/1,000,000th of a second, 1 μs and is low (0) for half that time and high (1) the other half. So the manufacturer might additionally specify a minimum pulse width of 0.5 μs. BS2-SX clock output is not a square wave; it consists of a 5.6 μs pulse repeating every 24 μs. Suppose a device's specs call for a max clock rate of 100kHz with a minimum pulse width of 8 μs. You'd be OK on the clock rate, but out-of-compliance with the minimum pulse width. You will probably never encounter such a problem, but it pays to be prepared.

TABLE 16-1	BS2-SX BAUDMODES AND SERIAL DATA RATES					
	DIRECT CONNECTION		THROUGH LINE DRIVER		TIMING ACCURACY	
DATA RATE	8 BITS, NO PARITY	7 BITS, EVEN PARITY	8 BITS, NO PARITY	7 BITS, EVEN PARITY	ACTUAL BAUD RATE	PERCENT TIMING ERROR
9600	16624	24816	240	8432	9615	+0.16%
19200	16494	24686	110	8302	19231	+0.16%
38400	16429	24621	45	8237	38462	+0.16%
57600	16407	24599	23	8215	58139	+0.93%
115200	16385	24577	1	8192	119048	+3.34%

communication. The baudmode value isn't the only potential source of error; all BS2-SX timing is controlled by a ceramic resonator, which has a tolerance of approximately ±1 percent.

Bear in mind the potential for timing errors if you design a project around the 115,200-bps capability. You'll have no problem communicating between two BS2-SX devices (since most of the error, the baudmode, is the same in each). Communicating with a PC at 115,200 bps via a very short cable should usually be OK, but it entails more risk of communication errors. Other devices will have varying ability to tolerate baud-rate-timing errors; check them out carefully or consider dropping back to 57,600.

In addition to changing baudmodes, the SX speedup affects serial timeouts,[5] reducing their units of time from 1 ms to 0.4 ms.

Digital-to-analog conversion: PWM, Freqout, DTMFout   PWM, the instruction that allows you to make an analog output from a stream of digital pulses, is only slightly changed by the SX speedup. You still specify 0-255 cycles of PWM output for a given duty cycle (also 0-255), but in the SX version each burst is 0.4 ms instead of 1 ms.

Freqout, which generates one or two sinewave tones for audio or musical purposes, gets a greatly expanded frequency range in the BS2-SX. It operates in units of 2.5 Hz over a range of 0 to 32,767 units, meaning it can output sine-wave audio from an ultralow 2.5 Hz to ultrahigh 82.917 kHz. One interesting benefit of this extended frequency range is that the BS2-SX can generate ultrasound for sonar applications (eliminating a couple of components from the project in Chapter 14, for example).

Tone durations are specified in units of 0.4ms (versus 1ms in the regular BS2).

DTMFout is basically unchanged, except that, like Freqout, its tone durations are specified in units of 0.4ms (versus 1ms in the regular BS2).

SCRATCHPAD MEMORY: FLEXIBLE TEMPORARY STORAGE

A standard BS2 has 32 bytes of RAM; 6 bytes are set aside for the I/O pins, leaving 26 bytes for use as variables in your programs.

5. Timeouts are optional timing values used with Serin or Serout to specify how long the Stamp should wait for data to be received (Serin) or for permission to send data (Serout).

BS2 programmers who run up against the 26-byte limit sometimes resort to swapping data in and out of the EEPROM program memory using the Write and Read instructions. This is a questionable practice, since EEPROM has a couple of big limitations: It can only be rewritten a few million times, and each Write takes several milliseconds to complete. It's no substitute for RAM, which has no write limitations, and no speed penalty.

Scratchpad memory is not a perfect substitute for RAM either, but it eliminates the two worst problems with the EEPROM trick (write cycles and speed). The BS2-SX has 64 bytes of scratchpad RAM. Although you can't use it to define additional variables (using the VAR directive), you can use it as temporary storage space for variables, or as a way of communicating between multiple programs.

Using the scratchpad is similar to using EEPROM. To write to scratch RAM, use the instruction PUT location, data where *location* is a value from 0-62 representing the byte of scratchpad RAM to be written, and *data* is a byte (0-255) value to be written there. Both *location* and *data* can be variables or constants.

Note that byte 63 of scratch RAM is not accessible to the Put instruction. The BS2-SX stores the number of the currently active program in byte 63. If your program needs to determine where it is loaded—program number 0-7—it can determine this by Getting byte 63.

To read from scratch RAM, use the instruction GET location, variable where *location* is a value from 0-63 representing the byte of scratchpad RAM to be read, and *variable* is a byte variable to which the stored value will be transferred.

Put and Get only deal with byte-sized data, but you can use PBASIC's variable modifiers to store other types of data in bytes. We'll look at this in a later example.

MULTIPLE PROGRAMS: MULTIPLE PERSONALITIES

The final SX enhancement is multiple program storage. On a standard BS2, your program resides in a 2kB EEPROM. The BS2-SX has a 16kB EEPROM that is divided into eight separate 2kB program memories numbered 0 through 7.

When a BS2-SX starts up (after power-on, reset, or a fresh program download) it runs program 0. In fact, unless program 0 instructs it otherwise, the SX will run *only* program 0, ignoring any other loaded programs.

Switching between programs requires using the Run instruction: Run progNumber where *progNumber* is a value from 0-7 representing the program number you wish to run.

If only it were that simple! There are actually quite a few housekeeping details you must observe in order to make this multiple-program stuff work properly. I'm going to provide some practical exercises later, but here are the main points:

■ Program 0 (the default, startup program) should contain a special comment listing all of the other programs it will Run (subordinate programs).

■ When a program Runs another, all variables and IO pin settings remain the same until or unless the new program changes them. And if both programs use the same variable names and sizes (VAR directives) in *exactly the same order,* variables will retain their names and values from one program to another. However, even the slightest change in the variable declarations and this won't necessarily hold true.

- When a program Runs another, the previous program's Gosub/Return markers are lost. You cannot put a Run instruction inside a subroutine and return to the correct spot when you later Run the original program. This unfortunately means that subordinate programs cannot be used as batches of subroutines to support a main program—at least not without some careful forethought and additional programming.

- Programs always start at their first line of code when Run by another program. If you need to run different portions of a program depending on where the Run instruction came from, you have to program the mechanics yourself.

- Finally, the multiple programs only run one at a time. This is not like *multitasking* in which processor time is divided up so that two or more programs can be run simultaneously (at reduced performance). Although the multiple SX programs execute only one at a time, they get the full performance of the processor.

Practical Examples of BS2-SX Programming

If you've programmed the standard BS2 before, most of the new capabilities of the BS2-SX should be easy to integrate into your existing knowledge. Just watch out for side effects of the fast processing speed and be alert to opportunities to use the new scratch RAM.

The multiple-program capability may throw you for a loop, however. It requires a considerable change in your programming style, unless you just happen to have an application in which you need multiple, separate programs loaded. In my experience, most Stamp programmers who run up against the program-size limitation of the standard BS2 want the ability to write *larger* programs, not necessarily *more* programs.

Our examples will therefore concentrate on making the multiple-program capability act more like a single big program that is broken up into manageable subprograms.

EXAMPLE 1: MOVING BETWEEN PROGRAMS

Working with multiple programs means working with multiple PBASIC program files. To avoid confusion, I recommend that you create a separate Windows folder to hold all programs that will be loaded into the BS2-SX for a given application. I'll refer to such a set of program files as a *project.*

To begin our first example project, I

- Opened the Windows Stamp editor,
- Hit Ctrl-N to begin a new file,
- Typed in the first program listing of Figure 16-1,
- And hit Ctrl-S to save the file.

Next I clicked the Create New Folder icon and named the folder Example1. I opened the new folder and saved the file under the name example1_pgm0, with the Save as Type selection set for BASIC Stamp 2SX files (*.bsx).

It's important that BS2-SX files get the correct file extension (.BSX). If they don't, the Stamp editor will flag SX-specific instructions like Run, Get, and Put as errors when you try to run the program. After all, the standard BS2 has no such instructions.

Then, without closing example1_pgm0, I hit Ctrl-N to get another program window. If you follow my steps, you'll see that the Stamp software automatically creates a tabbed window so that you can easily switch amongst open program files. I then typed in example1_pgm1 and saved it to the same folder (Example1) as a .BSX file.

Finally, one more stab at Ctrl-N to create another window, and I typed and saved example1_pgm2 in exactly the same way.

It's not necessary to name your programs as I have, but it does help keep things organized.

Next I selected Preferences from the Edit menu and clicked on the tab named Editor Operation, then set the Default BS2SX Download Mode to Modified and clicked OK. This setting ensures that all of the programs get loaded into the BS2-SX. There's also an All selection for Download Mode that forces all programs to download every time, whether they need to or not. It's much slower and rarely if ever necessary to use All.

If you've followed the same steps I did, you're ready to run. Hit Ctrl-R to run the programs. A Debug window pops up and text from each of the three programs appears there. This demonstrates that the BS2-SX can hop from one program to another.

And that's about all that the trio of programs does. Note that the Pauses in program 0 and program 2 are not really necessary—they just prevent the Debug printouts from whizzing by on the screen.

Check out the special comment at the beginning of example1_pgm0. It tells the Stamp software that this is a BS2-SX program, and lists the other programs that make up this project. Such a comment implies that this is program 0, and the other programs are loaded in order of their listing.

Within each of the three programs the Run instruction is used to jump to the next program. Run can send the BS2-SX to any program slot-even slots that don't *have* programs. For example, if you change the Run 1 to Run 4 in example1_pgm0, you won't get an error message, but the project won't work as expected.

Another item to notice is in example1_pgm2. Here we Get the contents of scratch RAM location 63. This location holds the number of the currently running program.

Before we go on to the next program, I'd like you to select Close from the File menu. If the software prompts you to save your file(s), do so. After all three files are closed, reopen example1_pgm0 (Ctrl-O) and notice what happens. Thanks to the special comment in that program, all three files are automatically opened together. This takes some of the strain out of project programming. (See Figure 16-1.)

EXAMPLE 2: VARIABLES AND SCRATCHPAD RAM

When you switch between programs, all memory locations retain their current contents. Remember that I/O pins are considered special types of memory, so they retain their states, too.

The tricky aspect of this is that individual programs may refer to the same piece of memory by different names. This may be helpful in some cases, but it's usually an opportunity to get yourself very confused. Unless you have a good reason to do otherwise, I recommend that you either keep all variable declarations (lines with the Var directive in them)

```
' PROGRAM: example1_pgm0
'{$STAMP BS2SX, example1_pgm1, example1_pgm2}
pause 1000
debug CR,"This is program 0.",CR
debug "Switch to program 1.",CR,CR
run 1


' PROGRAM: example1_pgm1
debug "This is program 1.",CR
debug "Switch to program 2.",CR,CR
run 2


' PROGRAM: example1_pgm2
pgmN   var    byte
get 63,pgmN
debug "This is program ",DEC pgmN,".",CR
debug "Back to program 0.",CR
pause 1000
run 0
```

Figure 16-1 Three short programs demo the Run instruction.

identical across all programs in a project, or use fixed-position variable names (b0, b1, b2… or w0, w1, w2…).

The programs of Figure 16-2 demonstrate how this works, and occasionally doesn't work. As with the first example, make a folder, type and save the programs, and run them.

The debug output shows that program 0 sets the variables length and width to 275 and 120, respectively. Program 1, which has identical variable declarations, multiplies length by width and gets the correct answer for area, 33,000. Program 2, whose variable declarations include an extra item, multiplies length by width and gets 0; the wrong answer!

This illustrates that while the memory contents remain after switching from one program to another, the variable names do not. Only if you keep the identical names, sizes, and sequence of variable declarations can you count on variable names and values being preserved properly.

The project also demonstrates the use of scratch RAM. Program 0 copies the values of length and width to the first four bytes of scratch. It uses the variable modifiers highbyte and lowbyte to split up the word (16-bit) variables into bytes. Program 2 Gets this data back into the variables length and width, and recalculates the area correctly. Of course, if program 2 then Ran program 1 or 0, the contents of width and length would not line up, and we'd get more errors. (See Figure 16-2.)

EXAMPLE 3: NAVIGATING INSIDE PROGRAMS

Although the multiple-program capability of the BS2-SX is a powerful feature, sometimes what you really want is to simply thin out the code in a single, large program. For example, if you find yourself bundling up a big block of instructions as a subroutine just to get it out of the way, you know what I mean. Instead of 96 lines of PBASIC code, you'd rather see something like Gosub bigRoutine.

Sadly, you can't use Gosub/Return between multiple programs. The return address is erased when you execute the Run instruction.

A related limitation is that Run always goes to the first line of the target program. There's no provision to specify which part of the program you wish to run.

It is possible to code around this limitation. The workaround uses the Branch instruction and a variable to go to specific points in the target program.

```
' PROGRAM: example2_pgm0
'{$STAMP BS2SX, example2_pgm1, example2_pgm2}
length        var    word
width         var    word
length= 275
width= 120
debug CLS,"length: ", DEC length
debug CR,"width: ", DEC width
put 0,width.highbyte
put 1,width.lowbyte
put 2,length.highbyte
put 3,length.lowbyte
run 1

' PROGRAM: example2_pgm1
length        var    word
width         var    word
pause 1000
debug CR,CR,"Program 1 says area= ", DEC (length*width)
run 2

' PROGRAM: example2_pgm2
area          var    word
length        var    word
width         var    word
pause 1000
area = length * width
debug CR,CR,"Program 2 says area = ", DEC area
debug CR,"That's not right!"
get 0,width.highbyte
get 1,width.lowbyte
get 2,length.highbyte
get 3,length.lowbyte
area = length * width
debug CR,CR,"Using data from scratch, area = ", DEC area
stop
```

Figure 16-2 Variables are preserved between programs, but watch out.

Figure 16-3 is our example. There are only two programs in this project, since each of the programs is longer than those in the previous examples.

If you enter and run this project, the debug screen will display "Curly Larry Moe Done." Looking at the code in Figure 16-3, you can see that program 0 had to selectively invoke the messages (msg1, msg2, and msg3) in program 1 and return to the proper points in its own code.

The key to this is the Branch instruction. Branch takes an index value and a list of labels used in the program. The labels are numbered 0,1,2,3... Branch jumps to the label corresponding to the value of index. So if index contains 2, Branch redirects the program to label number 2.

When example3_pgm0 starts up, all variables contain 0, so the initial Branch jumps to the 0th label, which is start. It loads the constant msg3_ (which has a value of 3) into toLabel and r1_ (value of 1) into retLabel, then Runs example3_pgm1.

The Branch instruction at the top of example3_pgm1 redirects the program to the label msg3 based on the value of toLabel (3). It prints "Curly" and Runs example3_pgm0.

```
' PROGRAM: example3_pgm0
'{$STAMP BS2SX, example3_pgm1}
toLabel       var   nib
retLabel      var   nib
msg1_ con     1
msg2_ con     2
msg3_ con     3
r1_   con     1
r2_   con     2
r3_   con     3

pause 1000
branch retLabel,[start,r1,r2,r3]

start:
toLabel= msg3_      ' "Gosub" message 3
retLabel= r1_       ' ..then "Return" to r1
run 1               ' Do it (hop to program 1).

r1:
toLabel= msg2_      ' "Gosub" message 3
retLabel= r2_       ' ..then "Return" to r1
run 1               ' Do it (hop to program 1).

r2:
toLabel= msg1_      ' "Gosub" message 3
retLabel= r3_       ' ..then "Return" to r1
run 1               ' Do it (hop to program 1).

r3:
debug CR,"Done."

' PROGRAM: example3_pgm1
toLabel       var   nib
retLabel      var   nib
msg1_ con     1
msg2_ con     2
msg3_ con     3
r1_   con     1
r2_   con     2
r3_   con     3

branch toLabel,[null,msg1,msg2,msg3]
null:
run 0

msg1:
debug "Moe",CR
run 0

msg2:
debug "Larry",CR
run 0

msg3:
debug "Curly",CR
run 0
```

Figure 16-3 Here's a technique using subroutines within programs.

The variable retLabel contains 1, so the Branch at the top of example3_pgm0 jumps to the label r1:. The project continues using this mechanism in which toLabel specifies the label to Branch to in example3_pgm1 and retLabel specifies the label to return to in example3_pgm0. Works like a charm.

Note that we could not define constants with the same names as the program labels (i.e., you can't have r1 con 1 in the same program as the label r1:. PBASIC won't allow it. So I added an underscore character to the names of the constants to satisfy the nitpicky software.

Also be aware that we're depending on identical variable declarations in both programs to ensure that toLabel and retLabel retain their correct contents between programs. Remember what happened in example2_pgm2 when we moved variables around? That's right—sorry, wrong number. Now imagine what havoc that would cause if the wrong number specified a destination or return Branch. This kind of programming requires you to stay on your toes.

EXAMPLE 4: BACKING UP YOUR VARIABLES

Throughout these examples, I've emphasized the need to use identical variable declarations so that multiple programs refer to the same variable data by the same names. The assumption is that all of the programs will be working with the same data in different ways.

What if one or two of the programs in a given project are doing something entirely different, with completely different data? And what if it's important that their variables be preserved intact even if another program is Run?

Scratchpad RAM and a little-known feature of BS2 variable memory can pull together to back up all of a program's variables before a program quits, then restore them the next time the program runs.

Figure 16-4 shows how backup/restore works. It requires the use of a single byte variable located at the end of variable memory. All programs in a project that use this technique must leave the last byte unused. If you're unsure, hit Ctrl-M in the Stamp editor to get a memory map, then look at REG12. The right-hand half of its bar, representing the final byte of Stamp memory, must be gray (unused).

In the example program (Figure 16-4), I've used the default byte name b25 to designate the last byte. Because variables organized by the automatic Var directive build from the bottom up, b25 will always be the last byte of memory used, so it's a safe choice for the backup scheme.

In example4_pgm0, subroutines backup and restore copy all data to and from variable memory to scratchpad RAM. They do this by treating the variable memory as a 25-byte array. An array is simply a set of consecutive variables that can be accessed by an index number. The usual way to work with an array is to create one like so: myArray var byte(10). The 10 bytes that make up myArray can be addressed by number, 0 through 9; for example myArray(6) = 211 stores the value 211 to the 7th byte of myArray. (Byte 6 is the 7th byte because the counting starts at 0.)

A useful loophole to the array concept is that you don't need to explicitly create an array in order to use the array mechanism. When PBASIC2 sees something like aVar(index) in a program, it takes these steps:

```
' PROGRAM: example4_pgm0
'{$STAMP BS2SX, example4_pgm1}
count1        var    word
count2        var    word
count3        var    word

pause 1000
gosub restore
count1 = count1 + 1
count2 = count2 + 2
count3 = count3 + 3
debug CR,CR,"Pgm0 variables: "
debug DEC count1,TAB,DEC count2,TAB,DEC count3,CR
gosub backup
run 1

backup:
  for b25 = 0 to 24
      put b25,b0(b25)
  next
return

restore:
  for b25 = 0 to 24
      get b25,b0(b25)
  next
return

' PROGRAM: example4_pgm1
oops1         var    word
oops2         var    word
oops3         var    word

pause 1000
debug CR,"Pgm1 variables: "
debug DEC oops1,TAB,DEC oops2,TAB,DEC oops3,CR
random oops1
random oops2
random oops3
debug "After random: "
debug DEC oops1,TAB,DEC oops2,TAB,DEC oops3,CR

run 0
```

Figure 16-4 Program saves/restores all variables to/from scratch RAM.

■ Determine the size of the variable aVar (bit, nib, byte, or word)
■ Count up in memory by *index* number of spaces corresponding to the variable size
■ Access the data located at the resulting offset location

So assuming that aVar is defined as a byte, aVar(3) refers to the byte located three bytes higher in variable memory than aVar.

Used indiscriminately without the proper array declaration, this could be a recipe for trouble. Most of the time, it's better to use arrays as prescribed in the Stamp manual. But our situation is a perfect opportunity to exploit this loophole; we want to start at byte 0 (b0) and backup or restore every byte up to b24 to or from scratchpad memory. And that's exactly what the For/Next loops in example4_pgm0 do.

If you type, save, and run the project you'll see proof: example4_pgm0 declares three variables to use as counters, while example4_pgm1 declares three overlapping variables that get randomized. Nonetheless, example4_pgm0 manages to preserve the contents of its variables even after example4_pgm1 has shredded them.

If you need further proof that the backup/restore mechanism works, try removing the line gosub backup from example4_pgm0 and rerun the project. The random data from example4_pgm1 ruins the count. (See Figure 16-4.)

Scratchpad RAM has a total of 63 usable bytes, so it is possible to apply the backup/restore techniques to two programs within a given project. Both programs could use the variable space without worrying about contamination from the other(s).

CHARGE-TRANSFER CAPACITANCE
SENSING (BS2-SX)

CONTENTS AT A GLANCE

Harnessing the power of the BS2-SX to a technique as old as the study of electricity itself, we can measure ultralow capacitance values for such applications as

- Sorting and salvaging unmarked capacitors
- Measuring moisture and humidity
- Detecting human touch
- Measuring liquid level inside a sealed vessel

In the process, we'll look at a powerful technique for storing calibration data as BS2-SX programs.

Measuring Capacitance

If we want to measure capacitance, the BS2 family's RCTime instruction is an obvious candidate for the job. It is capable of measuring the time required for an unknown capacitor to charge through a known resistance value. With a little math (see Appendix C), we can calculate the capacitance from the charge time.

RCTime does an admirable job with middling values of capacitance, but it requires some setup time,[1] which limits its usefulness with very small capacitance values in the picofarads (1×10^{-12} Farads). For example, from Appendix C we know that the time required to charge a capacitor in the preferred RCTime arrangement is calculated as

$$R \times C \times 1.204 = time$$

So if we have a 100k resistor and a 100 pF capacitor, the calculation is:

$$(100 \times 10^3) \times (100 \times 10^{-12}) \times 1.204 = 12 \text{ }\mu\text{s}$$

Twelve microseconds is far too short for RCTime to measure, given the minimum setup time of 214 μs for the BS2 and 86 μs for the BS2-SX. Cranking up the value of the resistor would help, but only to a point. In most cases we ignore the fact that there can be a very small leakage current of about 1μA in or out of a Stamp I/O pin in its input mode. But if we increased the resistor to 1M, that measly 1 μA would cause a 1-volt drop across the resistor, throwing off our calculations.

What to do? A common way of measuring small capacitance values is to build an oscillator whose frequency is dependent on a resistor-capacitor combination, then measure that frequency. That's a time-honored method and one I once experimented with to measure water level.[2] Works fine, but it's in my nature to look for simpler solutions. I found one.

The Charge-Transfer Technique

A simplistic way to think about a capacitor is as a container that holds a certain amount of electrical charge. Larger (higher-μF) capacitors hold more charge; smaller ones hold less. This leads to an interesting analogy: Say you wanted to determine the volume of a small cup. And suppose you have a 5-gallon bucket and access to a water faucet. You could determine the volume of the cup by filling it, dumping it into the bucket, and repeating until you had filled the bucket. The volume of the cup would be 5 gallons divided by the number of cupfuls needed to fill it.

A similar method can be used to measure capacitance. The small, unknown capacitor (the "cup") is charged up. It is then disconnected from power and reconnected across a larger, known-value capacitor (the "bucket") to which its charge is transferred. This process is repeated until the bucket capacitor is charged up. The number

---

1. Based on casual measurements, a standard BS2 needs about 214 μs to set up for RCTime; a BS2-SX, about 86 μs.
2. See "Measure Water Level without Getting Wet," *Nuts & Volts* Magazine, May 1997. A reprint of this article is available from the magazine web site at www.nutsvolts.com. Go to the Stamp Applications section of their ftp library and download column number 27, st_ap27.pdf.

of cup-to-bucket transfers required to fully charge the bucket is proportional to the cup-capacitor's value.[3]

Figure 17-1 shows a simplified charge-transfer circuit. Assuming that the accumulator capacitor is discharged when we start, the process goes like this: The switch is thrown to the left position to charge up the small, unknown capacitance. It is then thrown to the right to transfer that small charge to the larger accumulator capacitor. Each transfer increases the voltage across the accumulator cap by a small amount.

The goal of the process is to count the number of transfer cycles required to bring the voltage on the accumulator cap to some preset level. In the case of a purely digital device like the Stamp, we're not really measuring this voltage, just waiting for it to cross the logic threshold—the imaginary line that separates a 0 input from a 1.

As we'll see later, it's not necessary to use a switch *per se* to accomplish the transfer. We can take advantage of the three states of the Stamp's I/O pins (high, low, and input/open) to do the job with no external parts.

Implementing Charge Transfer with the BS2-SX

Only one component—the bucket or "accumulator" capacitor—is needed to perform charge-transfer measurements with the BS2-SX. Figure 17-2 shows the circuit. Figure 17-3 is a

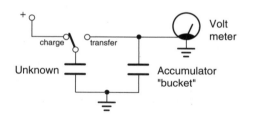

Figure 17-1 Simplified charge-transfer circuit.

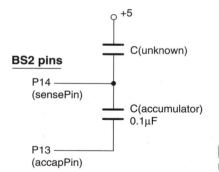

Figure 17-2 Circuit for charge-transfer measurements.

---

3. Charge transfer is a very versatile method for sensing humidity, liquid levels, human touch, etc. For information on commercial applications and more background on the theory behind charge-transfer applications, visit www.qprox.com (Quantum Research Corp. web site).

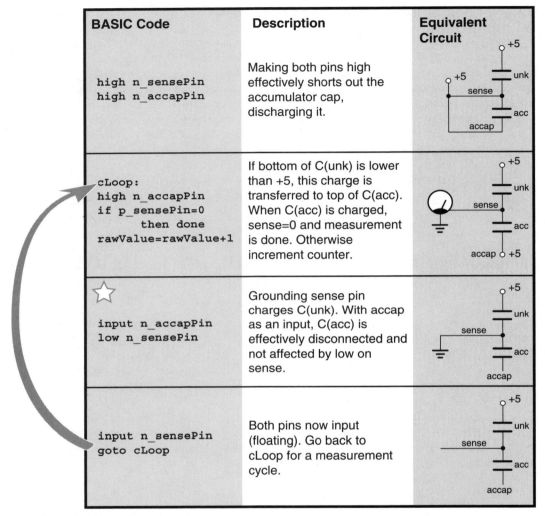

BASIC Code	Description	Equivalent Circuit
`high n_sensePin` `high n_accapPin`	Making both pins high effectively shorts out the accumulator cap, discharging it.	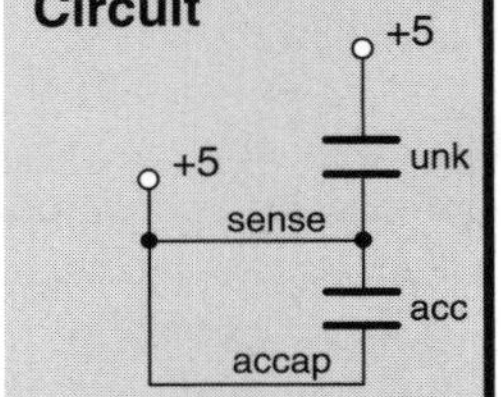

`cLoop:`
`high n_accapPin`
`if p_sensePin=0`
`      then done`
`rawValue=rawValue+1`
	If bottom of C(unk) is lower than +5, this charge is transferred to top of C(acc). When C(acc) is charged, sense=0 and measurement is done. Otherwise increment counter.	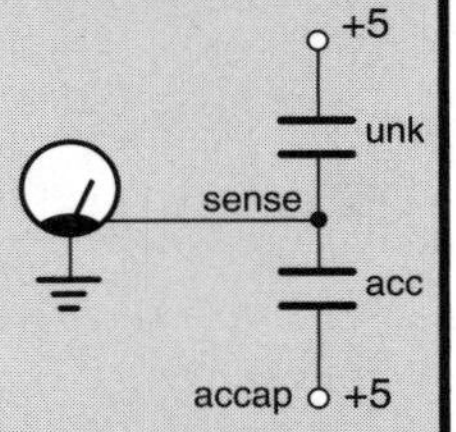

`input n_accapPin` `low n_sensePin`	Grounding sense pin charges C(unk). With accap as an input, C(acc) is effectively disconnected and not affected by low on sense.	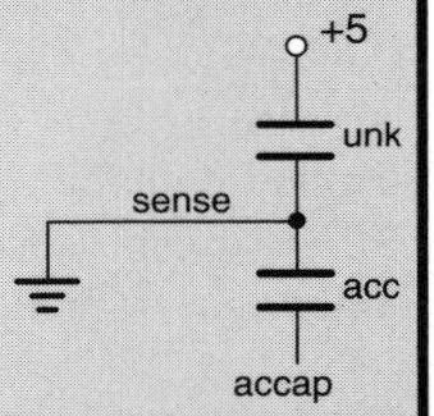
`input n_sensePin` `goto cLoop`	Both pins now input (floating). Go back to cLoop for a measurement cycle.	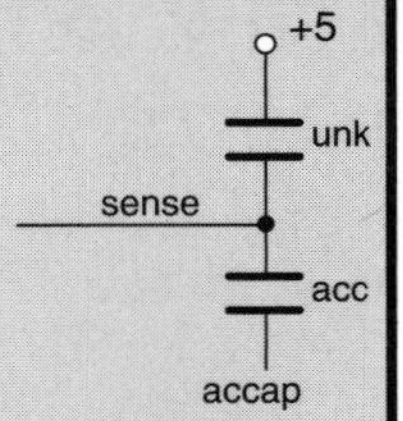

Figure 17-3 **How the charge-transfer code works.**

closeup look at the PBASIC code that drives the charge-transfer process, while 17-4 is the complete BS2-SX project.

I spent some time experimenting with charge transfer and found that it would respond in a reasonably repeatable way to a given value of unknown capacitor. But the response wasn't entirely linear, and I didn't feel like going through the hassle of trying to find an appropriate curve-fitting equation—the usual way of straightening out ill-behaved sensor data.[4]

Even if I had been ambitious enough to work out the math that would do the conversion, the whole process would have to be repeated in order to convert charge-transfer

4. See Parallax BS1 application note #7, Using a Thermistor. Thermistors are temperature-dependent resistors with a distinct curve to their temperature response. The app note outlines the process of converting this nonlinear response to an equation that can calculate the temperature. It's a useful technique, but not particularly pretty.

results to other units. For example, if the unknown capacitor were a humidity sensor or a liquid-level sensor, we would want the output in terms of percent or gallons, not capacitance.

I covered all bases by using a calibration table and interpolation to substitute storage space and simple arithmetic for curve-fitting math. The multiple program memories of the BS2-SX play into this perfectly. I put the main program in one memory area, and the calibration data with simple read/write routines in another.

Let's walk through the BS2-SX project and see how it works.

CHARGE-TRANSFER MECHANICS

It would be perfectly reasonable to build the charge-transfer circuit of Figure 17-1 and control it with the BS2-SX. A relay or solid-state switch could be used to control the charge/transfer function, and an I/O pin in input mode substituted for the meter to monitor the progress of the charging. The same I/O pin could also discharge the accumulator cap in preparation for another measurement cycle by going output low.

Following that line of thought, it might be possible to manipulate an I/O pin to eliminate the need for the switch. After all, the Stamp I/O pins are conceptually just a collection of switches that go to +5V (high), 0V (low), or open (input). See Figure 7-3 back in chapter 7 for a refresher on this concept.

Figure 17-2, the charge-transfer circuit, is deceptively simple. It doesn't convey how the Stamp pins change state to discharge the accumulator (acc) cap, charge the unknown (unk) cap, and transfer charge from unk to acc. For that, we have to look at the PBASIC program side-by-side with equivalent circuits, as shown in Figure 17-3.

The first step in 17-3 is to discharge the acc cap. This is accomplished by setting both I/O pins high. This puts +5V on either side of the acc cap, making for 0V difference across the cap. No voltage difference = no charge.

Now skip down to the third step (with the star). This is where the unk cap gets charged by a low on the sense pin. Although the top of the acc cap is also connected to the sense pin, the other end of the acc cap is effectively disconnected because its I/O pin is set to input. This means there's no complete circuit across the acc cap, so it's unaffected.

In the next step both pins are set to input in preparation for a charge-transfer cycle. The program goes to the label cLoop: and sets the bottom of the acc cap high. It then checks for a low at the top end of the acc cap. Remember that when the acc cap was discharged, both ends were high. As the cap charges up, the voltage at the sense pin drops.

If the level at the sense pin is still high (>1.5V), the routine increments a counter, rawValue and continues. When the level on sense is low (<1.5V), the program considers the acc cap to be charged, and the measurement process ends. The variable rawValue contains the number of transfer cycles required to charge up acc cap from unk cap, which is proportional to the value of unk cap. (See Figure 17-4.)

CALIBRATION AND INTERPOLATION

The charge-transfer routine returns a result in <<PROGRAM>>rawValue<</PROGRAM>> that is related to the value of the unknown capacitor, unk. Figure 17-5 shows that relationship as a graph of capacitance versus transfer cycles. As you can see, the greater the value of C(unk), the fewer transfer cycles required to charge C(acc). According to our cup-and-

```
' Program: CAPMTR01.BSX (Measure pF caps using charge transfer)
' This program uses a technique called charge transfer to measure
' the capacitance of low-value capacitors (0-220 pF). It
' demonstrates the enhanced capabilities of the BS2-SX,
' including speed, multiprogram chaining, and scratchpad RAM.
' A companion program, CAPDAT01, stores calibration data for
' use by this one. Since the BASIC code used in that program
' is quite small, it leaves lots of space for calibration
' points.
' This technique requires some ugly but straightforward
' programming. What we want to do is set up a Gosub/Return
' capability between separate programs. Gosub/Return doesn't
' work for this, and in fact even local, in-program Returns
' are lost by Running a different program. So each time we
' use code in the other program, we must:
'  - Indicate what we want to do (read or write data)
'      via rwFlag
'  - Set up the address and data for that action
'  - Record the point in this program that we wish to
'      return to (returnPtr), and ..
'  - Run the other program
' Upon returning from the other program, the first line of
' code executed here is a Branch instruction that jumps to
' the correct return point based on returnPtr.

'{$STAMP BS2SX, CAPDAT01.BSX}

rawValue      var   w0          ' No. of charge xfers of cap
interpVal     var   w1          ' Result of interpolating rawValue
newEntry      var   w1          ' Entry for new calibration value
tableValue    var   w2          ' Value read from EEPROM
tableEntry    var   w3          ' Entry for above value
iHiValue      var   w4          ' Value >= rawValue
iHiEntry      var   w5          ' Entry for above value
iLoValue      var   w6          ' Value <= rawValue
iLoEntry      var   w7          ' Entry for above value
index         var   w8          ' Counter for EEPROM loop(s)
maxEntry      var   w9          ' Highest address of EEPROM data

scaleFactor var   b20           ' Power-of-2 of calculated results
replyYN       var   scaleFactor ' Temporary use of var as YN input
returnProg    var   b21.lownib       ' Program to return to (0)
returnPtr     var   b21.highnib      ' Point in program to return
rwFlag        var   b22.bit0         ' 0=read; 1=write
resultBit     var   b22.bit1         ' True/False flag (0=False)
eeData        var   b23

STARTADDR     con   2           ' Starting address of data in EEPROM
DEBUG_PIN     con   16          ' Serial output to debug connector
DBBAUD        con   $F0         ' Debug baudmode
READ_         con   0
WRITE_        con   1
SCRATCH       con   0           ' Data exchange location in scratch RAM
ENTRYSIZE     con   0           ' Address of number of cal points in EEPROM

n_sensePin    con   14      ' Pin 14 senses charge transfer completion
p_sensePin    var   in14    ' Same, but ref'd by pin name, not #
n_accapPin    con   13      ' Ground side of accumulator cap
```

Figure 17-4 Complete BS2-SX capacitance-measurement program.

```
' Branch to correct return point in program based on value of
' returnPtr.
branch returnPtr, [mainMenu, r1,r2,r3,r4,r5,r6,r7,r8,r9,r10,r11,r12]


' This program uses Serin and Serout to communicate through the debug
' screen. Inserting this unnecessary Debug instruction causes the Stamp
' host software to automatically open a preconfigured Debug screen.
debug CLS
mainMenu:
  serout DEBUG_PIN,DBBAUD, [CLS,"CAPACITANCE METER",CR,"(C)alibrate (M)easure
(P)urge"]
  serin DEBUG_PIN,DBBAUD, [replyYN]
  replyYN.bit5 = 0      ' Change to capital letter
  if replyYN = "C" then getCal
  if replyYN = "M" then doMeasurement
  if replyYN = "P" then purgeCalTable
goto mainMenu


' SUBROUTINE: getCal=========================================
' Accept a new calibration point from the user by prompting for
' a capacitance value, measuring the number of charge-transfer
' cycles and writing the result into the calibration table of
' program 1 (CAPDAT).
getCal:
  serout DEBUG_PIN,DBBAUD, [CLS,"CALIBRATION",CR,"Cap value in pF: "]
  serin DEBUG_PIN,DBBAUD, [DEC newEntry]
  serout DEBUG_PIN,DBBAUD, [CR,CR,"Confirm ",DEC newEntry, "pF (Y/N) "]
  serin DEBUG_PIN,DBBAUD, [replyYN]
  replyYN.bit5 = 0 ' Change to capital letter
  if replyYN = "N" then doneCal
  gosub cTransfer
  goto ckForDupeCal
rFromCk:
  branch resultBit, [noDupe]
  serout DEBUG_PIN,DBBAUD, [CR,"Calibration point duplicates previous entry."]
  goto doneCal
noDupe:
  goto addCalPoint
rFromAdd:
  serout DEBUG_PIN,DBBAUD, [CR,"Calibrated: ",DEC rawValue, "=", DEC
newEntry,"pF"]
  serout DEBUG_PIN,DBBAUD, [CR,DEC maxEntry ," cal points."]
doneCal:
  gosub pressAny
goto mainMenu


' SUBROUTINE: doMeasurement=================================
' Take a charge-transfer measurement to get rawValue, then
' interpolate from entries saved in the pgm-1 calibration table.
doMeasurement:
  serout DEBUG_PIN,DBBAUD, [CLS,"Measuring...please wait."]
  gosub cTransfer
  goto interpolate
rFromInterp:
  serout DEBUG_PIN,DBBAUD, [CR,"Value= ", DEC interpVal,"pF"]
  serout DEBUG_PIN,DBBAUD, [CR,"Based on calibration points of "]
  serout DEBUG_PIN,DBBAUD, [DEC iLoEntry, "pF and ", DEC iHiEntry, "pF"]
  gosub pressAny
goto mainMenu
```

Figure 17-4   (*Continued*)

```
' SUBROUTINE: cTransfer======================================
' This routine transfers charge from the unknown, low-value
' capacitor into a larger accumulator cap until the accumulator
' cap is sufficiently charged to reach the Stamp's logic
' threshold. To take advantage of the 3.5V difference between
' 5V and the 1.5V logic threshold, the accumulator cap is
' wired so that it reads "1" when discharged and "0" when
' charged. When done, the variable rawValue contains the
' number of transfers required to charge the accumulator cap.
' This number is proportional to the capacitance of the
' unknown cap.
cTransfer:
  high n_sensePin       ' Make both ends of accumulator cap
  high n_accapPin       ' high—thus discharging it.
  pause 100             ' Wait..
  rawValue = 0          ' Clear count
cLoop:
  high n_accapPin       ' High bottom end of accumulator cap
  if p_sensePin = 0 then done
  rawValue = rawValue+1 ' Count a cycle
  input n_accapPin      ' Now let accumulator cap float
  low n_sensePin        ' Put a charge onto unknown C
  input n_sensePin      ' Let the pin float.
goto cLoop
done:
return


' SUBROUTINE: Interpolate===================================
' This routine accepts a number in the variable rawValue
' and searches through the data in EEPROM for values greater
' than and less than (but as close as possible to) this
' value. It stores these values (iHiValue and iLoValue)
' and their corresponding table entries (iHiEntry and iLoEntry).
' The "values" represent measurements taken by our system;
' the "entries" are actual calibration points for those
' measurements. E.g., a value of 6234 units might correspond to
' an entry of 273pF, while another value of 7011 to 338pF.
' The goal of interpolation is to answer a question like
' what # of pF would correspond to 6894? This is done by
' figuring what fraction of the difference between the values
' 6894 is, then applying that fraction to the difference between
' the calibration entries to find the intermediate value.

interpolate:
  index = ENTRYSIZE     ' Get the number of entries from CAPDAT.
  rwFlag = READ_ : returnProg = 0 : returnPtr = 1 : run 1
r1:
  get SCRATCH, maxEntry.lowByte
  get SCRATCH+1, maxEntry.highByte
  if maxEntry = 0 then rFromInterp   ' No cal data, so why bother.
  maxEntry = 1 + (maxEntry * 4)      ' Compute address of max entry.
  iHiValue = 65535     ' Preload with maximum possible value.
  iLoValue = 0         ' Prelaod with minimum possible value.
' The FOR loop below goes through the calibration table in CAPDAT
' (program 1) looking for data points that are as close as possible
' to the rawValue. When the loop is done, iHiValue will contain the
' closest cal table entry greater than rawValue and iLoValue will
' contain the closest entry less than rawValue. iHiEntry and iLowEntry
' will contain the corresponding calibration values in pF. This is
```

Figure 17-4 (*Continued*)

```
' the raw material for interpolation.
  for index = STARTADDR to maxEntry step 4
    returnPtr = 2 : run 1
r2:
    get SCRATCH,tableValue.lowByte
    get SCRATCH+1,tableValue.highByte
    get SCRATCH+2,tableEntry.lowByte
    get SCRATCH+3,tableEntry.highByte
    if NOT ((tableValue >= rawValue) AND ((tableValue-rawValue)<(iHiValue-
rawValue))) then skip01
    iHiValue = tableValue
    iHiEntry = tableEntry
skip01:
    if NOT ((tableValue <= rawValue) AND ((rawValue-tableValue)<(rawValue-
iLoValue))) then skip02
    iLoValue = tableValue
    iLoEntry = tableEntry
skip02:
  next          ' Next index in table search.
' Here is where interpolation actually happens. First we need to
' know whether entries rise when the raw value is bigger, or fall—
' In algebraic terms, whether the line has a positive or negative slope.
' If positive slope (higher raw -> higher entry) then we figure out what
' fraction of the difference between hi and lo values the raw value is,
' multiply that fraction by the difference between hi and lo entries,
' and add the result to the lo entry to get the interpolated answer.
' If negative slope, we follow the same basic procedure, but relative
' to the hi value and entry. The scaleFactor computation adds the powers
' of 2 that will be used in the multiplication part of the fraction and
' determines whether the multiplication will exceed 2 to the 15th power
' (32,768). If it will, the scaleFactor causes the value to be divided
' by a power of 2 (shifted right a number of times) that will prevent
' an overflow (result larger than 65,535—the largest number the Stamp
' can compute).
  if iHiEntry < iLoEntry then downSlope
  scaleFactor= NCD(rawValue-iLoValue) + NCD(iHiEntry-iLoEntry) min 15 - 15
  interpVal = (rawValue-iLoValue)*((iHiEntry-iLoEntry)>>scaleFactor)/((iHiValue-
iLoValue)>>scaleFactor)+iLoEntry
  goto interpDone
downSlope:
  tableValue= iLoEntry: iLoEntry = iHiEntry: iHiEntry = tableValue
  scaleFactor= NCD(rawValue-iLoValue) + NCD(iHiEntry-iLoEntry) min 15 - 15
  interpVal= iHiEntry-((rawValue-iLoValue)*((iHiEntry-
iLoEntry)>>scaleFactor)/((iHiValue-iLoValue)>>scaleFactor))
interpDone:
goto rFromInterp


' SUBROUTINE: addCalPoint======================================
' Takes the data in rawValue and newEntry and attempts to
' add them to the EEPROM data table at the address specified
' by index. Index and maxEntry are set to the correct values by
' ckForDupeCal, which must be called before addCalPoint.
' If the new point is not a duplicate/replacement, then
' maxEntry will be incremented.
addCalPoint:
  rwFlag=WRITE_ : returnProg=0
  returnPtr=5 : eeData=rawValue.lowByte : run 1
r5:
  index=index+1 : returnPtr=6 : eeData=rawValue.highByte : run 1
```

Figure 17-4 *(Continued)*

17

CHARGE-TRANSFER CAPACITANCE

```
r6:
   index=index+1 : returnPtr=7 : eeData=newEntry.lowByte : run 1
r7:
   index=index+1 : returnPtr=8 : eeData=newEntry.highByte : run 1
r8:
   if resultBit = 1 then doneAddCal
   index=ENTRYSIZE : returnPtr=9 : eeData=maxEntry.lowByte : run 1
r9:
   index=ENTRYSIZE+1 : returnPtr=10 : eeData=maxEntry.highByte : run 1
r10:
doneAddCal:
goto rFromAdd


' SUBROUTINE: ckForDupeCal=====================================
' Takes the data in rawValue and checks for a duplicate in the
' EEPROM data. Returns answer (0=NO; 1=YES) in the variable
' resultBit. In variable tableEntry, leaves the entry
' corresponding to the duplicated value.
ckForDupeCal:
   resultBit= 0
   index = ENTRYSIZE
   rwFlag = READ_ : returnProg = 0 : returnPtr = 3 : run 1
r3:
   get SCRATCH,maxEntry.lowByte
   get SCRATCH+1,maxEntry.highByte
   if maxEntry = 0 then addPoint
   for index = STARTADDR to (1+(maxEntry * 4)) step 4
      returnPtr = 4 : run 1
r4:
   get SCRATCH,tableValue.lowByte
   get SCRATCH+1,tableValue.highByte
   get SCRATCH+2,tableEntry.lowByte
   get SCRATCH+3,tableEntry.highByte
   if tableValue<>rawValue then ckNext ' If duplicate
   resultBit= 1 ' ..then resultBit= 1; else 0.
   goto ckDone
ckNext:
 next
addPoint:
 maxEntry = maxEntry + 1       ' If no dupe, new value
ckDone:     ' will be added to end of EEPROM list.
goto rFromCk


' SUBROUTINE: purgeCalTable=================================
' When you use this project for the first time, or if you want to
' change its calibration from say pF to gallons (capacitive
' liquid-level sensing), you need to delete old calibration
' data. The easiest way to do this is simply reset the
' number of calibration entries to 0. Old data will be ignored
' and then overwritten when new calibration points are added.
purgeCalTable
   serout DEBUG_PIN,DBBAUD,[CR,CR,"Confirm delete all calibration (Y/N) "]
   serin DEBUG_PIN,DBBAUD,[replyYN]
   replyYN.bit5 = 0      ' Change to capital letter
   if replyYN = "N" then r12
   index=ENTRYSIZE : returnPtr=11 : eeData=0 : run 1
r11:
   index=ENTRYSIZE+1 : returnPtr=12 : eeData=0 : run 1
```

Figure 17-4 (*Continued*)

```
r12:
goto mainMenu

' SUBROUTINE: pressAny=========================================
' Prompts user "Press any key..." and waits for press
pressAny:
  serout DEBUG_PIN,DBBAUD,[CR,CR,"Press any key to continue.."]
  serin DEBUG_PIN,DBBAUD,[replyYN]
return


' Program: CAPDAT01.BSX (Calibration utility for cap meter)
' This program serves consists mostly of EEPROM storage for
' capacitance calibration entries used by the charge-transfer
' capacitance-measurement program. When that program wants to
' read a value from EEPROM it sets rwFlag to 0, places the
' address in the variable index, and runs this program.
' Here, we branch on rwFlag to readEE, and read four bytes
' from EEPROM into addresses 0-3 of scratchpad RAM. (Reads
' are always 4 bytes, since that's what's most often needed,
' and the additional reads cost nothing but a little
' processing time.)
' When the cap meter wants to write to EEPROM, it sets rwFlag
' to 1, index to the EEPROM address, and eeData to the value
' to be stored, then runs this program. Here, we branch on
' rwFlag to writeEE and store eeData to index.
entrySize    data   word (1)        ' Number of calibration entries
interpTable  data   word (980)      ' Calibration data storage

index        var    w8              ' Counter for EEPROM loop(s)
returnProg   var    b21.lownib      ' Program to return to
returnPtr    var    b21.highnib     ' Point in program to
rwFlag       var    b22.bit0        ' 0=read; 1=write
eeData       var    b23

STARTADDR    con    2          ' Starting address of data in EEPROM
FROM_EE      con    0          ' Point in scratchpad RAM for data exchange

' If rwFlag = 0 then read; if 1 then write
branch rwFlag,[readEE,writeEE]

' Read 4 bytes located at index into scratchpad RAM
readEE:
  read index,eeData
  put FROM_EE,eeData
  read index+1,eeData
  put FROM_EE+1,eeData
  read index+2,eeData
  put FROM_EE+2,eeData
  read index+3,eeData
  put FROM_EE+3,eeData
run returnProg

' Write value in eeData to address in index
writeEE:
  write index,eeData
run returnProg
```

Figure 17-4   *(Continued)*

bucket analogy, that makes sense. But if you graphed the relationship of cup size to number of cups needed to fill a bucket, you'd get a straight line, not a curve like Figure 17-5.

The problem becomes how to translate a given number of transfer cycles to the corresponding capacitance. Suppose I asked you what capacitance corresponded to a rawValue of 2000 based on Figure 17-5. You would imagine (or sketch) a straight line up from 2000 on the horizontal axis until it touched the curve, then trace back to the vertical axis to locate a point about a quarter of the way between 20 and 40 pF. You'd mentally calculate that the difference between 20 and 40 is 20, and one quarter of 20 is 5, so that point on the chart is approximately 25 pF.

Now we're ready to define some terms: A *calibration table* is a list of raw measurements and their corresponding values in actual units. *Interpolation* is the process of finding an unknown point that lies between two other points based upon available information. When you look at a graph and estimate a value that's not specifically marked, you're *interpolating* between the marked data points.

Let's look at an example calculation. Table 17-1 shows the calibration data used to create Figure 17-5.

Suppose a new measurement of an unknown capacitance is 2511. According to the calibration table, what is the approximate value in pF? First step is to find the table entries closest to 2511:

- HiValue 2728 = 12 pF (HiEntry)
- RawValue 2511 = ?? pF
- LoValue 2367 = 18 pF (LoEntry)

Right away we know that 2511 corresponds to some value that is less than 18 but greater than 12 pF. But how much? The difference between HiValue and LoValue is 272 − =2367=361. And the difference between HiValue and RawValue is 2728 − 2511=217.

We can say that RawValue lies at a point 217/361 of the way between HiValue and

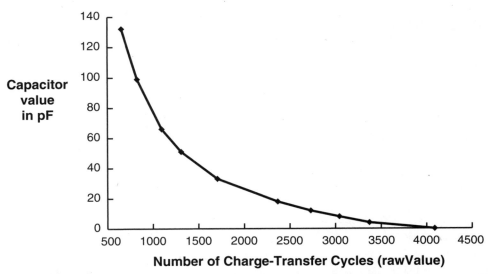

Figure 17-5 Graph of charge-transfer cycles for capacitors from 0 to 130 pF.

TABLE 17-1 CHARGE-TRANSFER CALIBRATION DATA

RAW VALUE	CALIBRATION ENTRY (PF)
4088	0
3371	4
3043	8
2728	12
2367	18
1706	33
1308	51
1095	66
827	99
654	132

LoValue. So its corresponding pF should be about 217/361 of the way between HiEntry (12 pF) and LoEntry (18 pF). The difference between those values is $18 - 12 = 6$. Multiply by the fraction; $6 \times 217/361 = 3.61$, and add to HiEntry; $3.61 + 12 = 15.61$ pF.

We added the 3.61 to HiEntry because the table shows that the pF entries increase as the input values decline. The program makes the same sort of decision, adding or subtracting as needed.

That's the procedure for interpolating between table entries. Note that if this calculation had been performed by a Stamp, the answer would be expressed as 15 pF, not 15.61 pF, due to the Stamp's integer-only math.

Using the Charge-Transfer Project

Set up the circuit of Figure 17-2 on a breadboard, making it convenient to temporarily add and remove capacitors to be measured. Gather up a batch of marked capacitors with values in the range of 1 to 220 pF to be used for calibration. The more the merrier, since ultimately the performance of the project will depend on having lots of closely spaced samples in its calibration table.

Copy the project from the CD-ROM and run it. A Debug screen will present you with three choices:

```
CAPACITANCE METER
(C)alibrate (M)easure (P)urge
```

The first thing you should do is press P to purge stray data from your calibration table. You don't want random data being interpreted as calibration points.

Before inserting any capacitor into the C (unknown) spot, press C to calibrate the unit for 0 pF. The lower the capacitance, the longer the charge-transfer process takes, so be prepared to wait several seconds.

Now repeat the calibration process on your collection of known capacitors. You don't have to go in any particular order, and it doesn't matter if you use the same capacitor twice.

Once you have a reasonable number of calibration values, you may check the project's performance by measuring one of the known capacitors. The result will be within 1-2 pF of the actual value.

OTHER APPLICATIONS OF CHARGE TRANSFER

Measuring low-value capacitors is a useful but slightly dull application. Fortunately, there are other applications for this capability:

- Measuring water level. Form a crude capacitor by attaching metal foil to the sides of a piece of thin plastic pipe (1/4″ or less). Since capacitance varies with the characteristics of the material between the conducting plates—the *dielectric*—you'll see a lower capacitance as water replaces air inside the pipe. Using this project's calibration/interpolation approach, you can calibrate in units of volume instead of pF. See the article referenced in footnote 2.
- Measuring humidity. A couple of closely spaced metal plates with an air gap form a capacitor. Since capacitance depends on the air dielectric, and the dielectric properties of air vary with humidity, you can get a rough estimate of humidity with this simple setup. There are also ready-made humidity sensors whose output is a variable capacitance.[5]

If you decide to try one of these alternative applications, you may want to change the value of the accumulator capacitor. For highest precision, the accumulator cap should be much higher in value than the capacitance to be measured. For fastest response, though, the accumulator should be as low in value as possible. This leaves you plenty of opportunities for fruitful experimentation. Thanks to the calibration-table approach used in the project, you won't have to jump through mathematical hoops of fire to make your changes—just provide a new set of calibration points and you're done.

---

5. A capacitance humidity sensor is available from Fascinating Electronics for less than $15 (at the time of this writing). Capacitance at 43% RH and 25°C is 122 pF ± 15%. Capacitance varies over a nominal range of 110 to 150 pF for a humidity change of 0 to 100% RH.

BASIC STAMP QUICK REFERENCE GUIDES

The Stamp user manuals provide excellent insight into the workings and uses of the PBASIC instruction sets. Of course, they take dozens of pages to do it. These guides distill PBASIC programming information into a compact format so that you can find essential answers quickly.

This super-condensed reference is too concise for novices. If you are a newcomer to PBASIC, browse the detailed writeups in the manual (provided on the CD-ROM that accompanies this book) and do the the BASIC Stamp Boot Camp exercises. Once you get the gist of PBASIC, you'll get the most benefit out of this reference.

BASIC Stamp 1 Language Reference

This section describes the BS1 dialect of PBASIC. Coverage of the BS2 begins later in this apppendix.

PROGRAM MEMORY (EEPROM)

A PBASIC program is stored in a 256-byte EEPROM. This kind of memory retains data with the power off, but is easily erased or rewritten. When the Stamp is powered up, it automatically runs the program stored in EEPROM.

Those 256 bytes hold about 75 PBASIC instructions. It's hard to pin down the exact number of instructions, because the STAMP host program squeezes each instruction into a compact token. The size of a given token depends more on the size and number of the instruction's arguments (data it processes; see summary below) than the instruction itself. For example, the High instruction takes a single three-bit argument (a number from 0 to 7) to specify a pin number. So its token consists of a sequence of bits to identify the instruction, High, plus three bits to specify the pin.

At the other end of the spectrum, take a typical serial-output instruction: SEROUT 0,N2400, ("Hello"). After identifying the instruction as Serout, the token must contain 3 bits for the pin number, 4 bits for the baudmode (N2400 is one of 16 predefined symbols), and 5 bytes (40 bits) for the message "Hello."

So that figure of 75 instructions assumes a mixture of big and small tokens. Your mileage may vary.

In spite of limited program space, many programs still have room left over to store data in the EEPROM. There are two ways to store data to EEPROM. If the data is determined before the program runs (such as tables of constants, like those used in the persistence-of-vision project) use the EEPROM directive. If the program itself will store the data, use the Write instruction. In either case, PBASIC accesses the data with the Read instruction.

Since the EEPROM directive is managed by the STAMP host software, you will get an error message if your data overlaps with space needed to store your program. Program storage begins at the highest EEPROM address (255) and works downward, while EEPROM-managed data storage starts at 0 (or an address you specify) and works upward. If the two try to store something to the same address, the STAMP host software stops and presents you with an error message.

When you use the Write instruction, it is up to your program to avoid overwriting itself. Fortunately, this is easy to do. Address 255 holds the address of the lowest EEPROM location used by the program. Your program just has to restrict Write instructions to addresses less than this value. For example, if you Read 255 and find that it holds a value of 100, you may write to locations 0–99.

Write has a couple of limitations. First, writing to EEPROM is slower than writing to RAM, requiring about 10 milliseconds (ms) versus about 0.5 ms. Second, EEPROM locations wear out after about 10 million writes. If you are tempted to replace RAM variables with EEPROM variables, make sure that you don't run afoul of these limitations.

DATA MEMORY (RAM)

The BS1 has 14 bytes of memory. Each byte has a predefined variable name: B0, B1, B2... B13. Pairs of bytes can be used as 16-bit words. For example, B0 and B1 can be accessed together as the single word variable W0. B0 is the low byte of W0, B1 is the high byte. In a similar way, you may use the individual bits of W0 as 16-bit variables, BIT0 through BIT15. This is a flexible arrangement, but it can be confusing. If you write to W0, the change will affect B0, B1, and all the bit variables. So if you are using W0, you should probably avoid using B0, B1, and the bit variables, unless you specifically want to use them as windows into portions of the larger variable. Table A-1 shows BS1 memory organization.

Note that W6 (bytes B12 and B13) should not be used in programs that contain Gosub instructions. Gosub stores return-address data in this location. This overwrites data stored in W6/B12/B13, and writing to this location while a Return is pending would prevent the subroutine from returning to the correct point in the program.

When a PBASIC program begins, all variables are cleared to 0s. Rely on this. Training in other programming languages may make you uneasy about not "explicitly initializing variables" (writing 0 or other value to them before use). In PBASIC, it's a waste of precious instruction space.

TABLE A-1 BS1 MEMORY MAP

WORDS	BYTES	BITS
W0	B0	Bit0-Bit7
	B1	Bit8-Bit15
W1	B2	
	B3	
W2	B4	
	B5	
W3	B6	
	B7	
W4	B8	
	B9	
W5	B10	
	B11	
*W6	B12	
	B13	

*Used by GOSUB

A

PBASIC-1 does not allow arrays, pointers, or other means of indirect addressing of variables. In other words, you cannot specify a variable by number rather than name. (The predefined names incorporate numbers, but you can't specify the 3 in B3 as the contents of a variable. PBASIC-1 won't let you.) If you desperately need arraylike capability, consider using the EEPROM Write and Read instructions. Just make sure that you can live with the limited number of writes (10 million total per byte) and the relatively slow write cycle. Or, if you mostly need to read RAM variables indirectly (selecting them with an index number), you can put them in a Lookup table, e.g., LOOKUP index, (b0,b1,b2,b3,b4,b5,b6,b7),result. If index contains 4, that instruction would place the contents of b4 into the variable result. Unfortunately, there's no counterpart for storing data to variables in the same manner.

INPUT/OUTPUT (I/O)

The BS1 has 8 input/output (I/O) pins. To understand the state of an I/O pin, you need to know two facts: the state of the pin (1 or 0) and the direction (input or output). That's exactly how PBASIC-1 represents its I/O pins in memory—as a 16-bit variable called Port consisting of a byte called PINS containing the I/O states, and another called DIRS containing the I/O directions (0=input; 1=output). Individual pins' states and directions can be accessed via bit variables PIN0, PIN1... and DIR0, DIR1... Table A-2 shows BS1 I/O variable organization.

When a PBASIC program starts, all variables are cleared to 0, including the I/O variables. This makes all 8 I/O pins inputs initially. Storing 1s to particular bits of DIRS, or executing certain instructions, such as HIGH, can change pins to output.

MATH AND LOGIC (LET)

PBASIC offers 14 math and logic operators that work on 16-bit values in any combination of variables and constants. If a value is smaller than 16 bits—a byte, for instance—it is internally converted to a 16-bit value by padding the missing higher-order bits with 0s.

Multiple operations can be combined into a single program step. Operations are performed strictly from left to right; you cannot use parentheses to alter the order of operations. This may require you to break complex operations up into multiple steps. Table A-3 lists BS1 math and logic operators.

Math and logic operations can be written with or without the instruction LET. For instance $x = 3*y + z$ and LET $x = 3*y + z$ produce the same results.

INSTRUCTION SUMMARY

Each instruction is accompanied by a brief description, followed by a table listing the instruction's arguments. Arguments are values that specify how the instruction should

TABLE A-2   BS1 I/O VARIABLES

WORD	BYTES	BITS
Port	Pins	Pin0-Pin7
	Dirs	Dir0-Dir7

TABLE A-3 BS1 MATH AND LOGIC OPERATORS

+	add
-	subtract
*	multiply (returns low word of result)
**	multiply (returns high word of result)
/	divide (returns quotient)
//	divide (returns remainder)
MIN	keep variable greater than or equal to value
MAX	keep variable less than or equal to value
&	logical AND
|	logical OR
^	logical XOR
&/	logical AND NOT
|/	logical OR NOT
^/	logical XOR NOT

work. In the instruction HIGH 3, the constant 3 is an argument that specifies which pin to make high.

In most cases, PBASIC instructions will accept a variable or constant of any size up to 16 bits (range of 0–65,535). Sometimes only part of this range is valid. Using High as an example again, PBASIC will permit you to specify an out-of-range value like HIGH 65535 without an error message. However, it will truncate (chop off) all but the lowest three bits in order to get a valid pin number (0–7). The ranges specified in this instruction summary are the valid choices for the instruction.

PBASIC instructions process three types of data: variables, constants, and labels. Variables and constants are usually interchangeable; such cases are listed as any type in the tables below. Labels are locations within a program and may not be interchanged with variables or constants.

BRANCH: Use an offset to pick a program address from a list and go to that address.

BRANCH	offset,	(address0,	address1,	...	addressN)
range/size:	0—N	—	—	—	—
type:	any	label	label	label	label
meaning, usage:	index # of label to go to; if greater than N (last label in list) continues with next instruction	address label to go to if offset is 0	address label to go to if offset is 1	more address labels	address label to go to if offset is N

BUTTON: Read button input with optional debounce and autorepeat. Go to a program address if specified conditions are met. Must be executed from within a loop.

BUTTON	pin,	downstate,	delay,	rate,	bytevariable,	targetstate,	address
range/size:	0—7	0,1	0—255	0—255	byte	0,1	—
type:	any	any	any	any	variable	any	label
meaning, usage:	I/O pin	state when button is pushed	loops before autorepeat; 0=no debounce or repeat; 255= debounce, no repeat	number of loops between repeats	must be cleared to 0 before 1st use; if multiple Buttons, use separate variables	0=goto address when not pressed; 1=goto address when pressed	address label to go to when target-state requirement is met

DEBUG: Show a variable or message in a window on the STAMP.EXE screen.

DEBUG	data	, ...
range/size:	any	
type:	variable or quoted text message	
meaning, usage:	DEBUG variable prints *variable = value*, where variable is the variable name and value is its contents. Modifiers before variable change display as follows: # (just value), $ (value in hex), % (value in binary), and @ (value as ASCII character). Text in quotes, like "message" will appear as typed; CR skips to next line; and CLS clears Debug window.	more entries, separated by commas

EEPROM: Store data into EEPROM during program downloading process.

EEPROM	location,	(data, data...)
range/size:	0—255	0—255
type:	constant	constant
meaning, usage:	optional EEPROM address at which to store first item in data list; subsequent items stored at address+1, address+2... If no location is specified, then first address = 0.	byte values to store in EEPROM. May be specified as a list of constants separated by commas, or as a quoted text string such as— "text string".

END: End program and put Stamp into low-power Sleep mode until reset.

FOR...NEXT: Create a loop that increments or decrements a variable and repeats until it is no longer between specified start and end values. The loop begins with the FOR instruction as outlined below and ends with NEXT. Instructions between FOR and NEXT execute each time the loop repeats. NEXT can stand alone, or can specify the indexVariable used by the corresponding FOR instruction.

FOR	indexVariable	=	start	TO	end	STEP	stepVal
range/size:	0—65535		0—65535		0—65535		0—65535
type:	variable		any		any		any
meaning, usage:	variable initially set to start value, then incremented or decremented until it passes end value		starting value for the indexVariable		ending value for indexVariable. If loop is incrementing, it ends when indexVariable is greater than end; if decrementing, less than end.	optional: If no step/stepVal specified then indexVariable increments by 1 each loop.	specifies value to add to indexVariable each loop. If preceded by minus (-) sign, value is subtracted each loop.

GOSUB: Store the location of the next instruction and go to a specified program address. Uses variable W6 internally to store next-instruction data. See Return.

GOSUB	address
range/size:	—
type:	label
meaning, usage:	address label to go to.

GOTO: Go to a specified program address.

GOTO	address
range/size:	—
type:	label
meaning, usage:	address label to go to.

HIGH: Make the specified pin output/high.

HIGH	pin
range/size:	0—7
type:	any
meaning, usage:	Pin's output and direction bits will be set to 1, making the pin output/high.

IF...THEN: Make comparison(s) and, if true, go to a specified program address.

IF	value1	?	value2	AND/OR value ? value	THEN	address
range/size:	0—65535	—	0—65535			—
type:	any	—	any			label
meaning, usage:	variable or constant compare to value2 via ?	comparison operator: = (equal) <> (not equal) > (greater than) < (less than) <= (less than or equal to) >= (greater than or equal to)	variable or constant compare to value1 via ?	optional additional comparisons in same pattern, connected by AND or OR. With AND both comparison must be true to go to THEN address. With OR, one or both comparisons can be true to go to THEN address.		address label to go to if comparison(s) true

INPUT: Make the specified pin an input.

INPUT	pin
range/size:	0—7
type:	any
meaning, usage:	Pin's direction bit will be cleared to 0, making the pin an input. Output bit is unaffected, but does not control pin state.

LET: Assign a value to a variable. See Math and Logic section.

LOOKDOWN: Find a value in a list and return the index number 0—N of the match.

LOOKDOWN	target,	(value0,	value1,	...	valueN),	result
range/size:	0—65535	0—65535	0—65535		0—65535	0—N
type:	any	any	any		any	variable
meaning, usage:	value to search for in list	list of values: may be a series of variables or constants separated by commas, or a quoted text string like "ABCDEFG." The first character is index 0, the second is 1, etc.				variable in which index of match value is stored. If no match, result is unchanged.

LOOKUP: Use an offset to pick a value from a list and store that value in a variable.

LOOKUP	offset,	(value0,	value1,	...	valueN),	result
range/size:	0—65535	0—65535	0—65535		0—65535	0—65535
type:	any	any	any		any	variable
meaning, usage:	index # of value to get	list of values: may be a series of variables or constants separated by commas, or a quoted text string like "ABCDEFG." The first character is index 0, the second is 1, etc.				receives value whose index is equal to offset. If offset it greater than N (last value in list) result is unchanged.

LOW: Make the specified pin output/low.

LOW	pin
range/size:	0—7
type:	any
meaning, usage:	Pin's output bit will be cleared to 0 and its direction bit set to 1, making the pin output/low.

NAP: Go into power-saving mode for a short time. See also Sleep.

NAP	period
range/size:	0—7
type:	any
meaning, usage:	length of power-down, computed as 2 raised to the power of period, times 18ms. Example: period = 3, 2^3 = 8, 8 x 18ms = 144ms, so NAP 3 powers down for 144ms.

OUTPUT: Make the specified pin an output.

OUTPUT	pin
range/size:	0—7
type:	any
meaning, usage:	Pin's direction bit will be set to1, making the pin an output. Output bit controls pin state.

PAUSE: Delay for a specified number of milliseconds.

PAUSE	time
range/size:	0—65535
type:	any
meaning, usage:	length of delay in milliseconds.

POT: Measure a variable resistance.

POT	pin,	scale,	result
range/size:	0—7	0—255	0—255
type:	any	any	variable
meaning, usage:	Pin connects to 5-50k variable resistor in series with 0.1μF capacitor to ground.	value used to scale resistance range to 0—255 range. Press ALT-P for help determining scale.	variable in which Pot result is stored.

PULSIN: Measures the edge-to-edge duration of a pulse in units of 10 microseconds.

PULSIN	pin,	state,	result
range/size:	0—7	0,1	0—65535
type:	any	any	variable
meaning, usage:	Pin's direction bit is cleared to 0, making the pin an input. Output bit is unaffected but does not control pin state.	triggering edge state: 0= 1-to-0 transition 1= 0-to-1 transition. Timing starts on state change; ends on opposite change	length of time between edges in units of 10μs. If no pulse occurs within 0.65535 seconds, or pulse is longer than this, result = 0.

PULSOUT: Generate a pulse by inverting pin state for time specified in units of 10 microseconds.

PULSOUT	pin,	time
range/size:	0—7	0—65535
type:	any	any
meaning, usage:	Pin's direction bit is set to 1, making the pin an output. Output bit is inverted for pulse duration, then returned to previous state.	duration of pulse in 10μs units

PWM: Output fast pulse-width modulation that can be filtered to DC voltage.

PWM	pin,	duty,	cycles
range/size:	0—7	0—255	0—255
type:	any	any	any
meaning, usage:	Pin's direction bit is set to 1, making it an output during PWM. When done, direction bit is reset to 0 (input).	duty cycle, where 0 to 100% is represented by 0 to 255 (0.39% per unit)	number of 5ms cycles of PWM output

RANDOM: Generate a pseudorandom number.

RANDOM	variable
range/size:	word
type:	variable
meaning, usage:	word variable to scramble in a pseudorandom pattern. Pattern is always the same, so starting with different values (seeds) in variable is necessary to simulate truly random results.

READ: Retrieve data stored in EEPROM.

READ	location,	data
range/size:	0—255	byte
type:	any	variable
meaning, usage:	EEPROM address	data retrieved from EEPROM

RETURN: Go to program line stored by most recent Gosub instruction.

REVERSE: Reverse the data direction of the specified pin.

REVERSE	pin
range/size:	0—7
type:	any
meaning, usage:	Pin's direction bit will be inverted: if 1 (output), changed to 0 if 0 (input), changed to 1

SERIN (1): Receive asynchronous serial transmission (e.g., RS-232)

SERIN	pin,	baudmode,	(variable)
range/size:	0—7	0—7	byte
type:	any	any	variable
meaning, usage:	Pin's direction bit is cleared to 0, making it an input.	baud rate and mode. Built-in constants select rates of 300, 600, 1200, or 2400 with true (T) or inverted (N) modes. Example: N2400 is inverted, 2400 baud. True modes are used with inverting line receivers; inverted for direct RS-232 input through a 22k resistor.	byte variable or list of byte variables in which to store serial data. May be preceded by # to convert text numbers like "123" into values. Multiple entries (with or w/o #) may be separated by commas. Variables may be mixed with qualifiers [see Serin (2)].

SERIN (2): Receive asynchronous serial and match against qualifiers

SERIN	pin,	baudmode,	(qualifier)
range/size:	0—7	0—7	byte
type:	any	any	any
meaning, usage:	Pin's direction bit is cleared to 0, making it an input.	baud rate and mode. Built-in constants select rates of 300, 600, 1200, or 2400 with true (T) or inverted (N) modes. Example: N2400 is inverted, 2400 baud. True modes are used with inverting line receivers; inverted for direct RS-232 input through a 22k resistor.	constant or list of constants to compare with incoming serial bytes. Holds until a match is achieved. Multiple constants may be separated with commas. Constants can be text in quotes like "PASSWORD." May be mixed with variables.

SEROUT: Transmit data via asynchronous serial (e.g., RS-232)

SEROUT	pin,	baudmode,	(data)
range/size:	0—7	0—15	byte
type:	any	any	any
meaning, usage:	Pin's direction bit is set to 1, making it an output.	baud rate and mode. Built-in constants select rates of 300, 600, 1200, or 2400 with true (T) or inverted (N) modes. Example: N2400 is inverted, 2400 baud. True modes are used with inverting line drivers; inverted for direct RS-232 ouput. Addition of "O" (for Open) causes output to be driven only to one supply rail; OT drives to +5V, ON drives to ground. Open modes used for networking.	byte or list of bytes to be transmitted. May be preceded by # to convert values into text numbers like "123." Multiple entries (with or w/o #) may be separated by commas. Constants may be text in quotes like "TEMPERATURE:." May be mixed with variables.

SLEEP: Go into power-saving mode for a specified time. See also Nap.

SLEEP	seconds
range/size:	0—65535
type:	any
meaning, usage:	length of power-down, in seconds rounded up to the nearest multiple of 2.304. Example: SLEEP 10 produces an 11.52-second (5 x 2.304) power-down period.

SOUND: Generate square-wave sound(s) of specified duration(s).

SOUND	pin ,	(note0,	duration0	,...)
range/size:	0—7	1—255	1—255	—
type:	any	any	any	—
meaning, usage:	Pin's direction bit is set to 1, making it an output.	relative frequency of tone. 0 is silent. Values from 1—127 are ascending tones; 128+ are white noises.	length of tone in units of 12ms	list of note/duration pairs continues, ending with closing paren.

TOGGLE: Make the specified pin an output and invert its state.

TOGGLE	pin
range/size:	0—7
type:	any
meaning, usage:	Pin's direction bit will be set to 1, making the pin an output. Its data bit will be the reverse of the former state; 1 changes to 0 and 0 to 1.

WRITE: Store data in EEPROM.

WRITE	location,	data
range/size:	0—255	byte
type:	any	any
meaning, usage:	EEPROM address. Care must be taken not to overwrite PBASIC program, whose lowest address may be Read from address 255. Only addresses less than that address should be used.	data to be stored in EEPROM

BASIC Stamp 2 Language Reference

This section describes the PBASIC dialect used in the standard BS2. The previous section covers the BS1; the next section covers the enhanced BS2-SX and -E.

PROGRAM MEMORY (EEPROM)

A PBASIC-2 program is stored in a 2048-byte EEPROM. This kind of memory retains data with the power off, but is easily erased or rewritten. When the Stamp is powered up, it automatically runs the program stored in EEPROM.

Those 2048 bytes hold about 500 PBASIC instructions. It's hard to pin down the exact number of instructions, because the Stamp host program squeezes each instruction into a compact *token*. The size of a given token depends more on the size and number of the instruction's arguments (the data it processes; see summary that follows) than the instruction itself. For example, the High instruction takes a single four-bit argument (a number from 0 to 15) to specify a pin number. So its token consists of a sequence of bits to identify the instruction, High, plus four bits to specify the pin.

At the other end of the spectrum, take a typical serial-output instruction: SEROUT 0,16780,["Hello"]. After identifying the instruction as Serout, the token must contain 4 bits for the pin number, 16 bits for the baudmode (the value 16780, which sets the speed and other serial parameters) and 5 bytes (40 bits) for the message "Hello."

So that figure of 500 instructions assumes a mixture of big and small tokens. Your mileage may vary.

You may also store data in unused EEPROM. There are two ways to store data to EEPROM. If the data is determined before the program runs (such as tables of constants) use the DATA directive. If the program itself will store the data, use the Write instruction. In either case, PBASIC can access the data with the Read instruction.

Since the DATA directive is managed by the STAMP host software, you will get an error message if your data overlaps with space needed to store your program. Program storage begins at the highest EEPROM address (2047) and works downward, while EEPROM-managed data storage starts at 0 (or an address you specify) and works upward. If the two try to store something to the same address, the Stamp host program stops and presents you with an error message.

DATA assigns a label to the starting address of the data it stores, making it easy for your program to locate the data. DATA automatically manages an address pointer that ensures that all data is stored in sequential EEPROM addresses starting at 0. You may also use DATA to set aside a portion of EEPROM for use by Write instructions. By doing so, you make the Stamp host program check to ensure that your desired storage area doesn't overlap the space used by your program. Here are a few examples of the DATA directive:

```
table     data    72,69,76,76,79  ' Store 5 bytes beginning at address "table"
string    data    "HELLO"    ' Store 5-byte beginning at address "string"
storage   data    (100)      ' Set aside 100 bytes beginning at "storage"
cleared   data    0(100)     ' 100 bytes, cleared to 0, starting at "cleared"
fixed     data    @10,"HELLO"     ' 5 bytes, beginning at (@) address 10
```

Once you have set aside space in the EEPROM, your program may store data in it by using the Write instruction. Write has a couple of limitations. First, writing to EEPROM is slower than writing to RAM, requiring as much as 10 milliseconds (ms) versus less than

0.3 ms. Second, EEPROM locations wear out after about 10 million writes. If you are tempted to replace RAM variables with EEPROM variables, make sure that you don't run afoul of these limitations.

To write to EEPROM that you have set aside with the DATA directive, use the DATA-assigned label as the starting address, and that label plus (n-1) as the ending address, where n is the number of storage bytes specified in the DATA directive. For example, if you have a subroutine called getaByte that acquires a byte of data, and you want to store 100 of those bytes in EEPROM, your code might resemble this:

```
n          con    100          ' Number of bytes to set aside.
storage    data   (n)          ' Set aside n bytes beginning at "storage"
topStor    con    storage + n-1 ' Highest address of storage area
EEadr      var    word         ' Address in EEPROM
theByte    var    byte         ' Byte returned by getaByte
FOR EEadr = storage to topStor ' For each byte of EEPROM set aside
   gosub getaByte              ' Obtain byte from subroutine in theByte
   WRITE EEadr,theByte         ' Store it to current address
NEXT                           ' Continue thru highest allocated address
' Mythical subroutine getaByte not shown.
```

DATA MEMORY (RAM)

The BS2 has 26 bytes of memory for variables. You assign names and sizes to variables with the VAR directive. The Stamp host software automatically arranges variables in memory. If you request more variable space than is available, the software responds with an error message.

VAR offers four sizes of variables, as shown in Table A-4 below:

The VAR directive also allows you to define variables that are alternative names for other variables or portions of other variables. In VAR directives, or within your program itself, you can refer to portions of variables by adding modifiers to variable names. For example, assuming that you have defined a 16-bit variable named myWord, the variable myWord.low-byte is the lower byte of myWord. Table A-5 lists the modifiers and their meanings.

The STAMP host program will respond with an error message if you use inappropriate modifiers; for example, if myByte is a byte variable, using myByte.bit15 will generate an error message because the bits of a byte are numbered 0–7.

Here are some example variable definitions with VAR:

```
onOff      var    bit          ' Single-bit variable (0 or 1)
count      var    nib          ' 4-bit variable (0-15)
myByte     var    byte         ' 8-bit variable (0-255)
bigNum     var    word         ' 16-bit variable (0-65535)
oddEven    var    count.bit0   ' Lowest bit of variable count.
```

TABLE A-4		
VAR TYPE	**SIZE**	**RANGE OF VALUE**
bit	1 bit	0,1
nib	4 bits (nibble)	0–15
byte	8 bits (byte)	0–255
word	16 bits (word)	0–65535

TABLE A-5

MODIFIER	MEANING
lowbyte	low byte of a word
highbyte	high byte of a word
byte0	low byte of a word
byte1	high byte of a word
lownib	low nibble of a word or byte
highnib	high nibble of a word or byte
nib0–3	numbered (low to high) nibbles of a word or byte
lowbit	low bit of a word, byte, or nibble
highbit	high bit of a word, byte, or nibble
bit0–15	numbered (low to high) bits of a word, byte, or nibble

VAR allows you to define arrays, groups of variables of the same size and sharing a single name, broken up into numbered cells. To create an array, just add the desired number of cells in parentheses to the size (bit, nib, byte, word) of the VAR directive. For example:

```
aString    var    byte(10)       ' An array consisting of 10 bytes.
```

To access one of the cells of an array, you must specify the variable name, and the cell number (0 to n-1) in parentheses. This makes it possible to efficiently perform an operation on a list of variables. The example below computes and displays the squares of the numbers 0 through 6.

```
n          con    7              ' Number of bytes for array
maxCell    con    n-1            ' Maximum cell # is n-1
square     var    byte(n)        ' An array of 7 bytes.
cell       var    nib            ' Cell # to work on.
FOR cell = 0 to maxCell          ' For each of the cells-
  square(cell) = cell * cell     ' Square # and store to cell
NEXT
debug "The first 7 squares are:",cr
FOR cell = 0 to maxCell          ' For each of the cells-
  debug DEC square(cell),cr      ' Display contents.
NEXT
STOP
```

INPUT/OUTPUT (I/O)

The BS2 has 16 input/output (I/O) pins. Each pin can be separately configured for input or output. Many of the input/output-oriented instructions automatically set the specified pin to the mode they require. For example, High sets a pin to output, while Pulsin sets a pin to input.

Three word variables represent the states of the I/O pins. Each bit of these variables corresponds to the like-numbered pin. For instance, bit 3 of OUTS is the output bit of pin P3. The variables are:

DIRS data direction, where 1 in a given bit makes that pin an output; 0, an input

OUTS output latches storing the output states (1 or 0) of the pins

INS read-only variable reflecting the states (1 or 0) of the pins

When a pin's DIRS bit is 1 (output), the state of the corresponding bit of OUTS controls the state of the pin, and therefore the state of the corresponding bit of INS. When a pin's DIRS bit is 0 (input), the state of its OUTS bit is ignored, and the states of the pin and its INS bit are controlled by whatever external circuit is connected to the pin.

The three I/O words are variables, and may be used in any way that variables are used, except that INS may not be written. Programs can access portions of these variables using either the variable modifiers of Table A-5, or additional built-in names listed in the manual.

MATH AND LOGIC

PBASIC-2 offers 14 math and logic operators that work on 16-bit values in any combination of variables and constants. If a value is smaller than 16 bits—a byte, for instance—it is internally converted to a 16-bit value by padding the missing higher-order bits with 0s.

Multiple operations can be combined into a single program step. Operations are performed from left to right by default, but you may use parentheses to reorganize the order of operation. PBASIC-2 performs operations in parentheses before other operations. If parentheses are nested, PBASIC-2 starts with the innermost parentheses and works outward.

Table A-6 lists BS2 math and logic operators.

TABLE A-6 BS2 MATH AND LOGIC OPERATORS

UNARY (ONE-ARGUMENT) OPERATORS

ABS	absolute value
SQR	square root
DCD	power-of-2 decode (sets nth bit of 16-bit word)
NCD	priority encode (returns power of 2, plus 1, of highest 1 bit)
-	negate (converts value to two's complement
~	complement (inverts—bitwise NOTs—all bits)
SIN	sine (returns two's complement, 8-bit sine of 0–255)
COS	cosine (returns two's complement, 8-bit cosine of 0–255)

BINARY (TWO-ARGUMENT) OPERATORS

+	add
-	subtract
*	multiply (returns low word of result)
**	multiply (returns high word of result)
/	divide (returns quotient)
//	divide (returns remainder)

TABLE A-6   CONTINUED

BINARY (TWO-ARGUMENT) OPERATORS

*/	multiply and return middle 16 bits of 32-bit result
MIN	keep variable greater than or equal to value
MAX	keep variable less than or equal to value
DIG	return decimal digit of value, 0-4 (e.g., 378 DIG 2 returns 3)
<<	Shift left
>>	Shift right
REV	mirror specified # of bits (e.g., %110101 REV 2 returns %10)
&	logical AND
I	logical OR
^	logical XOR
&/	logical AND NOT
I/	logical OR NOT
^/	logical XOR NOT

PBASIC-2 does not allow the LET keyword. To put the value y+z into x, you write "x=y+z" but not the "LET x=y+z" form used in older dialects of BASIC.

INSTRUCTION SUMMARY

Each instruction is accompanied by a brief description, followed by a table listing the instruction's *arguments*. Arguments are values that specify how the instruction should work. In the instruction HIGH 3, the constant 3 is an argument that specifies which pin to make high.

In most cases, PBASIC-2 instructions will accept a variable or constant of any size up to 16 bits (range of 0–65,535). Sometimes only part of this range is valid. Using High as an example again, PBASIC-2 will permit you to specify an out-of-range value like HIGH 65535 without an error message. However, it will truncate (chop off) all but the lowest four bits in order to get a valid pin number (0–15). The ranges specified in this instruction summary are the valid choices for the instruction.

PBASIC instructions process three types of data: variables, constants, and labels. Variables and constants are usually interchangeable; such cases are listed as any type in the tables below. Labels are locations within a program and may not be interchanged with variables or constants.

BRANCH: Use an offset to pick a program address from a list and go to that address.

BRANCH	offset,	[address0,	address1,	...	addressN]
range/size:	0—N	—	—	—	—
type:	any	label	label	label	label
meaning, usage:	index # of label to go to; if greater than N (last label in list) continues with next instruction	address label to go to if offset is 0	address label to go to if offset is 1	more address labels	address label to go to if offset is N

BUTTON: Read button input with optional debounce and autorepeat. Go to a program address if specified conditions are met. Must be executed from within a loop.

BUTTON	pin,	downstate,	delay,	rate,	bytevariable,	targetstate,	address
range/size:	0—15	0,1	0—255	0—255	byte	0,1	—
type:	any	any	any	any	variable	any	label
meaning, usage:	I/O pin	state when button is pushed	loops before autorepeat; 0=no debounce or repeat; 255=debounce, no repeat	number of loops between repeats	must be cleared to 0 before 1st use; if multiple Buttons, use separate variables	0=goto address when not pressed; 1=goto address when pressed	address label to go to when target-state requirement is met

COUNT: Count the number of cycles occurring on a pin during a specified period of time.

COUNT	pin,	period,	result
range/size:	0—15	0-65535	0—65535
type:	any	any	variable
meaning, usage:	Pin's direction bit is cleared to 0, making the pin an input. Output bit is unaffected but does not control pin state.	number of milliseconds during which cycles are to be counted	number of cycles counted during the specified period. Cycles must be less than 125kHz (square wave), or at least 4µs between transitions.

DEBUG: Show a variable or message in a window on the PC Debug screen.

DEBUG	modifier	data	, ...
range/size:	—	any	
type:	—	variable or quoted text message	
meaning, usage:	Debug modifiers determine how a value will be shown on the screen. No modifier means, "print the ASCII character corresponding to this value." The basic modifiers are: DEC Print value as decimal number HEX Print value as hexadecimal number BIN Print value as binary number See the Serin listing for options like I and S that can display values with indicators (% for binary or $ for hex) and two's complement sign (- if negative). Debug can also display arrays as strings using the modifier STR followed by the name of the array. Following the array with \n specifies how many characters to display. The modifier REP byte\n causes Debug to output a given byte repeated n times.		more entries, separated by commas

DTMFOUT: Output a telephone touch tone or list of tones.

DTMFout	pin,	ontime,	offtime,	[tone, tone, tone...]
range/size:	0—15	0-65535	0-65535	0—15
type:	any	any	any	any
meaning, usage:	Pin's direction bit is set to 1, making it an output.	optional: number of milliseconds to emit tone; if omitted, default is 200ms	optional: number of milliseconds silence after tone; if omitted, default is 50ms	list of tones to send; numbers correspond to the phone keypad—values 0–9 send the digits, 10 is star (*), 11 is pound (#), and 12–15 are the (normally unused) 4th-column tones.

END: End program and put Stamp into low-power Sleep mode until reset.

FOR...NEXT: Create a loop that increments or decrements a variable and repeats until it is no longer between specified start and end values. The loop begins with the FOR instruction as outlined below and ends with NEXT. Instructions between FOR and NEXT execute each time the loop repeats. NEXT can stand alone, or can specify the indexVariable used by the corresponding FOR instruction.

FOR	indexVariable	=	start	TO	end	STEP	stepVal
range/size:	0—65535		0—65535		0—65535		0—65535
type:	variable		any		any		any
meaning, usage:	variable initially set to start value, then incremented or decremented until it passes end value		starting value for the indexVariable		ending value for indexVariable. If loop is incrementing, it ends when indexVariable is greater than end; if decrementing, less than end.	optional: If no step/stepVal specified then indexVariable increments by 1 each loop.	specifies value to add to indexVariable each loop. If end is less than start, PBASIC-2 automatically makes stepVal negative.

FREQOUT: Output one or two sinewave tones for a specified duration.

Freqout	pin,	duration,	freq1	,freq2
range/size:	0—15	0—65535	0—32767	0—32767
type:	any	any	any	any
meaning, usage:	Pin's direction bit is set to 1, making it an output.	number of milliseconds to emit tone(s)	frequency in hertz (Hz) of first tone	optional: frequency in hertz (Hz) of 2nd tone

GOSUB: Store the location of the next instruction and go to a specified program address. May be nested four deep. See Return.

GOSUB	address
range/size:	—
type:	label
meaning, usage:	address label to go to.

GOTO: Go to a specified program address.

GOTO	address
range/size:	—
type:	label
meaning, usage:	address label to go to.

HIGH: Make the specified pin output/high.

HIGH	pin
range/size:	0—15
type:	any
meaning, usage:	Pin's output and direction bits will be set to 1, making the pin output/high.

A

IF...THEN: Make comparison(s) and, if true, go to a specified program address.

IF	value1	?	value2	AND/OR/XOR value ? value	THEN	address
range/size:	0—65535	—	0—65535	—		—
type:	any	—	any	—		label
meaning, usage:	variable or constant compare to value2 via ?	comparison operator: = (equal) <> (not equal) > (greater than) < (less than) <= (less than or equal to) >= (greater than or equal to)	variable or constant compare to value1 via ?	optional additional comparisons in same pattern, connected by AND or OR or XOR. With AND both comparison must be true to go to THEN address. With OR, one or both comparisons can be true to go to THEN address. With XOR, one (but not both) of the comparisons must be true. NOT may also be used to invert the outcome of a comparison.		address label to go to if comparison(s) true

INPUT: Make the specified pin an input.

INPUT	pin
range/size:	0—15
type:	any
meaning, usage:	Pin's direction bit will be cleared to 0, making the pin an input. Output bit is unaffected, but does not control pin state.

LOOKDOWN: Find a value in a list and return the index number 0—N of the match. Optionally uses a comparison operator to find first item that makes the comparison true.

LOOKDOWN	target,	?	[value0,	value1,	...	valueN],	result
range/size:	0—65535	—	0—65535	0—65535		0—65535	0—N
type:	any	—	any	any		any	variable
meaning, usage:	value to search for in list	optional comparison operator; if not specified, = is default: = (equal) <> (not equal) > (greater than) < (less than) <= (less than or equal to) >= (greater than or equal to)	list of values: may be a series of variables or constants separated by commas, or a quoted text string like "ABCDEFG". The first character is index 0, the second is 1, etc.				variable in which index of match value is stored. If no match, result is unchanged.

LOOKUP: Use an offset to pick a value from a list and store that value in a variable.

LOOKUP	offset,	[value0,	value1,	...	valueN],	result
range/size:	0—65535	0—65535	0—65535		0—65535	0—65535
type:	any	any	any		any	variable
meaning, usage:	index # of value to get	list of values: may be a series of variables or constants separated by commas, or a quoted text string like "ABCDEFG". The first character is index 0, the second is 1, etc.				receives value whose index is equal to offset. If offset it greater than N (last value in list) result is unchanged.

LOW: Make the specified pin output/low.

LOW	pin
range/size:	0—15
type:	any
meaning, usage:	Pin's output bit will be cleared to 0 and its direction bit set to 1, making the pin output/low.

NAP: Go into power-saving mode for a short time. See also Sleep.

NAP	period
range/size:	0—7
type:	any
meaning, usage:	length of power-down, computed as 2 raised to the power of period, times 18ms. Example: period = 3, 2^3 = 8, 8 x 18ms = 144ms, so NAP 3 powers down for 144ms.

OUTPUT: Make the specified pin an output.

OUTPUT	pin
range/size:	0—15
type:	any
meaning, usage:	Pin's direction bit (DIRS) will be set to 1, making the pin an output. Output bit (OUTS) controls pin state.

PAUSE: Delay for a specified number of milliseconds.

PAUSE	time
range/size:	0—65535
type:	any
meaning, usage:	length of delay in milliseconds.

PULSIN: Measures the edge-to-edge duration of a pulse in units of 2 microseconds.

PULSIN	pin,	state,	result
range/size:	0—15	0,1	0—65535
type:	any	any	variable
meaning, usage:	Pin's direction bit is cleared to 0, making the pin an input. Output bit is unaffected but does not control pin state.	triggering edge state: 0= 1-to-0 transition 1= 0-to-1 transition. Timing starts on state change; ends on opposite change	length of time between edges in units of 2µs. If no pulse occurs within 0.131 seconds, or pulse is longer than this, result = 0.

PULSOUT: Generate a pulse by inverting pin state for time specified in units of 2 microseconds.

PULSOUT	pin,	time
range/size:	0—15	0—65535
type:	any	any
meaning, usage:	Pin's direction bit is set to 1, making the pin an output. Output bit is inverted for pulse duration, then returned to previous state.	duration of pulse in 2µs units

A

APPENDIX A

PWM: Output fast pulse-width modulation that can be filtered to DC voltage.

PWM	pin,	duty,	cycles
range/size:	0—15	0—255	0—255
type:	any	any	any
meaning, usage:	Pin's direction bit is set to 1, making it an output during PWM. When done, direction bit is reset to 0 (input).	duty cycle, where 0 to 100% is represented by 0 to 255 (0.39% per unit)	number of 1ms cycles of PWM output

RANDOM: Generate a pseudorandom number.

RANDOM	variable
range/size:	word
type:	variable
meaning, usage:	word variable to scramble in a pseudorandom pattern. Pattern is always the same, so starting with different values (seeds) in variable is necessary to simulate truly random results.

RCTIME: Measures time until pin is in specified state, usually to measure RC timing.

RCTIME	pin,	state,	result
range/size:	0—15	0,1	0—65535
type:	any	any	variable
meaning, usage:	Pin's direction bit is cleared to 0, making the pin an input. Output bit is unaffected but does not control pin state.	state of pin that stops timer and ends the instruction 0= timing ends on 0 1= timing ends on 1	time in 2µs units from beginning of instruction until pin matches state; if pin does not match state within 0.131ms, returns 0.

READ: Retrieve data stored in EEPROM.

READ	location,	data
range/size:	0—2047	byte
type:	any	variable
meaning, usage:	EEPROM address	data retrieved from EEPROM

RETURN: Go to program line stored by most recent Gosub instruction (return from subroutine).

REVERSE: Reverse the data direction of the specified pin.

REVERSE	pin
range/size:	0—15
type:	any
meaning, usage:	Pin's direction bit will be inverted: if 1 (output), changed to 0 if 0 (input), changed to 1

SERIN: Receive asynchronous serial transmission (e.g., RS-232).

SERIN	rPin{\fPin},	baudmode,	{plabel,}	{time,}	{tlabel,}	[inputData]
range/size:	0—15	0—65535	—	0—65535	—	see below
type:	any	any	label	any	label	see below
meaning, usage:	rPin is changed to input and used to receive serial data; optional fPin changes to output to act as a flow-control flag (0=OK to send, 1=wait with noninverted baudmodes, opposite for inverted)	bits 0–12 are bit time minus 20µS bit13 is parity: 0= 8 bits no parity 1= 7 bits even parity bit14 is polarity 0= noninverted 1= inverted	optional program label to go to in event of a parity error; used only when bit13 of baudmode is 1	optional time in ms to wait for serial data to arrive; if no data arrives in time, program goes to tlabel	optional program label to go to in event that data does not arrive in specified time; used only when time is specified	input data may consist of individual byte(s) to be stored in variable(s) or array, or any data to be pattern-matched, or a combination, separated by commas; see options below

NUMERIC CONVERSION OPTIONS FOR SERIN

Serin modifiers can automatically convert text representations of numbers into values to be stored in variables. For example, if the code is *DEC2 variable* and the text bytes "17" are received, the value 17 (%10001 or $11) would be stored to the variable. You can construct numeric conversion modifiers from the elements listed in Table A-7. For example ISHEX3 variable would accept only 3-digit hex values indicated by the $ symbol, and, if preceded by the minus sign (−), would convert them to two's complement negative numbers.

STRING COLLECTION AND SEQUENCE MATCHING OPTIONS FOR SERIN

Serin can automatically acquire strings of bytes, or compare sequences of bytes to a string and wait until it receives a match. Table A-8 summarizes the options.

OPTIONS FOR SEROUT

The same modifiers used by Serin to interpret incoming text bytes as values may be used by Serout to convert values into text in various formats. See the Serin listing. Serout offers options for sending a string or a repeating sequence of bytes, as shown in Table A-9.

TABLE A-7 NUMERIC CONVERSION MODIFIERS

TYPE OF NUMBER	INDICATOR (require $ or % before hex, binary numbers)	SIGN (if minus (−) precedes number, convert to 2's complement)	NUMBER BASE	NUMBER OF DIGITS
Decimal (no indicator)	not used	S	DEC	1–5
Hexadecimal ($)	I	S	HEX	1–4
Binary (%)	I	S	BIN	1–16

TABLE A-8 STRING COLLECTION MODIFIERS

MODIFIER	ACTION
STR bytearray \L {\E}	Input a string of length L into an array. If specified, end character E causes the string input to end before length L. Remaining bytes filled with 0s (zeros).
WAIT (value,value,...) WAIT ("text")	Wait for byte sequence, which may be quoted text, up to six bytes long.
WAITSTR bytearray	Wait for byte sequence matching a string stored in array. The end of the array-string is marked by a byte containing 0 (zero).
WAITSTR bytearray\L	Wait for byte sequence matching a string of length L bytes stored in an array.
SKIP L	Ignore L bytes of serial input.

SEROUT: Transmit data via asynchronous serial (e.g., RS-232).

SEROUT	tPin{\fPin},	baudmode,	{time,}	{tlabel,}	[outputData]
range/size:	0—15	0—65535	0—65535	—	see below
type:	any	any	any	label	see below
meaning, usage:	tPin is changed to output and used to send serial data; optional fPin changes to input to read the state of the receiver's flow-control flag (0=OK to send, 1=wait with noninverted baudmodes, opposite for inverted)	bits 0–12 are bit time minus 20µS bit13 is parity: 0= 8 bits no parity 1= 7 bits even parity bit14 is polarity 0= noninverted 1= inverted bit15 determines whether tPin is always driven (0) or left open in one state (1) for networking	optional time in ms to wait for permission to send via fPin	optional program label to go to in event that permission to send (fPin) does not arrive in specified time	output data may consist of individual byte(s), arrays, or constants, such as quoted text; modifiers may convert numeric values to text or automatically send arrays as strings—see below

TABLE A-9   SEROUT STRING-CREATION MODIFIERS

MODIFIER	ACTION
STR bytearray {\n}	Send a string from bytearray until byte=0. If optional \n is used, send n bytes.
REP byte\n	Send a string consisting of byte repeated n times.

SHIFTIN: Clock in data synchronously.

SHIFTIN	dPin	cPin	mode	[variable {\n}]
range/size:	0—15	0—15	0—3	0—65535
type:	any	any	any	variable
meaning, usage:	dPin is changed to input and used to capture incoming data bits in sync with pulses on cPin; mode determines when dPin is captured	cPin is changed to output and pulsed to shift bits out of connected synchronous serial device	mode sets clock-to-data timing and order of data bits with predefined constants as follows: MSBPRE (0): start with msb; get data before clock pulse. LSBPRE (1): start with lsb; get data before clock pulse. MSBPOST (2): start with msb; get data after clock pulse. LSBPOST (3): start with lsb; get data after clock pulse.	variable stores the incoming data bits; n is an optional number of bits to be shifted (default is 8)

SHIFTOUT: Clock out data synchronously.

SHIFTOUT	dPin	cPin	mode	[outputData {\n}]
range/size:	0—15	0—15	0—1	0—65535
type:	any	any	any	any
meaning, usage:	dPin is changed to output and used to send data bits in sync with pulses on cPin; mode determined when dPin has valid data	cPin is changed to output and pulsed to shift bits out of dPin into connected synchronous serial device	mode sets the order of data bits with predefined constants as follows: LSBFIRST (0): start with lsb. MSBFIRST (1): start with msb.	outputData is the data to be shifted out; n is an optional number of bits to be shifted (default is 8)

SLEEP: Go into power-saving mode for a specified time. See also Nap.

SLEEP	seconds
range/size:	0—65535
type:	any
meaning, usage:	length of power-down, in seconds rounded up to the nearest multiple of 2.304. Example: SLEEP 10 produces an 11.52-second (5 x 2.304) power-down period.

STOP: End program until reset without putting Stamp into low-power mode.

TOGGLE: Make the specified pin an output and invert its state.

TOGGLE	pin
range/size:	0—15
type:	any
meaning, usage:	Pin's direction bit will be set to 1, making the pin an output. Its data bit will be the reverse of the former state; 1 changes to 0 and 0 to 1.

WRITE: Store data in EEPROM.

WRITE	location,	data
range/size:	0—2047	byte
type:	any	any
meaning, usage:	EEPROM address. Care must be taken not to overwrite PBASIC program, which is stored at address 2047 working downward. Data directive may be used to set aside "safe" EEPROM storage.	data to be stored in EEPROM

XOUT: Send an X-10 powerline control command.

XOUT	mPin	zPin	[house\keyCommand {\cycles}]
range/size:	0—15	0—15	0—15
type:	any	any	any
meaning, usage:	mPin is changed to output and used to send the X-10 signals (modulation) to the powerline interface.	zPin is changed to input and used to get the zero-crossing timing from the powerline interface.	house is the X-10 house code with 0-15 representing codes A-P. keyCommand is an X-10 command from the following list of built-in constants: unitOn, unitOff, unitsOff, lightsOn, dim, bright cycles is the number of times the key should be sent.

BASIC Stamp 2-SX and -E Language Reference

This section describes the PBASIC dialect used in the BS2-SX and BS2-E models. The previous section covers the standard BS2. Throughout this section I will refer to both the BS2-SX and -E as "BS2-SX" since the devices are the same except for speed. Where the speed difference is significant, I will point out that difference.

PROGRAM MEMORY (EEPROM)

Like the standard BS2 described in the previous section, a BS2-SX program is a block of 2048 bytes, comprising approximately 500 PBASIC-2 instructions. Unlike the standard BS2, which can hold only one program, the BS2-SX can store up to eight programs at a time. Programs are numbered 0 through 7. When the BS2-SX starts up after power-on or reset, program 0 runs. Program 0 may switch to another program using the syntax Run `programNumber`.

When the BS2-SX switches programs using Run, all variables retain their current states. Programs that use identical variable declarations or fixed-location variables can refer to the same data by the same variable names. However, when Run executes, any Gosub/Return pointers are lost. In other words, a Run instruction within a subroutine cannot automatically find its way back to that location. The program must explicitly record any return-to location, then Goto or Branch to that location when it is again Run. See Chapter 16 for examples.

Also like the standard BS2, the BS2-SX can use program memory for data storage. All of the standard BS2 features for this purpose are available; see the previous section. However, a program can access EEPROM data only within its own segment of EEPROM. For instance, program 0 cannot Read or Write data in program 1's memory space. You may work around this limitation by writing a simple program to access EEPROM within its program segment. See the capacitance-measuring project in Chapter 17 for an example.

DATA MEMORY (RAM)

Just like the BS2, the BS2-SX has 26 bytes of memory for variables. All aspects of variable assignment and usage are identical to the standard BS2; see the previous section.

SCRATCHPAD MEMORY (RAM)

The BS2-SX has a block of 64 bytes of scratchpad RAM, a new feature not shared by the standard BS2. Scratchpad RAM is accessed using two instructions:

- Get address,byteVariable
- Put address,byteValue

where address is the location in scratch RAM (0-63 for Get; 0-62 for Put). Only byte-sized data is supported, so Get should have a byte variable as the destination for the data it fetches from scratch RAM, and Put should use a byte value (variable or constant) for the data it stores. Using data of different sizes will not generate an error, but the data will be padded with 0s if too small, or truncated if too large.

Note that address 63 of the scratchpad is special—it always contains the currently running program number (0-7) and cannot be overwritten by Put.

Like normal data memory, the scratchpad is cleared (to all 0s) at startup and reset. Scratchpad RAM retains its data when Run is used to switch programs, making it suitable for communicating between programs, or for backing up data for use upon return from another program. See Chapter 16 for examples.

INPUT/OUTPUT (I/O)

I/O programming for the BS2-SX is identical to the standard BS2 as described in the previous section. The only difference is that each BS2-SX I/O pin, when set to output, can source and sink up to 30mA; standard BS2 I/Os source 20mA max and sink 25mA max. (Source means act as the + supply for a load whose other side is grounded; sink means act as the ground connection for a load whose other side is connected to Vdd.)

MATH AND LOGIC

All math and logic instructions for the BS2-SX are identical to the standard BS2 set described in the previous section.

INSTRUCTION SUMMARY

Most BS2-SX instructions are identical to their counterparts in the standard BS2. The following instructions are unique to the BS2-SX and BS2-E:

- Get (store data to scratchpad RAM)
- Put (retrieve data from scratchpad RAM)
- Run (switch to another program)

In addition, the BS2-SX operates 2.5 times faster than the standard BS2. This affects the timing of a number of instructions. The BS2-E supports Get, Put, and Run, but operates at the same speed as the original BS2.

The listing that follows indicates the timing differences for the -SX and -E models.

BRANCH: Use an offset to pick a program address from a list and go to that address.

BRANCH	offset,	[address0,	address1,	...	addressN]
range/size:	0—N	—	—	—	—
type:	any	label	label	label	label
meaning, usage:	index # of label to go to; if greater than N (last label in list) continues with next instruction	address label to go to if offset is 0	address label to go to if offset is 1	more address labels	address label to go to if offset is N

BUTTON: Read button input with optional debounce and autorepeat. Go to a program address if specified conditions are met. Must be executed from within a loop.

BUTTON	pin,	downstate,	delay,	rate,	bytevariable,	targetstate,	address
range/size:	0—15	0,1	0—255	0—255	byte	0,1	—
type:	any	any	any	any	variable	any	label
meaning, usage:	I/O pin	state when button is pushed	loops before autorepeat; 0=no debounce or repeat; 255= debounce, no repeat	number of loops between repeats	must be cleared to 0 before 1st use; if multiple Buttons, use separate variables	0=goto address when not pressed; 1=goto address when pressed	address label to go to when target-state requirement is met

COUNT: Count the number of cycles occurring on a pin during a specified period of time.

COUNT	pin,	period,	result
range/size:	0—15	0-65535	0—65535
type:	any	any	variable
meaning, usage:	Pin's direction bit is cleared to 0, making the pin an input. Output bit is unaffected but does not control pin state.	number of units during which cycles are to be counted. Units are 0.4ms for –SX; 1ms for –E.	number of cycles counted during the specified period. For the SX, cycles must be less than 312.5kHz or at least 1.6μs between transitions. For –E, 125kHz max or 4μs min.

DEBUG: Show a variable or message in a window on the PC Debug screen.

DEBUG	modifier	data	, ...
range/size:	—	any	
type:	—	variable or quoted text message	
meaning, usage:	Debug modifiers determine how a value will be shown on the screen. No modifier means, "print the ASCII character corresponding to this value." The basic modifiers are: DEC Print value as decimal number HEX Print value as hexadecimal number BIN Print value as binary number See the Serin listing for options like I and S that can display values with indicators (% for binary or $ for hex) and two's complement sign (- if negative). Debug can also display arrays as strings using the modifier STR followed by the name of the array. Following the array with \n specifies how many characters to display. The modifier REP byte\n causes Debug to output a given byte repeated n times.		more entries, separated by commas

DTMFOUT: Output a telephone touch tone or list of tones.

DTMFout	pin,	ontime,	offtime,	[tone, tone, tone...]
range/size:	0—15	0-65535	0-65535	0—15
type:	any	any	any	any
meaning, usage:	Pin's direction bit is set to 1, making it an output.	optional: number of units to emit tone; if omitted, default is 200. Units are 0.4ms (-SX) or 1ms (-E).	optional: number of units silence after tone; if omitted, default is 50. Units are 0.4ms (-SX) or 1ms (-E).	list of tones to send; numbers correspond to the phone keypad—values 0–9 send the digits, 10 is star (*), 11 is pound (#), and 12–15 are the (normally unused) 4th-column tones.

END: End program and put Stamp into low-power Sleep mode until reset.

FOR...NEXT: Create a loop that increments or decrements a variable and repeats until it is no longer between specified start and end values. The loop begins with the FOR instruction as outlined below and ends with NEXT. Instructions between FOR and NEXT execute each time the loop repeats. NEXT can stand alone, or can specify the indexVariable used by the corresponding FOR instruction.

FOR	indexVariable	=	start	TO	end	STEP	stepVal
range/size:	0—65535		0—65535		0—65535		0—65535
type:	variable		any		any		any
meaning, usage:	variable initially set to start value, then incremented or decremented until it passes end value		starting value for the indexVariable		ending value for indexVariable. If loop is incrementing, it ends when indexVariable is greater than end; if decrementing, less than end.	optional: If no step/stepVal specified then indexVariable increments by 1 each loop.	specifies value to add to indexVariable each loop. If end is less than start, PBASIC-2 automatically makes stepVal negative.

FREQOUT: Output one or two sinewave tones for a specified duration.

Freqout	pin,	duration,	freq1	,freq2
range/size:	0—15	0—65535	0—32767	0—32767
type:	any	any	any	any
meaning, usage:	Pin's direction bit is set to 1, making it an output.	number of units to emit tone(s). Units are 0.4ms (-SX) or 1ms (-E).	frequency in units of first tone. Units are 2.5Hz (-SX) or 1Hz (-E).	optional: frequency in units of 2nd tone. Units are 2.5Hz (-SX) or 1Hz (-E).

GET: Retrieve data stored in scratchpad RAM (BS2-SX and –E only).

GET	location,	data
range/size:	0—63	byte
type:	any	variable
meaning, usage:	Scratch RAM address. Location 63 always holds the current program number (0-7)	data retrieved from scratchpad

GOSUB: Store the location of the next instruction and go to a specified program address. May be nested four deep. See Return.

GOSUB	address
range/size:	—
type:	label
meaning, usage:	address label to go to.

GOTO: Go to a specified program address.

GOTO	address
range/size:	—
type:	label
meaning, usage:	address label to go to.

HIGH: Make the specified pin output/high.

HIGH	pin
range/size:	0—15
type:	any
meaning, usage:	Pin's output and direction bits will be set to 1, making the pin output/high.

IF...THEN: Make comparison(s) and, if true, go to a specified program address.

IF	value1	?	value2	AND/OR/XOR value ? value	THEN	address
range/size:	0—65535	—	0—65535	—		—
type:	any	—	any	—		label
meaning, usage:	variable or constant compare to value2 via ?	comparison operator: = (equal) <> (not equal) > (greater than) < (less than) <= (less than or equal to) >= (greater than or equal to)	variable or constant compare to value1 via ?	optional additional comparisons in same pattern, connected by AND or OR or XOR. With AND both comparison must be true to go to THEN address. With OR, one or both comparisons can be true to go to THEN address. With XOR, one (but not both) of the comparisons must be true. NOT may also be used to invert the outcome of a comparison.		address label to go to if comparison(s) true

A

APPENDIX A

INPUT: Make the specified pin an input.

INPUT	pin
range/size:	0—15
type:	any
meaning, usage:	Pin's direction bit will be cleared to 0, making the pin an input. Output bit is unaffected, but does not control pin state.

LOOKDOWN: Find a value in a list and return the index number 0—N of the match. Optionally uses a comparison operator to find first item that makes the comparison true.

LOOKDOWN	target,	?	[value0,	value1,	...	valueN],	result
range/size:	0—65535	—	0—65535	0—65535		0—65535	0—N
type:	any	—	any	any		any	variable
meaning, usage:	value to search for in list	optional comparison operator; if not specified, = is default: = (equal) <> (not equal) > (greater than) < (less than) <= (less than or equal to) >= (greater than or equal to)	list of values: may be a series of variables or constants separated by commas, or a quoted text string like "ABCDEFG". The first character is index 0, the second is 1, etc.				variable in which index of match value is stored. If no match, result is unchanged.

LOOKUP: Use an offset to pick a value from a list and store that value in a variable.

LOOKUP	offset,	[value0,	value1,	...	valueN],	result
range/size:	0—65535	0—65535	0—65535		0—65535	0—65535
type:	any	any	any		any	variable
meaning, usage:	index # of value to get	list of values: may be a series of variables or constants separated by commas, or a quoted text string like "ABCDEFG". The first character is index 0, the second is 1, etc.				receives value whose index is equal to offset. If offset it greater than N (last value in list) result is unchanged.

LOW: Make the specified pin output/low.

LOW	pin
range/size:	0—15
type:	any
meaning, usage:	Pin's output bit will be cleared to 0 and its direction bit set to 1, making the pin output/low.

NAP: Go into power-saving mode for a short time. See also Sleep.

NAP	period
range/size:	0—7
type:	any
meaning, usage:	length of power-down, computed as 2 raised to the power of period, times 18ms. Example: period = 3, 2^3 = 8, 8 x 18ms = 144ms, so NAP 3 powers down for 144ms.

OUTPUT: Make the specified pin an output.

OUTPUT	pin
range/size:	0—15
type:	any
meaning, usage:	Pin's direction bit (DIRS) will be set to1, making the pin an output. Output bit (OUTS) controls pin state.

PAUSE: Delay for a specified number of milliseconds.

PAUSE	time
range/size:	0—65535
type:	any
meaning, usage:	length of delay in milliseconds.

PULSIN: Measures the edge-to-edge duration of a pulse in units of 2 microseconds.

PULSIN	pin,	state,	result
range/size:	0—15	0,1	0—65535
type:	any	any	variable
meaning, usage:	Pin's direction bit is cleared to 0, making the pin an input. Output bit is unaffected but does not control pin state.	triggering edge state: 0= 1-to-0 transition 1= 0-to-1 transition. Timing starts on state change; ends on opposite change	length of time between edges in units. If no pulse occurs within 65535 units, or pulse is longer than this, result = 0. Units are 0.8µs (-SX) or 2µs (-E).

PULSOUT: Generate a pulse by inverting pin state for time specified in units of 2 microseconds.

PULSOUT	pin,	time
range/size:	0—15	0—65535
type:	any	any
meaning, usage:	Pin's direction bit is set to 1, making the pin an output. Output bit is inverted for pulse duration, then returned to previous state.	duration of pulse in units of 0.8µs (-SX) or 2µs (-E).

PUT: Store data in scratchpad RAM (BS2-SX and –E only).

PUT	location,	data
range/size:	0—62	byte
type:	any	any
meaning, usage:	Scratchpad address.	data to be stored in scratch RAM

PWM: Output fast pulse-width modulation that can be filtered to DC voltage.

PWM	pin,	duty,	cycles
range/size:	0—15	0—255	0—255
type:	any	any	any
meaning, usage:	Pin's direction bit is set to 1, making it an output during PWM. When done, direction bit is reset to 0 (input).	duty cycle, where 0 to 100% is represented by 0 to 255 (0.39% per unit)	number of time units of PWM output. Units are 0.4ms (-SX) or 1ms (-E).

RANDOM: Generate a pseudorandom number.

RANDOM	variable
range/size:	word
type:	variable
meaning, usage:	word variable to scramble in a pseudorandom pattern. Pattern is always the same, so starting with different values (seeds) in variable is necessary to simulate truly random results.

A

APPENDIX A

RCTIME: Measures time until pin is in specified state, usually to measure RC timing.

RCTIME	pin,	state,	result
range/size:	0—15	0,1	0—65535
type:	any	any	variable
meaning, usage:	Pin's direction bit is cleared to 0, making the pin an input. Output bit is unaffected but does not control pin state.	state of pin that stops timer and ends the instruction 0= timing ends on 0 1= timing ends on 1	time in units from beginning of instruction until pin matches state; if pin does not match state within 65535 units, returns 0. Units are 0.8µs (-SX) or 2µs (-E).

READ: Retrieve data stored in EEPROM.

READ	location,	data
range/size:	0—2047	byte
type:	any	variable
meaning, usage:	EEPROM address	data retrieved from EEPROM

RETURN: Go to program line stored by most recent Gosub instruction (return from subroutine).

REVERSE: Reverse the data direction of the specified pin.

REVERSE	pin
range/size:	0—15
type:	any
meaning, usage:	Pin's direction bit will be inverted: if 1 (output), changed to 0 if 0 (input), changed to 1

RUN: Execute another program. (BS2-SX and –E only)

RUN	programNumber
range/size:	0—7
type:	any
meaning, usage:	program number to run. Program invoked by run always starts at the first line of code, but all variables and I/O pin settings remain intact.

SERIN: Receive asynchronous serial transmission (e.g., RS-232).

SERIN	rPin{\fPin},	baudmode,	{plabel,}	{time,}	{tlabel,}	[inputData]
range/size:	0—15	0—65535	—	0—65535	—	see below
type:	any	any	label	any	label	see below
meaning, usage:	rPin is changed to input and used to receive serial data; optional fPin changes to output to act as a flow-control flag (0=OK to send, 1=wait with noninverted baudmodes, opposite for inverted)	bits 0–12 are bit timing; $(2.5 \times 10^6/\text{baud}) - 20$ for -SX or $(1 \times 10^6/\text{baud}) - 20$ for -E bit13 is parity: 0= 8 bits no parity 1= 7 bits even parity bit14 is polarity 0= noninverted 1= inverted	optional program label to go to in event of a parity error; used only when bit13 of baudmode is 1	optional time to wait for serial data to arrive; if no data arrives in time, program goes to tlabel. Time units: 0.4ms for -SX; 1ms for -E	optional program label to go to in event that data does not arrive in specified time; used only when time is specified	input data may consist of individual byte(s) to be stored in variable(s) or array, or any data to be pattern-matched, or a combination, separated by commas; see options below

NUMERIC CONVERSION OPTIONS FOR SERIN

Serin modifiers can automatically convert text representations of numbers into values to be stored in variables. For example, if the code is *DEC2 variable* and the text bytes "17" are received, the value 17 (%10001 or $11) would be stored to the variable. You can construct numeric conversion modifiers from the elements listed in Table A-10. For example, ISHEX3 variable would accept only 3-digit hex values indicated by the $ symbol, and, if preceded by the minus sign (-), would convert them to two's complement negative numbers.

STRING COLLECTION AND SEQUENCE MATCHING OPTIONS FOR SERIN

Serin can automatically acquire strings of bytes, or compare sequences of bytes to a string and wait until it receives a match. Table A-11 summarizes the options.

TABLE A-10   NUMERIC CONVERSION MODIFIERS

TYPE OF NUMBER	INDICATOR (require $ or % before hex, binary numbers)	SIGN (if minus (-) precedes number, convert to 2's complement)	NUMBER BASE	NUMBER OF DIGITS
Decimal (no indicator)	not used	S	DEC	1–5
Hexadecimal ($)	I	S	HEX	1–4
Binary (%)	I	S	BIN	1–16

TABLE A-11   STRING COLLECTION MODIFIERS

MODIFIER	ACTION
STR bytearray \L {\E}	Input a string of length L into an array. If specified, end character E causes the string input to end before length L. Remaining bytes filled with 0s (zeros).
WAIT (value,value,...) WAIT ("text")	Wait for byte sequence, which may be quoted text, up to six bytes long.
WAITSTR bytearray	Wait for byte sequence matching a string stored in array. The end of the array-string is marked by a byte containing 0 (zero).
WAITSTR bytearray\L	Wait for byte sequence matching a string of length L bytes stored in an array.
SKIP L	Ignore L bytes of serial input.

SEROUT: Transmit data via asynchronous serial (e.g., RS-232).

SEROUT	tPin{\fPin},	baudmode,	{time,}	{tlabel,}	[outputData]
range/size:	0—15	0—65535	0—65535	—	see below
type:	any	any	any	label	see below
meaning, usage:	tPin is changed to output and used to send serial data; optional fPin changes to input to read the state of the receiver's flow-control flag (0=OK to send, 1=wait with noninverted baudmodes, opposite for inverted)	bits 0–12 are bit timing; $(2.5 \times 10^6/\text{baud}) - 20$ for -SX or $(1 \times 10^6/\text{baud}) - 20$ for -E bit13 is parity: 0= 8 bits no parity 1= 7 bits even parity bit14 is polarity 0= noninverted 1= inverted bit15 determines whether tPin is always driven (0) or left open in one state (1) for networking	optional time to wait for permission to send via fPin. Units are 0.4ms for –SX, 1ms for -E	optional program label to go to in event that permission to send (fPin) does not arrive in specified time	output data may consist of individual byte(s), arrays, or constants, such as quoted text; modifiers may convert numeric values to text or automatically send arrays as strings—see below

OPTIONS FOR SEROUT

The same modifiers used by Serin to interpret incoming text bytes as values may be used by Serout to convert values into text in various formats. See the Serin listing. Serout offers options for sending a string or a repeating sequence of bytes, as shown in Table A-12.

TABLE A-12 SEROUT STRING-CREATION MODIFIERS

MODIFIER	ACTION
STR bytearray {\n}	Send a string from bytearray until byte=0. If optional \n is used, send n bytes.
REP byte\n	Send a string consisting of byte repeated n times.

SHIFTIN: Clock in data synchronously.

SHIFTIN	dPin	cPin	mode	[variable {\n}]
range/size:	0—15	0—15	0—3	0—65535
type:	any	any	any	variable
meaning, usage:	dPin is changed to input and used to capture incoming data bits in sync with pulses on cPin; mode determines when dPin is captured	cPin is changed to output and pulsed to shift bits out of connected synchronous serial device	mode sets clock-to-data timing and order of data bits with predefined constants as follows: MSBPRE (0): start with msb; get data before clock pulse. LSBPRE (1): start with lsb; get data before clock pulse. MSBPOST (2): start with msb; get data after clock pulse. LSBPOST (3): start with lsb; get data after clock pulse.	variable stores the incoming data bits; n is an optional number of bits to be shifted (default is 8)

SHIFTOUT: Clock out data synchronously.

SHIFTOUT	dPin	cPin	mode	[outputData {\n}]
range/size:	0—15	0—15	0—1	0—65535
type:	any	any	any	any
meaning, usage:	dPin is changed to output and used to send data bits in sync with pulses on cPin; mode determined when dPin has valid data	cPin is changed to output and pulsed to shift bits out of dPin into connected synchronous serial device	mode sets the order of data bits with predefined constants as follows: LSBFIRST (0): start with lsb. MSBFIRST (1): start with msb.	outputData is the data to be shifted out; n is an optional number of bits to be shifted (default is 8)

SLEEP: Go into power-saving mode for a specified time. See also Nap.

SLEEP	seconds
range/size:	0—65535
type:	any
meaning, usage:	length of power-down, in seconds rounded up to the nearest multiple of 2.304. Example: SLEEP 10 produces an 11.52-second (5 x 2.304) power-down period.

STOP: End program until reset without putting Stamp into low-power mode.

TOGGLE: Make the specified pin an output and invert its state.

TOGGLE	pin
range/size:	0—15
type:	any
meaning, usage:	Pin's direction bit will be set to 1, making the pin an output. Its data bit will be the reverse of the former state; 1 changes to 0 and 0 to 1.

WRITE: Store data in EEPROM.

WRITE	location,	data
range/size:	0—2047	byte
type:	any	any
meaning, usage:	EEPROM address. Care must be taken not to overwrite PBASIC program, which is stored at address 2047 working downward. Data directive may be used to set aside "safe" EEPROM storage.	data to be stored in EEPROM

XOUT: Send an X-10 powerline control command.

XOUT	mPin	zPin	[house\keyCommand {\cycles}]
range/size:	0—15	0—15	0—15
type:	any	any	any
meaning, usage:	mPin is changed to output and used to send the X-10 signals (modulation) to the powerline interface.	zPin is changed to input and used to get the zero-crossing timing from the powerline interface.	house is the X-10 house code with 0-15 representing codes A-P. keyCommand is an X-10 command from the following list of built-in constants: unitOn, unitOff, unitsOff, lightsOn, dim, bright cycles is the number of times the key should be sent.

APPENDIX B

NUMBERING SYSTEMS, INTEGER MATH, AND BOOLEAN LOGIC

These three topics—numbers, math, and logic—go to the very heart of programming. To program well, or even just to understand well-written programs, you will have to come to grips with these subjects. This is not difficult stuff; much of it is just a second look at subjects covered in the second and third grades. So let's go back (possibly way back) to those days when you were just beginning your acquaintance with numbers.

Numbering Systems

Suppose I take a piece of chalk and make a series of marks on a blackboard, like so:

* * * * * * * * * * * *

How many marks did I make? Most people would reply, "twelve" or "a dozen" or write the number "12." A smart-aleck might write the Roman numeral "XII." A programmer would have even more options: "C hex" or "1100 binary."

Would any of these responses change the actual number of marks? No, of course not. The way we express a number doesn't matter; a number is a concept that is independent of how it is expressed.

Let's apply this principle to computers like the BASIC Stamps. When you write a Stamp program, you can specify numbers in any of three different numbering systems—decimal, hexadecimal, or binary.

Since people are comfortable with decimal numbers and representations don't matter, why bother with hexadecimal or binary? While decimal numbers are usually the best way to express quantities (time, volts, distance, counts), hexadecimal and binary are useful for visualizing the states of individual bits. Let's use the Stamp circuit in Figure B-1 as a point of discussion. Knowing that a 1 on a particular Stamp pin will light the connected LED, what would be the state of the LEDs after this short program?

```
dirs = 255      ' Set all pins to output.
pins = 170      ' Write 170 to the pins.
```

The decimal number 170 provides no clue to the state of the individual bits. What about this example:

```
dirs = %11111111      ' Set all pins to output.
pins = %10101010      ' Write %10101010 to the pins.
```

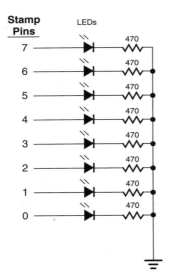

Stamp Pins   LEDs

7
6
5
4
3
2
1
0

470 (each)

Figure B-1 LEDs light when corresponding pin is 1.

TABLE B-1 BINARY, DECIMAL, AND HEXADECIMAL NUMBERS

BINARY	DECIMAL	HEXADECIMAL
0000	0	0
0001	1	1
0010	2	2
0011	3	3
0100	4	4
0101	5	5
0110	6	6
0111	7	7
1000	8	8
1001	9	9
1010	10	A
1011	11	B
1100	12	C
1101	13	D
1110	14	E
1111	15	F

Using a binary number[1] makes the states of the pins clear as day. Starting from the left, pin 7 is 1, pin 6 is 0, pin 5 is 1...and pin 0 is 0. The kicker is that the binary number %10101010 is equal to the decimal number 170. The two program fragments produce the exact same results.

The only drawback to using binary numbers in bit-oriented programs is length. A 16-bit number is by definition a 16-digit number in binary notation. Fortunately, there's a shorthand for binary in the form of hexadecimal (hex) numbers. The hexadecimal system is named for the fact that it uses 16 different symbols to represent decimal values of 0 through 15 and binary values of 0000 through 1111.

Each digit of a hex number represents four binary digits (bits). If you learn the bit patterns for the 16 hex digits, you can express long binary numbers in a compact way. Table B-1 lists the hex numbers and their binary and decimal equivalents.

Here's the previous program example using hexadecimal numbers:

```
dirs = $FF     ' Set all pins to output.
pins = $AA     ' Write $AA (%10101010) to the pins.
```

1. PBASIC knows that you mean a binary number when you precede it with the percent sign (%), and it recognizes hexadecimal numbers by the prefix $. The choice of these symbols is arbitrary, and has nothing to do with percentages or money!

DIGITS, WEIGHTS, AND SIGNIFICANCE

In the decimal numbering system, we use just 10 symbols—0,1,2,3,4,5,6,7,8,9—to represent vast quantities. How? By grouping as many symbols as necessary to express a number. Working from right to left, the significance (value or weight) of each digit increases by a factor of ten. Let's analyze the number 123. It really means $(1 \times 100) + (2 \times 10) + (3 \times 1)$. The position of a digit determines its relative weight. The 1 in 123 is really 100 because of its position in the "hundreds place" of the number.

Here are the weights of the first 5 decimal digits:

digit:	4	3	2	1	0
weight:	10,000	1000	100	10	1
	(10^4)	(10^3)	(10^2)	(10^1)	(10^0)

A pattern emerges—the weight of a digit is 10 raised to the power of the digit position, starting with 0 at the right and working upward by 1 with each digit to the left.

The same scheme applies to binary and hex numbers. In binary, we have two symbols, 0 and 1. The value of each digit[2] of a binary number is determined by its position in the number. Here are the weights of the first 16 binary digits (bits):

bit:	15	14	13	12	11	10	9	8	7	6	5	4	3	2	1	0
weight:	2^{15}	2^{14}	2^{13}	2^{12}	2^{11}	2^{10}	2^9	2^8	2^7	2^6	2^5	2^4	2^3	2^2	2^1	2^0

Powers of 2:

2^0	1	2^8	256
2^1	2	2^9	512
2^2	4	2^{10}	1024
2^3	8	2^{11}	2048
2^4	16	2^{12}	4096
2^5	32	2^{13}	8192
2^6	64	2^{14}	16,384
2^7	128	2^{15}	32,768

Now for hexadecimal, whose digits each represent four bits:

digit:	3	2	1	0
weight:	4096	256	16	1
	(16^3)	(16^2)	(16^1)	(16^0)

---

2. The term bit was coined by combining binary and digit.

CONVERSIONS

To convert binary or hex to decimal, take the decimal equivalent of each digit and multiply it by the decimal weight for that digit. Add up all of the results, and you get the decimal equivalent. For example, take the binary number %110110.

$$0 \times 2^0 = \quad 0$$
$$1 \times 2^1 = \quad 2$$
$$1 \times 2^2 = \quad 4$$
$$0 \times 2^3 = \quad 0$$
$$1 \times 2^4 = \quad 16$$
$$1 \times 2^5 = \quad \underline{32}$$

$$\text{TOTAL} = 54$$

The same applies to hex numbers. Let's try $2B16:

$$6 \times 16^0 = \quad\quad 6$$
$$1 \times 16^1 = \quad\quad 16$$
$$B\ (11) \times 16^2 = \quad\quad 1816$$
$$2 \times 16^3 = \quad\quad \underline{8192}$$

$$\text{TOTAL} = \quad\quad 11{,}030$$

Converting decimal to binary or hex is harder, since you're not accustomed to doing arithmetic in those number systems. When the going gets tough, the smart pawn the job off on somebody else. You can use the Stamp[3] to do numerical conversions for you. To convert the number 1776 to hex and binary, just use the debug instruction like so:

```
BS1:              BS2:
w0 = 1776         w0 = 1776
debug $w0         debug Hex ? w0
debug %w0         debug BIN ? w0
```

The Stamp will respond with $06F0 and %0000011011110000, the hex and binary representations of 1776 decimal.

Numbers as Text and Text as Numbers

I began this appendix by emphasizing that a number is a number, regardless of how it's represented; 170 decimal is %10101010 is $AA. But there's a common situation in which a number is something else, and that's when it is used to represent text.

When you type a letter with a word processor, you're probably aware that the text is stored as a sequence of bytes, first in the computer's memory, then on a disk. And you've seen that a byte is an 8-bit number that can represent values from 0 to 255. But how does

---

3. Many scientific calculators also do conversions. Doing these conversions manually is not hard, but simple errors in arithmetic can become serious bugs in your programs. While you should understand the basis of the number systems, I firmly believe that minds are for thinking and computers are for computing. Therefore I encourage you to offload the arithmetic onto a calculator if you can!

a number represent text? You could take a stab in the dark and say that perhaps someone took the letters of the alphabet and assigned each a serial number—A=1, B=2, C=3...Z=26. They also needed to set aside some numbers for punctuation marks, lowercase letters, and even numbers. That's right, numbers to represent numbers!

If you thought about the problem long enough, you would realize that these number-symbols should also include some special characters, like a carriage return to end a line or paragraph and a backspace to erase mistakes. In the earliest days of computing and digital communications, the powers that be went through this very process, and worked out a system called ASCII (American Standard Code for Information Interchange). The ASCII character set uses seven bits (128 possible values) to represent the most commonly used letters, numbers, punctuation, and control codes. The arrangement of symbols in ASCII is pretty clever. Programs can alphabetize text by simply comparing and sorting bytes. To convert lowercase to uppercase, merely clear bit 5 of a byte (which has the effect of subtracting 32—the difference between ASCII codes for upper- and lowercase letters).

In Stamp applications involving serial communication you may experience some confusion over when is a number a number and when is it text? It's really quite simple. If you ask Serout to send a variable or constant, and you don't ask for it to be transmitted as text, the instruction will send one byte with that value. However, if you want Serout to create a text representation of a number, you can use a prefix to tell the instruction how to format that text. An example:

```
BS1:                                 BS2:
b0 = 76                              b0 = 76
serout 0,N2400,(b0)      ` byte      serout 0,$418D,[b0]       ` byte
serout 0,N2400,(# b0)    ` text      serout 0,$418D,[DEC b0]   ` text
```

Those examples send the contents of variable b0, first as a single byte set to a value of 76 (the letter "L"), then as a text representation of the number 76, consisting of two bytes with values of 55 and 54 (ASCII symbols for "7" and "6"). (Table B-2.) Serin works the same way.

Math with Fixed-size Integers

The calculator I'm holding has an eight-digit display, but it can easily deal with numbers larger than 99,999,999. It can also handle fractional values like 0.0375 and negative values like −47. Calculators have been around so long now that we take these capabilities, known as floating-point math, for granted.

The Stamps aren't calculators. They cannot deal with numbers larger than 16 bits (65,535 max) or numbers to the right of the decimal point. The BS2 can handle negative numbers to a limited degree, but it's up to you, the programmer, to understand the limitations of the Stamp's modest math abilities.

You can understand some of the Stamps' math quirks through an example. Think of a car's odometer with only five digits, like the one in Figure B-2. When the car passes 99,999 miles, it's apparently reborn; the odometer returns to zero. The reason is that the odometer's mechanical counter is supposed to perform a carry-the-one operation to display 100,000, but it has no sixth digit to carry the one into. So the one is lost and we're left looking at the lower five digits of 100,000–00000.

TABLE B-2   THE ASCII CHARACTER SET

NAME/FUNCTION (KEYBOARD COMBINATION)	ASCII CODE DECIMAL	HEX	NAME/FUNCTION (KEYBOARD COMBINATION)	ASCII CODE DECIMAL	HEX
null (control-@)	0	00	data line escape (control-P)	16	10
start of heading (control-A)	1	01	device control 1 (control-Q)	17	11
start of text (control-B)	2	02	device control 2 (control-R)	18	12
end of text (control-C)	3	03	device control 3 (control-S)	19	13
end of xmit (control-D)	4	04	device control 4 (control-T)	20	14
enquiry (control-E)	5	05	neg acknowledge (control-U)	21	15
acknowledge (control-F)	6	06	synchronous idle (control-V)	22	16
bell (control-G)	7	07	end of xmit block (control-W)	23	17
backspace (control-H)	8	08	cancel (control-X)	24	18
horizontal tab (control-I)	9	09	end of medium (control-Y)	25	19
line feed (control-J)	10	0A	substitute (control-Z)	26	1A
vertical tab (control-K)	11	0B	escape (control-[)	27	1B
form feed (control-L)	12	0C	file separator (control-\)	28	1C
carriage return (control-M)	13	0D	group separator (control-])	29	1D
shift out (control-N)	14	0E	record separator (control-^)	30	1E

*NOTE: The control codes have no standardized screen symbols. Rather than appear on the screen (or paper, in the case of a printer), the control codes are meant to instruct the terminal or printer how to present data. For example, a line-feed character moves the cursor or print head to the next line of the screen or paper. The keyboard combinations listed indicate how to enter these non-printing codes from the keyboard; for example, to send a Bell character, hold down the Control (Ctrl) key and press G. The BASIC Stamp host software understands many of these control codes and will correctly process them in its Debug screen.

TABLE B-2 (CONTINUED)

PART 2: PRINTING CHARACTERS (NUMBERS, LETTERS, AND COMMON PUNCTUATION)

CHR	DEC	HEX	CHR	DEC	HEX	CHR	DEC	HEX	CHR	DEC	HEX
sp	32	20	@	64	40	`	96	60	p	112	70
!	33	21	A	65	41	a	97	61	q	113	71
"	34	22	B	66	42	b	98	62	r	114	72
#	35	23	C	67	43	c	99	63	s	115	73
$	36	24	D	68	44	d	100	64	t	116	74
%	37	25	E	69	45	e	101	65	u	117	75
&	38	26	F	70	46	f	102	66	v	118	76
'	39	27	G	71	47	g	103	67	w	119	77
(	40	28	H	72	48	h	104	68	x	120	78
)	41	29	I	73	49	i	105	69	y	121	79
*	42	2A	J	74	4A	j	106	6A	z	122	7A
+	43	2B	K	75	4B	k	107	6B	{	123	7B
,	44	2C	L	76	4C	l	108	6C	|	124	7C
-	45	2D	M	77	4D	m	109	6D	}	125	7D
.	46	2E	N	78	4E	n	110	6E	~	126	7E
/	47	2F	O	79	4F	o	111	6F	del	127	7F

NOTE: sp is the Space character; del is Delete.

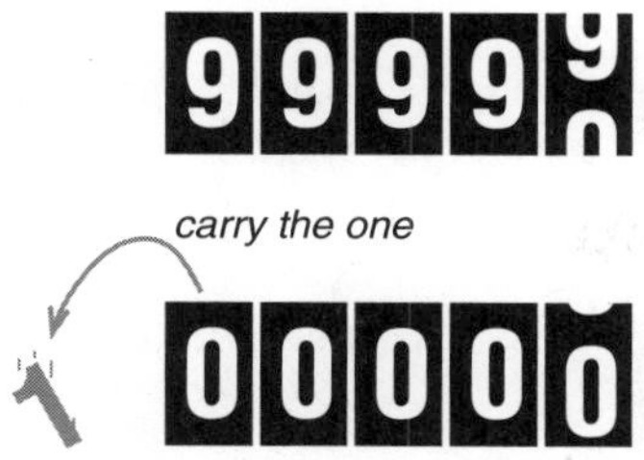

carry the one

Figure B-2 With a fixed number of digits, large values roll over to zero.

%1111111111111111
+   %1
————————————————————
%10000000000000000

Figure B-3 Adding 1 to 65535 yields a 17-bit number.

Something similar goes on with the Stamp. The largest variables in PBASIC are words, which are made up of 16 bits. The largest number they can hold is %1111111111111111, which is equivalent to 65,535 decimal. If we add 1 to that number, we cause a series of carry-the-one operations that ripple up from right to left, as shown in Figure B-3. The one that is carried out of bit 15 has no place to go, and we end up looking at the lowest 16 bits of a 17-bit number: %0000000000000000.

This applies to every kind of math operation. If a result exceeds the size of the variable, the additional digits (bits) will simply be lost. One of a programmer's primary duties is to make sure that the variables he or she selects for a given job are the right size. In debugging programs, it's smart to look for mismatches in size. Here's a specimen of just such a bug:

```
BS1:                 BS2:
  symbol a = b0        a var byte
  symbol b = b1        b var byte
  symbol c = b2        c var byte
  a= 13                a = 13
  b= 100               b= 100
  c= a * b             c= a * b
  debug c              debug ? c
```

The program assigns three variables, then multiplies 13 by 100, and stores it in one of the variables. Then it uses debug to show you the result: 20. Given our previous discussion, you know what happened; the Stamp correctly calculated that 13 * 100 = 1300, then stored the result into a byte. In the process, the upper eight bits of 1300 were lost.[4] To fix the program, make the following small change:

```
BS1:                 BS2:
  symbol c = w1        c var word
```

4. Here's where your scientific calculator can help. Make it convert 1300 to binary. The calculator will probably show you the lower byte of the result, and you'll have to press a key to view the upper byte. At any rate, the lower byte will be %00010100, which is 20 decimal. Now look at the upper byte: %00000101; that's the part that was lost when the Stamp stored the value into a byte. Now convert 1300 to hexadecimal. You'll get $0514. Since each hex digit represents eight bits, $05 is the upper byte and $14 is the lower. In decimal, $14 is 20 (the result the Stamp showed).

NEGATIVE NUMBERS

So far we have concentrated on what happens when numbers are larger than the variables used to store them. What happens when they are smaller? For example, what is the result of this program:

```
BS1:                          BS2:
  w0 = 0 - 1                    w0 = 0 - 1
  debug w0      ' decimal       debug ? w0       ' decimal
  debug %w0     ' binary        debug BIN ? w0   ' binary
```

Debug announces that 10 − 11 is 65535, not the −1 we expected. The odometer analogy helps us to understand this as well. What if we started rolling the odometer backwards, as some unscrupulous used car dealers do? If we dialed the digits down to 00000 and kept going, the next number to show would be 99999. In other words, if we subtract 1 mile from 0 miles, we get 99999, not −1.

The same applies to the Stamp, but we're dealing with 16 binary digits instead of 5 decimal digits.

There are two ways to deal with this: (1) Restrict your programs to only positive values; or (2) use math tricks and Stamp features to fake negative-number capabilities. Let's discuss choice 2 in more detail.

One way to define a negative number is as a value subtracted from zero. In the example above, we showed that − 1 is equal to 0 − 1, which in the Stamp's 16-bit math comes out to %1111111111111111. Taking this further, we could define −2 as 0 − 2, which is %1111111111111110.

The negative numbers follow a predictable pattern known as two's complement. Other than subtracting from 0, you can get the negative of a number by inverting all of its bits (changing 1s to 0s and 0s to 1s), and adding 1 to the result. Using 2 as an example: 2 is equivalent to %0000000000000010. Inverting gives us %1111111111111101, and adding 1 yields %1111111111111110—the same result we got by subtracting 2 from 0.

This gives us a clue as to how to convert negative numbers into a sign (+ or −) and a value. We just have to recognize that a number is negative, then undo the two's complement.

Recognizing a negative number turns out to be easy. If bit 15, the highest bit of a 16-bit number, is 1, the number is negative. So bit 15 is the sign bit. Bits 14 through 0 are the magnitude of the number—just a fancy way to say the absolute value or distance from zero on the number line.

If the sign bit is 0, then the number is positive and needs no special treatment in order for the Stamp to perform calculations with it and display it properly.

If the sign bit is 1, then the number is negative. To display it correctly, we must precede it with a minus sign (−), clear bit 15 to 0, subtract 1 from bits 14 through 0, then invert bits 14 through 0. Figure B-4 shows the process.

Using two's complement, a 16-bit number can represent values from −32767 to +32767. A convenient property of two's complement numbers is that they work correctly as negative numbers in most math operations. For example, 152 + (−7) should equal 145. If you take the two's complement of 7 and add it to 152, you do indeed get 145. Addition, subtraction, and multiplication all work fine with two's complement (provided that the result doesn't exceed the range of ±32,767 for a 16-bit word).

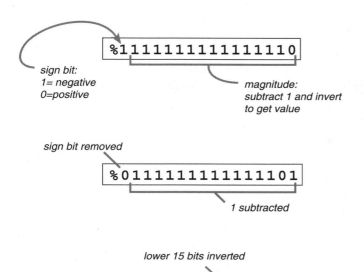

sign bit:
1= negative
0=positive

magnitude:
subtract 1 and invert
to get value

sign bit removed

1 subtracted

lower 15 bits inverted

answer: %1111111111111110 = –2 in two's complement

Figure B-4 Converting two's complement negative value to absolute value.

Division does not work correctly with two's complement, but there's an easy fix: prior to division, convert the two's complement negative number to its positive value and make note of the sign. If one of the values involved in division is negative, then the result is negative (and should be changed back to two's complement format by inverting and incrementing). If both values are negative, the result is positive, and can be used as-is. There's a good example of this in the BS2 documentation on the accompanying CD-ROM under Runtime Math and Logic in the writeup on the division operator.

Here are example programs that display two's complement negative numbers. They count down 10, 9, 8...0, –1, –2... Note that the BS2 has a built-in feature for dealing with negative numbers; SDEC in the Debug instruction stands for "signed decimal" and it automatically performs the procedure described above.

```
BS1:                              BS2:
  for b0 = 0 to 100                 for b0 = 0 to 100
    w1 = 10 - b0                      w1 = 10 - b0
    if w1 > 32767 then neg            debug SDEC w1,cr
    debug #w1, cr                   next
  continue:
  next
  neg:
    w1 = w1 & $7FFF-1 ^ $7FFF
    debug "-", #w1,cr
  goto continue
```

Don't worry if you don't understand all of the symbols used in the BS1 program. We'll talk about them in the next section.

Boolean Logic

Computers and microcontrollers are built out of thousands or millions of logic gates. These units take one or two binary inputs and produce a single binary output. The output depends on the inputs and the logic function of the gate. There are four basic types of gates, called NOT, AND, OR, and XOR, that can be combined to perform almost any conceivable function.

Figure B-5 shows the four fundamental gates and their outputs for various combinations of inputs (known as a truth table). As you can see, the gates' names reflect their operation; the output of a NOT gate is not the same as its input; an OR gate outputs a 1 if one or the other (or both) of its inputs is 1; an AND gate outputs a 1 only if one input and the other is 1; an XOR (exclusive-OR) gate outputs 1 if one input or the other (but not both) is 1.

Stamp programs can apply the fundamental logic operations to any variable. You saw this in the previous section on negative numbers; the BS1 program applied AND (&) and XOR operators to a word variable to convert a two's complement negative number to its positive counterpart.

The gates depicted in the figure apply their logic to one or two bits to produce a single output. The PBASIC logic operators apply the logic to all of the bits that make up a variable. Let's use a short program to illustrate this:

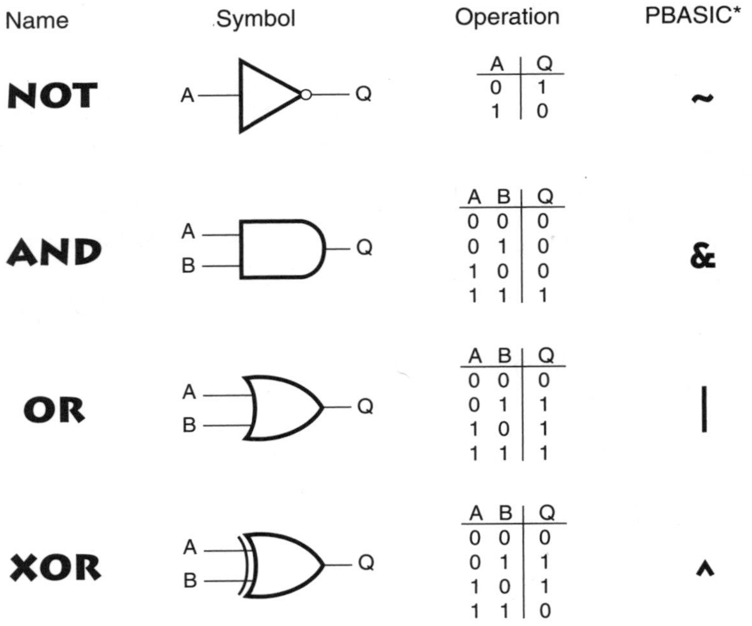

Name	Symbol	Operation	PBASIC*
NOT	A ▷o— Q	A | Q 0 | 1 1 | 0	~
AND	A, B ⟩— Q	A B | Q 0 0 | 0 0 1 | 0 1 0 | 0 1 1 | 1	&
OR	A, B ⟩— Q	A B | Q 0 0 | 0 0 1 | 1 1 0 | 1 1 1 | 1	|
XOR	A, B ⟩— Q	A B | Q 0 0 | 0 0 1 | 1 1 0 | 1 1 1 | 0	^

*The NOT operator is available only in PBASIC2. In PBASIC1, you may XOR a value with a constant containing all 1s to get the same result. See text.

Figure B-5 The fundamental four logic gates.

```
BS1:
symbol A = %11110000
symbol B = %00110011
symbol Q = b0     Q var byte
Q= A & B            ' Q = A AND B
debug "A AND B= ", #%Q,cr
Q= A | B            ' Q = A OR B
debug "A OR B= ", #%Q,cr
Q= A ^ B            ' Q = A XOR B
debug "A XOR B= ", #%Q
```

```
BS2:
A con %11110000
B con %00110011

Q= ~ A                  ' NOT A
debug "NOT A= ",BIN8 Q,cr
Q= A & B                ' Q = A AND B
debug "A AND B= ",BIN8 Q,cr
Q= A | B                ' Q = A OR B
debug "A OR B= ",BIN8 Q,cr
Q= A ^ B                ' Q = A XOR B
debug "A XOR B= ",BIN8 Q
```

SO WHAT?

Now that you have seen what the logic operators do, you may be wondering what they're good for. Each operator solves one or more common programming problems, as follows.

NOT: There's nothing very tricky about NOT; it just inverts the states of all bits. For example, if you had LEDs connected to pins 0 through 7 of the BS2, you could turn any lit LEDs off and dark LEDs on as a group by the instruction OUTL = ~ OUTL. XOR can be used as a selective form of NOT, so the BS1 doesn't even have a NOT operator. BS1 programmers simply XOR a variable with a constant containing all 1s instead.

AND: The important characteristic of AND is that there's only one way to get a 1 in the result, and that's by having a 1 in that bit position of both input values. Another way of looking at it is that ANDing an unknown bit with 1 faithfully copies the bit, while ANDing with 0 always yields 0. So programmers say that AND is useful for masking or stripping selected bits of a variable by clearing them to 0.

Here's a for-instance—say you needed to know the state of bit 4 of a variable called bitFlags. You would AND bitFlags with a constant containing a 1 in bit 4 and 0s in all other bit positions. If bitFlags AND %00010000 is 0, then bit 4 of bitFlags is 0. If bitFlags AND %00010000 is not 0, then bit 4 of bitFlags is 1.

Another application of AND is to limit the value of a variable to some smaller number of bits. For example, you have a byte variable called counter. Each time a user presses a button, 1 is added to counter. But you don't want a full 8-bit count, just the lower 5 bits (range of 0 to 31). Without AND, you would have to use an IF/THEN instruction like this:

```
counter = counter+1
IF counter < 32 then OK
 counter = 0
OK: ' program continues
```

Using AND gives us the same result with fewer instructions:

```
counter = counter+1 & %11111   ' Hold to 5 bits.
```

OR: Where AND selectively clear bits to 0s, OR selectively sets bits to 1s. For example, suppose your program was working with bytes that represented letters of the alphabet in standard ASCII characters (Table B-2). For whatever reason, you need to convert all incoming text to lowercase. Looking at the ASCII chart, you notice that the lowercase letters have the same codes as uppercase, but bit 5 is set to 1 for the lowercase version and cleared to 0 for uppercase. To convert a text byte to lowercase you'd simply OR it with a constant with bit 5 set to 1:

```
textByte = textByte | %00100000 ' Force to lowercase.
```

XOR: This is an interesting and versatile operator. XOR is most often used to selectively invert bits. XORing a bit with 1 inverts (NOTs) the bit; 0 leaves the bit unchanged. That's why the BS1 gets along without a NOT operator. Its designers figured that users would simply XOR a value with a constant containing all 1s.

An obvious application for XOR is to implement a Toggle instruction[5] that works on bits in memory. If you needed to invert bit 6 of the variable bitFlags, you would write:

```
bitFlags = bitFlags ^ %01000000 ' Invert bit 6.
```

Another use for XOR is to detect whether or not two values are alike. If you XOR two identical values, the result is always 0. For example, %10111110 ^ %10111110 equals 0. You can understand this in a commonsense way—if a 1 inverts the corresponding bit of the number it's XORed with, and two values have 1s in all the same places, then all of those 1s will be changed to 0s. All of the 0s will be left alone. So the result is a number containing all 0s.

Of course, the Stamp has no problem comparing values using an instruction like IF x <> y THEN... But it's not straightforward to use If/Then to compare individual bits within a larger byte or word variable. You would have to use AND to strip off the bits that didn't interest you before using If/Then.

Summary

It's possible to write PBASIC programs that work based on a casual understanding of the instructions. But to really extract every last drop of performance from PBASIC, you need a solid understanding of integer math and logic.

A good way to further your understanding is to take what you've learned here and apply it to the application programs. Scan the listings for logic operators and analyze them to see how and why they work. Pretty soon you'll be using them with confidence in your own programs.

5. Toggle only works on the I/O pins. It sets a specified pin to output and inverts its state.

ELECTRONIC CALCULATIONS AND STAMP-RELATED CIRCUITS

CONTENTS AT A GLANCE

Earlier you saw how Ohm's Law lets you calculate important values in electrical circuits. In this appendix we'll apply Ohm's Law and other electronics principles to a variety of common circuits. We'll also look at some common Stamp-related circuits, including switches, LEDs, transistors, relays, pots, and power supplies.

Ohm's Law Calculations

In electronics, we're concerned with three important characteristics of a circuit: potential (volts), current (amperes), and resistance (ohms). Ohm's law defines the relationship between these characteristics with three formulas:[1]

$$\text{Volts} = \text{Amperes} \times \text{Ohms}$$

$$\text{Amperes} = \text{Volts/Ohms}$$

$$\text{Ohms} = \text{Volts/Amperes}$$

Figure C-1 shows a circuit we can use to discuss Ohm's Law further. The figure also shows the Ohm's Law memory triangle; cover the symbol for the value you want to calculate, and the other symbols show the formula. For example, cover Ω (ohms) and you're left with V/A, the formula for calculating ohms.

Check your understanding: If the battery is 9 volts and the resistor is 1500 ohms (1.5k), what is the current through the circuit? Cover the A in the triangle; $A = V/\Omega$. Plug in the numbers; $A = 9/1500$. Calculate the result; $9/1500 = 0.006$ amperes. With small numbers like this, it's common to express the answer in thousandths of an ampere, milliamperes or mA for short. Using scientific notation, $1 \text{ mA} = 1 \times 10^{-3}$ amperes. So $9/1500 = 0.006$ amperes or 6 mA.

Resistances in Series and Parallel

When resistors are connected end-to-end, that's called a series connection. Side-by-side is called parallel. In using Ohm's Law on circuits with series and/or parallel resistances, we think of the individual resistances as a single, total resistance. See Figure C-2.

RESISTANCES IN SERIES

In a series circuit, resistances just add up. In the figure, if R1 is 470Ω and R2 is 2.2k then Rtotal is $470 + 2200 = 2670$ (2.67k). This holds for as many resistors as are connected in series.

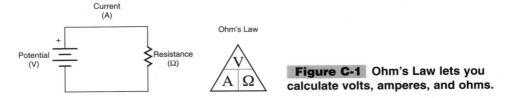

Figure C-1 Ohm's Law lets you calculate volts, amperes, and ohms.

1. This book uses the unit symbols in the Ohm's Law formulas, while many texts use the traditional symbols of physics. Same meaning, different letters:

E electromotive force   V voltage
I current   A amperes
R resistance Ω ohms

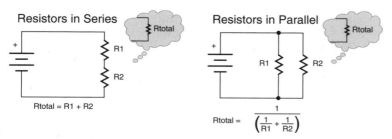

Figure C-2 Calculating resistances in series and parallel.

TABLE C-1 RULES FOR RESISTORS IN SERIES AND PARALLEL		
	SERIES CIRCUIT	**PARALLEL CIRCUIT**
Adding resistors makes the total resistance...	increase	decrease
Voltage across individual resistors...	calculated in accordance with Ohm's Law	is the same
Current through individual resistors...	is the same	calculated in accordance with Ohm's Law

RESISTANCES IN PARALLEL

In a parallel circuit, the total resistance is one over the sum of the reciprocals of the individual resistances. An example will clear up the math jargon. In the figure, if R1 is 470Ω and R2 is 2.2k, we proceed like so:

- $1/R1 = 1/470 = 0.002128$
- $1/R2 = 1/2200 = 0.0004545$
- $1/R1 + 1/R2 = 0.002128 + 0.0004545 = 0.0025825$
- Rtotal $= 1/0.0025825 = 387.26\Omega$

Most scientific calculators have a reciprocal (1/x) function key that makes this calculation easy: Enter 470; press 1/x; press +; enter 2200; press 1/x; press =; press 1/x.

The formula for parallel resistances holds for as many resistors as are connected in parallel.

RULES OF THUMB AND SHORTCUTS

The formulas for series and parallel resistors lead to the rules of thumb shown in Table C-1. There's a shortcut for figuring Rtotal in parallel circuits whose resistors are all the same value: Rtotal = R/number of resistors. For example, two 10k resistances in parallel have an Rtotal of 5k. Five 100k resistors in parallel total 20k.

CALCULATIONS WITH RTOTAL

Once you have calculated Rtotal you can use Ohm's Law on the circuit. In our series example, Rtotal is 2.67k; if the battery is 6V, the current is 6/2670 = 2.25mA. Since the resistors are in series, that current flows through both of them. We can use the current to calculate the voltage across each resistor. For R1 the voltage is 470 × 0.00225 = 1.06V. For R2 the voltage is 2200 × 0.00225 = 4.94V. Notice that the individual voltages add up to 6V—the battery voltage.

Why should you care about these voltages? We'll see later that knowing the voltage at particular points in a circuit can be important to designing Stamp circuits that read switches, and in understanding voltage dividers.

As a hands-on example, set up a circuit like the example. Calculate the voltages across the resistors, then check your calculations with a voltmeter. Figure C-3 shows how.

Circuits with Resistors in Series and Parallel

The idea of combining resistors into a single Rtotal applies to circuits with any combination of series and parallel resistors. Take Figure C-4 for instance. To use Ohm's Law on the circuit, we must calculate a single Rtotal. The first step in doing that is to combine parallel resistors R2 and R3. Then we can simply add the series combination of R1 to the total of R2 and R3.

Let's try an easy example. Suppose R1 is 4.7k and R2 and R3 are both 10k. The total of R2 and R3 is 5k (see the rule of thumb for identical resistors in parallel above). Rtotal = R1 + R2,3 = 4.7k + 5k = 9.7k.

You can apply this same technique to circuits of any size and complexity. Solve one part of the circuit, then redraw the circuit substituting a single resistor for the solved portion. Keep simplifying the circuit until you arrive at the answer. Beginning electronics books are filled with examples you can work to become proficient at this technique.

Series Resistors as a Voltage Divider

Figure C-3 showed how two resistors in a series circuit cause voltage drops that can be calculated with Ohm's Law and measured with a voltmeter. There's another way to look at a pair of resistors in series—as a voltage divider. Figure C-5 shows a shortcut formula for determining the voltage across points A and B.

BLACK PROBE	RED PROBE	METER READING
A	B	4.94V
A	C	6.00V
B	C	1.06V

Figure C-3 Measuring voltage drops.

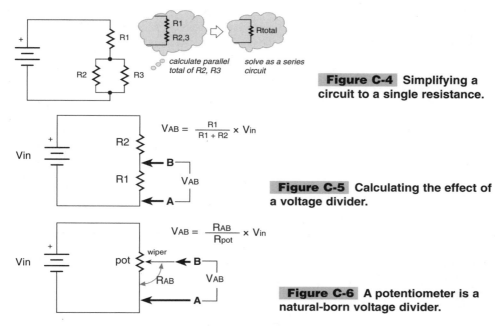

Figure C-4 Simplifying a circuit to a single resistance.

$$V_{AB} = \frac{R1}{R1 + R2} \times V_{in}$$

Figure C-5 Calculating the effect of a voltage divider.

$$V_{AB} = \frac{R_{AB}}{R_{pot}} \times V_{in}$$

Figure C-6 A potentiometer is a natural-born voltage divider.

Let's use the formula on the values from Figure C-3. R1 is 2200, R2 is 470, and Vin is 6V. That's 2200/(2200 + 470) × 6 = 2200/2670 × 6 = 0.824 × 6 = 4.94V, the same result we got using Ohm's Law. A convenient aspect of the voltage-divider view of a series circuit is that we don't have to know the input voltage in advance. We can calculate R1/(R1+R2) and just multiply that value by any Vin that comes along.

Potentiometers as Adjustable Voltage Dividers

A potentiometer is a resistor with three leads. The end leads, or legs are fixed at either end of a piece of resistive material whose resistance increases with its length. The third lead, called the wiper, can be moved along the length of the resistive material. This makes for a variable resistance between the legs and the wiper.

You can look at a pot as two resistors in series with a connection in the middle, like our previous voltage-divider examples. And that's one of the important uses for pots—as adjustable voltage dividers.

Because the pot's total resistance—the R1+R2 part of our previous voltage-divider example—is always the same, it's even easier to calculate the voltage between the wiper and the lower leg, as Figure C-6 shows.

If you connect a 1k to 100k pot and a 9V battery as shown in the figure, you can use a voltmeter to see the voltage-divider effect. Connect the circuit and adjust the pot while observing the meter.

One final note: There are two basic types of pots, linear taper and audio taper. The taper refers to the relationship of the pot setting and the resistance. With a linear taper pot, the relationship is strictly proportional; move the control shaft a quarter turn, and the resistance between the wiper and the lower leg changes by one quarter. Audio taper pots' resistance changes gradually at first, then much more rapidly. This is meant to match the ear's response to changes in sound intensity for volume-control applications. Unless otherwise specified, pots used as anything other than volume controls should be the linear taper variety.

Digital Switch Input

Stamps often need to sense the state of a switch or pushbutton. Figure C-7 shows two circuits that are commonly used for this purpose. Many newcomers to digital electronics wonder how to select the right value for the pullup or pulldown resistors used in these circuits. (The terms pullup and pulldown come from the schematic convention of putting the +supply at the top of the drawing and ground at the bottom, hence up to + and down to ground.)

In Stamp circuits, any value from 1k to 100k will do just fine. Here's the reasoning: When the switch is open, the resistor connects the Stamp pin to +5V or ground. In input mode, Stamp pins leak about 1 microampere (one-millionth of an ampere, abbreviated μA) into or out of connected circuitry. Using resistance and current, we can calculate a voltage drop: 100k × 1μA = 0.1V. So a 100k resistor to +5V would hold the Stamp pin at 4.9V and the same resistor to ground (0V) would hold the pin at 0.1V.

What happens when the switch is closed? You can think of the circuit as a voltage divider consisting of R1 and a tiny resistance (less than 0.1Ω) for the switch. There's almost no voltage drop across the switch, so it is as if the Stamp pin were now connected to ground or +5V (Figure C-7 left and right, respectively).

The other consequence of closing the switch is that current flows through R1 between ground and +5V. That current isn't doing anything useful, but it is placing a load on the power supply. That's why we avoid pullup/down resistors smaller than 1k. For instance, a 100Ω pullup or pulldown would waste 50mA (5V/100 = 0.05 = 50mA) during switch closure—that's enough current to shut down the Stamp's built-in power supply!

That being the case, why not use the largest resistor value possible? Since 100k only produces a 0.1V drop, why not 1 million ohms (1M) or more? It's true that on paper a 1M resistor would work. As a pullup it would hold the Stamp pin at 4V, and pull down to 1V. Since 1.4V is the threshold between 1 and 0 inputs, those voltages seem acceptable.

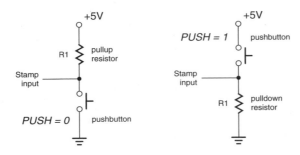

Figure C-7 Getting digital input from a switch or button.

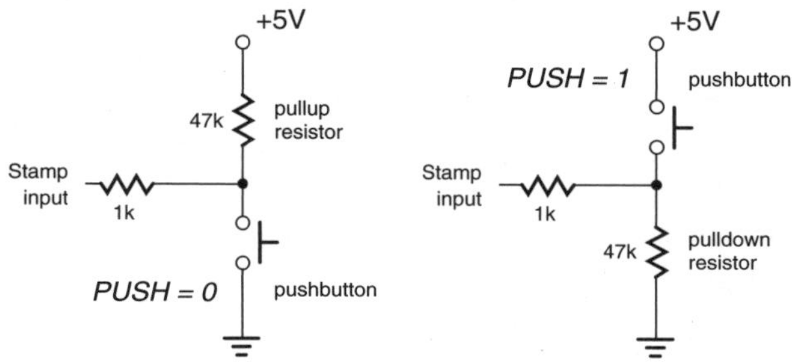

Figure C-8 An improved switch-input circuit.

However, in real-world applications, noise pickup on the wires to the switch might cause the Stamp to see a false input occasionally.

SWITCHES IN THE REAL WORLD

There are a couple of real-world complications that Figure C-7 is not designed to handle—static electricity and program bugs. Your body stores static charges like a capacitor, then discharges them when you come into contact with a conductor. For example, when you press the button in Figure C-7, one route the static discharge could take is straight from the switch terminal into the Stamp input. The Stamp has internal protection against static, but it can be overwhelmed by a direct hit.

The other hazard comes from program bugs. Suppose you have a switch connected as shown in the left side of Figure C-7, and instead of being in input mode, the Stamp pin is set for output because of a mistake in programming. And as luck would have it, the pin is output-high, connected through the Stamp to +5V.

What happens when the switch closes? For the sake of discussion, say that the switch presents a 1Ω resistance to ground; Ohm's Law says that the Stamp will be asked to output 5V/1Ω = 5 amperes of current. That's 250 times the 20mA that Stamp pins are rated for, and 100 times the 50mA capacity of the built-in voltage regulator. At best, pressing the switch will cause the Stamp to shut down. At worst, the Stamp may be damaged.

Adding a resistor between the Stamp input as shown in Figure C-8 helps harden the circuit against the real world. The 1k resistor lessens the impact of static discharges. It also limits short-circuit current in the event of a program bug to an acceptable 5mA.

WHY ARE PULLUPS PREFERRED?

Unless there's a good reason to do otherwise, use the pullup version of the switch circuit labeled "PUSH = 0" in Figures C-7 and C-8 above. Why prefer pullups? Remember that the threshold between inputs of 1 and 0 is around 1.4 to 1.5 volts. When an input is pulled up to 5 volts, it sees a 1. In order for electrical noise to cause an error (make the input see a 0 instead), it would have to cause a −3.5-volt change at the input pin.

Now consider a pulldown. The input pin is at 0 volts, seeing a digital 0. In order to cause an error (input of 1), noise would only have to raise the voltage at the pin by +1.5 volts.

So a pulled-up input is 2 volts farther away from the 1-to-0 crossover point than a pulled-down input. That extra 2 volts constitutes an extra margin of safety against false inputs resulting from electrical noise. That's why pullups are more commonly used in switch circuits.

LED Output

Figure C-9 shows how a Stamp pin can light an LED. Let's see how to calculate an appropriate value for R1. First we need a little background information.

LEDs are diodes—components that conduct in only one direction. Their positive connection, the anode, is represented schematically by the arrowhead, and their negative connection, the cathode, by the heavy line. An LED will conduct and light up only when + is connected to the anode and − to the cathode.

LEDs have a fixed voltage drop of approximately 1.5 to 2.3V. That is, when the LED is connected so that it conducts, its voltage drop (called the forward voltage and abbreviated VF) is more or less the same regardless of the amount of current flowing through it. This is different from a resistor, whose voltage drop varies with current.

The amount of light an LED gives off depends on the amount of current passing through it. At about 1mA, most LEDs barely glow; at 20mA they're pretty bright; and at some higher current, say 50mA or more, an LED may begin to overheat internally, shortening its life and reducing its light output.

LED characteristics match up nicely with the properties of the Stamp pins. In an output-high situation (Figure C-9, left), current through a Stamp pin should be 20mA or less; output-low current (Figure C-9, right) should be 25mA or less.

With this information, we need to calculate an appropriate value for R1 to light the LED. The general formula is:

$$R = (V_{SUPPLY} - V_F)/I_{LED}$$

where   R is the resistor value
$\quad\quad$ V_{SUPPLY} is the supply voltage
$\quad\quad$ V_F is the LED forward voltage
$\quad\quad$ I_{LED} is the desired LED current

You may not always know these values precisely, so here are some rules of thumb: In a Stamp circuit, use 5V for V_{SUPPLY}; V_F is usually about 1.6V for red LEDs and 2V for green and yellow LEDs; and most LEDs are decently bright with an I_{LED} of 10mA.

For a red LED, resistor R1 would be $(5 - 1.6)/0.01 = 340\Omega$. The closest standard resistor value is 330Ω. We can turn the formula around and calculate the current given this resistor:

$$I_{LED} = (V_{SUPPLY} - V_F)/R$$

That works out to $(5–1.6)/330 = 10.3$mA; close enough, especially considering that our calculations are based on approximations.

Switching Bigger Loads with Transistors

For microcontrollers, the Stamps can switch pretty large amounts of current. Stamp pins can source 25 mA and sink 20mA.[2] That's enough to light an LED or even turn on a sensitive relay. But it won't light any but the tiniest light bulb, activate a solenoid, or spin a motor. For those things, the Stamps need some help.

2. Source means provide the path for current to flow from + into a circuit; sink means provide the path for current to flow out of the circuit to ground. To keep these terms straight, just think, "current sinks into the ground."

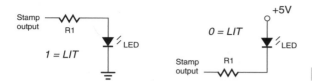

Figure C-9 Lighting an LED.

An easy way to increase the Stamp's current-handling ability is by adding a transistor switch. A transistor is a semiconductor device that acts like a couple of diodes. One diode is formed by the base and emitter, the other by the collector and emitter. A small current flowing through the base-emitter (BE) diode can cause a much larger current flow through the (CE) collector-emitter diode. The ratio of BE current to CE current varies, but it's generally in the range of 10 to 1000 times.

To switch a load with a transistor requires one of the hookups shown in Figure C-10. Designing a transistor switch boils down to (1) deciding whether to switch to the + supply rail or ground and selecting a PNP or NPN transistor accordingly; (2) determining how much current the load requires and further narrowing your choice of transistor to handle that current; and (3) calculating an appropriate value for the base resistor.

You generally won't go wrong if you figure your base resistor to pass 1/10th the load current through the BE diode. Like other diodes, BE has a forward voltage drop—about 0.6 volts. So you have to subtract that voltage from the voltage into BE in order to use Ohm's Law to calculate the resistance needed for a given current. Here's an example: We want to control a 100 mA (0.1-ampere) load. One-tenth of 100mA is 10mA, our desired BE current. Voltage out of a Stamp pin is close to 5 volts. Subtracting 0.6V for BE, we get 4.4V. By Ohm's Law R = V/A, so R = 4.4/.01 = 440 ohms. The closest standard value is 470 ohms—done.

Figure C-10c shows a combination of NPN and PNP transistors. This type of arrangement is useful when you wish to switch the +supply ("high side") as in a normal PNP switch, but the voltage to be switched is greater than the Stamp's operating voltage (5V). If you were to use a plain PNP transistor in such a case, the transistor would be switched on all the time because of current flowing from +supply down to the +5V level of the Stamp pin.

For Figure C-10c, calculate the base resistor for the PNP as described above. Then calculate the base resistor for the NPN figuring on a load current 1/10th the PNP load current.

Whole books have been written on the subject, but that's switching transistors in a nutshell. When you choose a transistor for such an application, look for parts marked "general purpose" or "switching transistor."

Finally, if your load is a relay or solenoid, use the circuit shown in Figure C-11. The diode across the load protects the transistor from spikes generated when the load is turned off.

Calculating Resistor-capacitor (RC) Timing

Many Stamp capabilities and applications depend on resistor-capacitor timing. The instructions Pot (BS1) and RCtime (BS2) are designed to measure resistance in terms of the time required for a fixed, known capacitance to charge or discharge through a variable, unknown resistance.

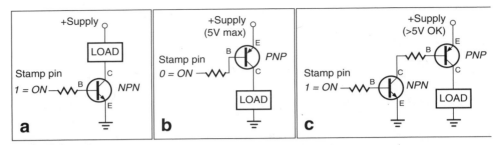

Figure C-10 Switching high-current loads with transistors.

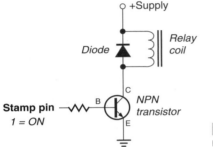

Figure C-11 Protecting a transistor from relay kickback.

In addition, the BS2 sonar project presented in this book uses an RC circuit to gradually increase the sensitivity of the circuit that detects sonar echoes. The longer it takes for an echo to return, the farther it has traveled, and the fainter it will be. It makes sense for the detector's sensitivity to increase and offset the decrease in strength of the signal.

RC timing is relatively easy to understand. A capacitor stores electrical charge. A resistor is a component that restricts the flow of electrical current. So the time required to fill (charge) or empty (discharge) a capacitor depends on the resistance that completes the charge/discharge circuit. The basic formula is:

$$t = R \times C$$

where t is time in seconds, R is resistance in ohms, and C is capacitance in farads.

The time that t refers to is that required for the capacitor to reach 63 percent charge or discharge, as measured by the voltage across it. The value t is also referred to as tau (the Greek symbol τ) or as the RC time constant.

An example: a $0.1\mu F$ capacitor will be charged through a 10k resistor from a 5-volt supply. How long would it take for the voltage across the cap to reach 63 percent of 5 volts (3.15 volts)? Calculate $(10 \times 10^3) \times (0.1 \times 10^{-6}) = 1 \times 10^{-3}$ or 1 millisecond.

That same time constant applies to both charging and discharging capacitors. Figure C-12 shows generalized graphs of voltage versus time for charging and discharging capacitors. The graphs are not a straight-line relationship. The voltage across the cap changes rapidly at first, then gradually slows down as it nears full or empty. The 63-percent point that the RC time constant gives us lies just about at the end of the fast initial charge/discharge stage.

Being able to calculate the RC time constant is a step in the right direction, but in Stamp applications, we often need a more specific timing value. In particular, for the BS2's

RCtime instruction, we need to know how long it will take for the voltage across a capacitor to go from 5V to 1.5V, the BS2's logic threshold. The threshold is the boundary between an input seeing a 0 (less than 1.5V) and a 1 (greater than 1.5V).

RCtime is typically used with a circuit like that of Figure C-13. Initially, the BS2 I/O pin is set output/high briefly to establish a 5V level at the junction between the capacitor and resistor. In the figure, we represent output/high by a closed switch from +5V inside the Stamp. When the RCtime instruction executes, it opens the switch by converting the I/O pin to input. RCtime counts the time interval required for the voltage to drop from 5V to 1.5V.

The general formula for calculating the time required for a change in the voltage across a capacitor is:

$$\text{time} = -\tau \left[\ln \left(\frac{V_{final}}{V_{final}} \right) \right]$$

where τ is the RC time constant, ln is the natural-logarithm function, Vinitial is the starting voltage, and Vfinal is the ending voltage.

Let's calculate the time required for the voltage at the junction of the 0.1µF cap and 5k resistor to go from 5V to 1.5V. Take the natural log of 1.5/5V (ln is a key on most scientific calculators; perform the division, then press the ln key). The result is −1.204. Now multiply by R × C, $(5 \times 10^3) \times (0.1 \times 10^{-6}) = 0.5 \times 10^{-3}$ to get -601.2×10^{-6}. Reverse the sign (−) and there's the answer: 601.2 × 10-6. So it would take about 600 microseconds for that RC circuit to go from 5V to 1.5V. That would be 300 RCtime units of 2µs each. Now we have a ballpark estimate of what values RCtime would return for a given set of component values.

Bear in mind that this is just an estimate; components can vary from their marked values by a percentage tolerance, typically 10 to 20 percent for capacitors and 5 to 10 percent

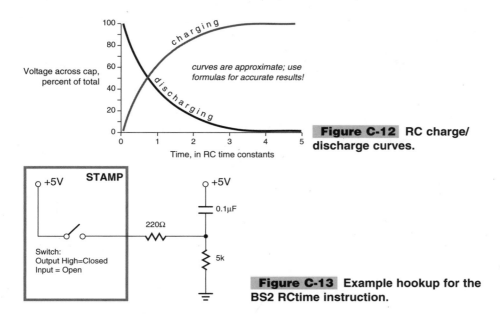

curves are approximate; use formulas for accurate results!

Voltage across cap, percent of total

Time, in RC time constants

Figure C-12 RC charge/discharge curves.

Figure C-13 Example hookup for the BS2 RCtime instruction.

STAMP

+5V

Switch:
Output High=Closed
Input = Open

220Ω

+5V

0.1µF

5k

for resistors. And there's a hidden voltage divider formed by the 220Ω series resistor and the 5k resistor to ground. This means that Vinitial can never reach 5V; it maxes out at 5V × 5000/5220 = 4.79V. (See the previous discussion of voltage dividers.) Even that value isn't perfect; the BS2's internal switch introduces a little more resistance, and its input voltage threshold can vary from the 1.5V rule of thumb. All in all, we have to be content with an estimate.

Using an External Voltage Regulator

One of the things that make the Stamps so easy to use are their built-in voltage regulators. You can directly connect 6 to 15Vdc to the Stamps' power input (Vin) and that onboard regulator will provide a steady 5Vdc to the Stamp circuitry.

The Stamps' regulator can deliver up to 50 mA to 150 mA of current, and the Stamps themselves require 6 mA to 60 mA. You can often use excess capacity of the onboard regulator to power circuitry.

Some applications need more current than the built-in reg. In these cases, you are supposed to connect a huskier 5Vdc supply directly to the Stamp's +5V (Vdd) pin and leave Vin unconnected. While you can purchase a regulated 5Vdc power supply, it's more economical to use an IC called a 3-terminal linear regulator to roll your own. Figure C-14 shows a couple of circuits that will do the job.

The first circuit uses the common 7805 regulator. I've drawn part rather than represented it schematically in order to show you how it's hooked up. (The schematic symbol for a three-terminal regulator is not very enlightening anyway—just a rectangle labeled in, ground, and out.)

The TO220 case of the 78L05 has a large metal tab with a hole in it. This allows you to attach a heat sink—a chunk of metal, usually aluminum, that helps cool the compo-

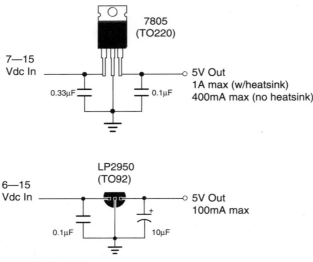

Figure C-14 External 5-volt regulators.

nent. With a heat sink installed, the 7805 can provide as much as 1000mA (1A) of current. Without it, figure on approximately 400mA maximum. The 7805 is internally protected against overheating; if you ask it for more current than it can deliver, it will shut down rather than self-destruct.

In addition to the current drawn by your circuit, the 7805 uses as much as 5mA for its own purposes. It's not a particularly good choice for Stamp circuits that make extensive use of the Sleep and Nap instructions to save power. Even with the Stamp knocked out, the regulator is draining your batteries dry.

The second circuit uses an LP2950. This regulator is more closely related to the one used in the Stamp, in that it uses very little current (75 microamperes) itself, making it more compatible with Sleep and Nap. It is not designed to be used with a heatsink (although you could probably improvise one) and puts out a maximum of 100mA.

With any voltage regulator, take care to connect the voltage input correctly, with + to the input and ground to ground. You can destroy a voltage regulator in a heartbeat by reversing these connections.

If you want to operate your Stamp projects off household AC power, use a commercial AC adapter. Radio Shack and Jameco (Appendix E) carry a wide variety of these devices. Pick one with a DC output that's in the lower end of the voltage range for your regulator, and rated for more than the amount of current you plan to draw.

For example, suppose you are using an LP2950 and expect to draw close to 100mA. You find two adaptors, one rated for 12Vdc at 100mA and another for 9Vdc at 200 mA; which do you pick? The better choice would be the 9Vdc model. The lower the input voltage to the regulator, the cooler it runs.

Building an Efficient Switching Power Supply

One of the Stamp's claims to fame is its low current draw. It can operate continuously from an alkaline 9V battery for days on end. However, the standard 9V battery supply is somewhat inefficient. Here's why:

The Stamp operates from 5V; the built-in regulator reduces 9V to 5V by wasting the excess energy as heat. As the 9V battery wears out, its output voltage drops while its internal resistance rises. When the output voltage drops below 5V, the voltage regulator can no longer actively regulate voltage to the Stamp; when it drops a little more, the Stamp stops working.

9V batteries consist of a stack of six 1.5V cells packed into a small case. Cylindrical flashlight batteries (such as AA, C, and D sizes) are individual cells, basically cans of electrically active chemicals. There's a lot more active content in these larger, single-cell batteries.

So, it would be nice if the Stamp could operate from, say, three AA cells instead of a 9V battery. Unfortunately, the voltage—4.5V—is too low for the 5V Stamp to operate reliably.

Figure C-15 provides a solution to this dilemma. It's a switching voltage regulator. It steps up the 4.5V from three series-connected AA cells to the 5V that the Stamp requires. It can continue to operate down to 2V, so it will squeeze pretty much every last drop of available energy out of the batteries.

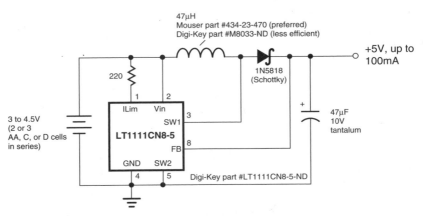

Figure C-15 **Switching power supply provides steady 5Vdc from 2 or 3 batteries.**

It's different from the Stamp's built-in regulator in a couple of respects. First, it takes an input voltage that is less than the output voltage. It steps up the voltage by storing energy in an inductor, which is basically an electromagnet. When current is run through an inductor, it creates a magnetic field. When current is removed, the field collapses, and the energy stored in the field is returned as a high-voltage spike across the inductor. A switching supply creates and captures these spikes to develop a higher output voltage.

In order to maintain a steady 5V output, the switching-regulator IC actively manages the on/off switching of current into the inductor. This is very efficient, since it does not require throwing away excess energy as heat (except tiny amounts due to small internal resistances in the circuit).

Enough theory. If you construct this circuit yourself, wire it as neatly as possible on a solder-type breadboard. A waffle board (plug-in board) won't do, since its small imperfections (contact resistance, small capacitance between socket rows) can prevent the regulator from working properly. Alternatively, you may purchase an inexpensive kit of parts with custom circuit board from Medonis Engineering (Battery Booster 5; see Appendix E for contact info).

WHY BOTHER?

Let's analyze some real-world figures to see how the switching power supply can eventually pay for itself. Table C-2 compares the likely performance of 9V and AA batteries running a 10-mA load.

TABLE C-2 BATTERY COST COMPARISON			
	*ESTIMATED LIFE AT 10MA	*TYPICAL COST ($)	COST PER OPERATING HOUR (CENTS)
9V Battery	47 hours	1 × 2.22 = 2.22	4.7
AA Cell (x3)	140 hours	3 × 0.57 = 1.71	1.2

*Figures based on Digi-Key specs/prices for alkaline batteries

Assuming that the switching power supply costs you $15, at a savings rate of 3.5 cents per operating hour, you get your money back in reduced battery use in just 429 operating hours, or three AA battery changes, or less than 3 weeks of continuous operation.

BUG-HUNTING GUIDE

This Bug-Hunting Guide is a feature that may be unique to this book. I'm hoping that others will imitate it. The idea is this: In any programming language or programmable gadget there are certain features that:

- Are prone to misunderstanding
- Are unlike similar features of other systems
- Have limited protection against mistakes

■ Can create problems that are difficult to trace
■ Are uncorrected flaws in the language or device itself

(Yes, I realize that I have just described the C programming language.)

As a programmer works with a system, he or she discovers these bug-o-genic properties, kills the bugs they caused, and files away the knowledge for future reference. This appendix is a rogue's gallery of the aspects of PBASIC1 and PBASIC2 that I've found to be at the root of common programming problems. I'm not suggesting that when you have a buggy program that you will immediately find the answer here. But when you get really stuck, browsing this list is a productive alternative to, say, breaking your keyboard over your knee.

Pin number versus pin variable (BS1 and BS2)—Instructions like LOW, HIGH, and TOGGLE take a pin number (0–7 for BS1, 0–15 for BS2) as an argument. Programmers sometimes confuse a pin name with the pin number—for example, using PIN4 (BS1) or OUT4 (BS2) in place of plain old 4. Pin names are actually bit variables containing 0 or 1, depending on the state of the pin, so this is a bug. It's especially common to make this mistake when you've renamed the pin variables.

16-bit intermediate results (BS2 only)—If you get unexpected results with a runtime expression or If...Then that uses the logical not (~) of some value, you've run into this aspect of PBASIC2. For example, If...Then regards any nonzero value as meaning true. So 1 is true, and not 1 is also true. See, not 1 computes a 16-bit result by padding 1 with 15 leading zeros. Not 1 produces an intermediate 16-bit result of %1111111111111110. Clearly, that's not equal to zero, so If...Then regards it as true. In If...Then instructions, always base decisions on a comparison and always use word-form logic operators (see the If...Then writeup). In other math/logic expressions, try to anticipate side effects of 16-bit results. If you get bugs, break your expression into smaller pieces, writing the intermediate results to the right-size variable at each step. For example, in this code we want to determine the highest bit containing 0. The NCD operator will give us the highest bit containing 1, so we have to invert (~) the value first. But the 16-bit intermediate result produces a bug:

```
x= %0111            ' Fourth bit is 0.
debug ? NCD ~ x     ' Show NCD of NOT x (should be 4; is 16!)
x= ~ x              ' Fix the bug by breaking into steps.
debug ? NCD x       ' Now we get correct result (4).
```

The reason this works is that in the process of writing ~ x back into the nibble variable x, PBASIC2 trims the 16-bit value it computed down to 4 bits.

Fixed variables (BS2 only)—A fixed variable can overlap an allocated variable, causing data meant to be written to one variable to show up in another. If you mix fixed and allocated variables, do so at your peril, and only with a very good reason.

Failed output (BS1 and BS2)—Pins must be in the output state for data written to them to actually show up at the corresponding pins. Instructions like LOW, HIGH, and TOGGLE automatically set the specified pin to output; instructions like pins = 255 (BS1) and OUTS = 255 (BS2) do not.

Arrays (BS2 only)—If your program supplies an out-of-range index value (subscript) to an array, it will access RAM outside the array, and possibly corrupt the contents of other variables.

Alias variables (BS1 and BS2)—When you assign two different names to the same location in RAM for the purpose of clarity, you may also be inviting a hard-to-trace bug. Values left by usage under one name may interfere with use under the other name.

GOTO versus GOSUB (BS1 and BS2)—It's easy to forget that a piece of code is a subroutine, i.e., code meant to be entered with GOSUB and exited by RETURN. If your program executes a GOTO to a subroutine, it does not store the proper return address. When it reaches the RETURN instruction at the end of the subroutine, it goes to some unexpected point in the program, causing a bug. This mistake is especially common with IF/THEN instructions, because neither PBASIC1 nor PBASIC2 supports IF condition THEN GOSUB some_routine.

Multiple Button instructions (BS1 and BS2)—If you use more than one Button instruction in a program, each requires its own unique byte variable. If you assign the same byte variable to more than one Button instruction, all kinds of unpredictable interactions between the buttons will result.

Missing ground wire (all circuits)—When you connect a signal from one system to another, you must also connect the grounds together. For instance, if you use the Stamp to control a servo, the Stamp has a power source (e.g., 9V battery) and the servo has a power source (4.8–6V battery). A classic mistake is to connect the Stamp to the servo's signal pin without connecting the battery grounds (-) together.

Mysterious failure of known-good program (BS1 and BS2)—The actual cause of this problem is open to speculation, but the cure is simple; recopy STAMP.EXE or STAMP2.EXE to your hard drive. Apparently the program can be corrupted in a subtle way that leaves it seeming fully functional, but has actually trashed its downloading function.

Erratic behavior with relays, solenoids, motors (BS1 and BS2)—Applications involving relays, solenoids, or motors can behave erratically when electrical noise from those devices causes the Stamp to reset. To determine if this is the case, substitute "quiet" loads like LEDs for the relays/solenoids/motors and see if the problem clears up. If it does, you need look at your power supply to ensure that it can drive the loads and that ground connections are as short as possible. In extreme cases, you may need to shield the Stamp with grounded foil or a grounded metal box.

Can't communicate with PC serial port (BS2)—The downloading connector on BS2 carrier boards is a convenient way to connect to a PC for communication. However, you must disconnect the BS2's ATN line (as described in the BS2 temperature-logger project) during communication, and reconnect it for programming. Another problem encountered with newer PCs is the 16550 UART's built-in buffers. Disable these when setting up a terminal program for communication with the Stamp.

String limitations (BS2 only)—The BS2 has some features associated with Serin, Serout, and Debug that treat arrays as strings. Don't get carried away; PBASIC2 doesn't support all string-handling functions. For instance, you cannot assign a quoted string to an array. The line myArray = "Hello" will not work. And you can't compare strings; if myArray = "Hello" then... also will not work, even if myArray contains "Hello."

Let reset do its job (BS1 and BS2)   Parallax tech support says some customers get the idea that the Reset pin (pin 6 on the BS1-IC and pin 22 on the BS2-IC) should be externally pulled high. Don't do it! This defeats, and may damage, the internal brownout-detect circuit. There are only two valid connections for reset, a switch to ground (e.g., pushbutton reset), or the input to another digital device (e.g., reset pin of another microcontroller). If you aren't using reset for either of these purposes, leave it unconnected.

Comparisons always positive (BS2)—The BS2 has a limited ability to treat 16-bit numbers as signed, two's complement negative numbers when formatting Serout and Debug statements. And most integer math except division works OK on these numbers. But comparisons don't work. For example:

```
myWord = -10
if myWord < 0 then isNegative
```

That code will not go to the label isNegative, because the value –10 is represented in two's complement form, which to the comparison operator looks like 65526.

Max and Min (BS1 and BS2)—In math expressions, watch out for expressions involving max 65535 and min 0. If the result of some calculation is larger than 65535, the Stamp truncates it, resulting in a number smaller than 65535. Same with min 0; numbers less than 0 turn up in two's complement form, resulting in large numbers. So max 65535 and min 0 do not work.

Weird Debug result (BS2)—In order to display the value of a variable in the Debug window, you must use a formatting modifier, as in debug ? myWord or debug DEC myWord. If you don't use a modifier, as in debug myWord, you get the ASCII character represented by the value stored in myWord. This is a little confusing, since the BS1's Debug operator defaults to decimal, and will give you a correct output without a formatting modifier. If you switch between the two flavors of PBASIC, watch out for this one.

Missing Debug characters (Windows Host Software)—If debug instructions sometimes print incorrectly, it may be due to Windows' management of the Serial-Port buffer. See Chapter 5 for step-by-step instructions for turning off the buffer.

SOURCES FOR ELECTRONIC

COMPONENTS

One of the biggest surprises that await beginners in electronics is how hard the shopping can be. There really is no such thing as one-stop shopping for electronic components. This should not be such a surprise, really. Novice cooks quickly learn that many interesting recipes call for ingredients that are not on the supermarket shelves. So they learn how to shop. And so must you.

The projects in this book list dealers' part numbers for the more unusual components. Generic items like resistors, capacitors, switches, and batteries are not listed by part number, since they can be obtained almost anywhere.

The best way to approach electronics shopping is to obtain as many catalogs as possible before putting together a shopping list. Then, when you are ready to buy parts for a project, start with the hard-to-find items. Check the catalog of the source for those parts to see whether you can order the generic parts from them, too.

Many electronics dealers now have catalogs on the Internet, allowing you to search their databases for a specific part. However, don't pass up the print versions of their catalogs. You can learn a lot about electronics by reading the catalog listings. For example, want to know the difference between ceramic and polyester capacitors? Browse the listings for each in the Digi-Key catalog. You'll notice that the ceramic caps are generally less expensive, but have looser tolerances ($\pm20\%$ versus $\pm5\%$) than polyester. See, you learned something.

Dealers can help with your education in other ways, too. If you're not familiar with a component that you are ordering for a project, request a data sheet. These are generally either free or inexpensive (typically $1 with purchase). Parts dealers also frequently sell manufacturers' data books filled with design information for next-to-nothing.

If you work at an electronics-oriented business, your purchasing department may be able to get data books or even samples free from commercial distributors. Freebies have become scarce, though, as scamming (noncustomers begging samples) has become prevalent.

The list below is admittedly incomplete, but includes a source for every component used in every project listed in this book. Each listing includes an Internet address (if known), toll-free phone (if applicable), and regular phone (if known, for readers outside the U.S.).

The vendors listed all do business by mail order, but don't forget electronics stores, like the huge Radio Shack chain. They carry a large variety of LEDs, resistors, capacitors, wire, tools, project boxes, switches, batteries, and semiconductors. The author lives in a town of 50,000 people and has three Radio Shack stores within a couple of miles.

Component Sources (Listed Alphabetically)

Arrick Robotics, www.robotics.com, 871-571-4528—Stamp-based ARobot kit, plus robotics and PC-based motion control for industrial and academic applications.

AWC Electronics, www.al-williams.com/awce/index.htm, 281-334-4341, PAK-series chips for math, I/O, pulse-timing, and PC-keyboard input.

Decade Engineering: www.decadenet.com/, 503-743-3194—Video-overlay display systems (BOB-series modules).

Digi-Key, www.digikey.com, 1-800-344-4539, 218-681-6674—BASIC Stamps, electronic components, LCDs (nonserial), keypads, switches, test equipment, tools and materials for electronics.

Digital Products Company, www.digitalproductsco.com, 916-985-7219—telephone line simulators.

EME Systems, www.emesystems.com, 510-848-5725—Data loggers and sensors.

ExpressPCB, www.expresspcb.com, offers free printed-circuit-board layout software and an inexpensive circuit-board manufacturing service for hobbyists and engineering prototypes.

Fascinating Electronics, www.fascinatingelectronics.com, 800-683-5487—humidity sensor, weather station, PC experimenter interface.

Hosfelt Electronics, www.hosfelt.com, 1-800-524-6464, 614-264-6464—Electronic components, LCDs (nonserial), switches, test equipment, tools and materials for electronics.

Jameco Electronic Components, www.jameco.com, 1-800-831-4242, 415-592-8097—BASIC Stamps, electronic components, LCDs (serial and nonserial), switches, test equipment, prototyping supplies, connectors, computers and peripherals, RS-485 adapters for PCs, modems, ac adapters, tools and materials for electronics.

JDR Microdevices, www.jdr.com, 1-800-538-5000—BASIC Stamps, electronic components, serial LCDs, switches, connectors, test equipment, computers and peripherals, modems, tools and materials for electronics.

Lynxmotion, www.lynxmotion.com, 309-382-1816—mobile and stationary robot kits, servos, Stamps in kit form.

Medonis Engineering, www.medonis.com, 503-860-1980—robotics products, switching power supply kit.

Mondo-tronics, The Robot Store, www.robotstore.com, 800-374-5764—Robotic kits and components, servos, electronics, hand tools, and information resources.

Mouser Electronics, www.mouser.com, 1-800-346-6873, 817-483-5712—Electronic components, keypads, switches, test equipment, tools and materials for electronics.

Oak Tree Systems, www.oaktreesystems.com, 734-604-8700—Serial I/O coprocessors for up to 32 outputs, model railroad controllers.

Parallax Inc., www.parallaxinc.com, 1-888-512-1024, 916-624-8333—BASIC Stamps and accessories, PIC and SX programming devices and chips, X-10 control devices, servos and servo controllers, serial LCDs, application kits for interfacing Stamps and peripheral chips.

Scott Edwards Electronics, Inc., www.seetron.com, 520-459-4802—Serial LCDs, BS2 data collection proto board, serial servo controller.

Selmaware, www.selmaware.com/, StampPlot and StampPlot Lite PC software for graphing Stamp-collected data.

Solutions Cubed, www.solutions-cubed.com, 916-891-8045—accessories for the BASIC Stamps, consulting services for Stamp users developing commercial products.

Tower Hobbies, www.towerhobbies.com, 800-637-4989, 217-398-3636—servos and R/C models.

SUGGESTED READING

CONTENTS AT A GLANCE

Suggested Reading

Parallax Web pages—This is the first stop in your Stamp journey. The Web page provides commercial information (product descriptions and pricing) as well as access to valuable reference material (application notes, user-manual updates, source code examples, latest host software, etc.); visit www.parallaxinc.com.

Parallax also hosts a site specifically aimed at educational uses of the BASIC Stamps, www.stampsinclass.com. The Stamps in Class program produces texts and workbooks based on Stamps; the curriculum currently includes fundamental electronics and programming, Earth measurements, industrial control and robotics, with more on the way.

Stamps discussion group—Parallax sponsors a lively discussion group whose central topic is the BASIC Stamp. Users roam far afield, though, discussing fundamental electronics, programming, sensors, robotics, etc. Preview or join the group at www.egroups.com/group/basicstamps.

Nuts and Volts Magazine—*Nuts and Volts* is home to the monthly *Stamp Applications* column, dedicated to tips, techniques, and projects for Stamp users.

The magazine also hosts an excellent column on robotics, and regular feature articles on programming, the Internet, electronics, and other high-tech topics. Back issues of Stamp Applications are available through the magazine's Web site. [Phone 1-800-783-4624 or 909-371-8497, www.nutsvolts.com]

List of Stamp Applications (LOSA)—Parallax' Swedish distributor regularly solicits descriptions of Stamp users' projects to compile this list. Walking and rolling robots, veterinary medical equipment, factory automation, games, demos of electronic principles, ham-radio accessories, and high-tech art all are just a few of the hundred or so projects described in LOSA. Best of all, most LOSA listings include an Internet address where you can find schematics, program listings, and even photos. [www.hth.com/losa/]

Microcontroller Projects with BASIC Stamps—Takes you on a whirlwind tour of the Stamps, electronics, and programming with projects ranging from reading buttons to building a data logger. Based on Al Williams' popular Stamp Project of the Month feature on the Internet. Includes textbook style exercises to check your progress. [Williams, Al 1999, CMP Books.]

BASIC Stamp: An Introduction to Microcontrollers—Aimed primarily at engineers and advanced hobbyists, with sophisticated and ingenious circuits and techniques. [Kuhnel, Claus and Zahnert, Klaus, 1997, Butterworth Heinemann Books.]

The Art of Electronics—No single book encompasses every possible subject in the field of electronics, but this 1000+ page volume comes very close. If you had to pick one book to use as a self-taught course and as a reference to components, circuits, and design techniques, this would be the one. [Horowitz, Paul and Winfield Hill. 1989. *The Art of Electronics.* 2d ed, Cambridge University Press.]

Mastering Serial Communication, Second Edition—This book is a great starting point for novices who need to understand serial communication basics. Explains serial framing, protocols, signaling, and programming with examples in BASIC, C, and assembly language. [Gofton, Peter W. 1994. *Mastering Serial Communication.* 2d ed. Alameda, CA: Sybex.]

Programming and Customizing the PIC Microcontroller—The title may sound familiar—this book is another in the same series. Once you've exhausted the potential of BASIC Stamps (impossible!) you may want to step up to the next level and program PICs directly in assembly language. This book is your launchpad. [Predko, Myke. 1998. *Programming and Customizing the PIC Microcontroller.* TAB, McGraw-Hill]

Encyclopedia of Electronic Circuits, volumes 1 through 6—Although most of the circuits featured in these books are designed to work without the assistance of a microcontroller, they can be great inspiration for Stamp-based projects. For example, suppose you're building a Stamp-based alarm of some sort. You'll find all kinds of alarm sensors for detecting everything from intruders to flooded basements. On the output end, you'll also find all sorts of alarm noisemakers. [Graf, Rudolph F. and William Sheets. 1996. *Encyclopedia of Electronic Circuits,* volume 6. TAB, McGraw-Hill]

Bebop to the Boolean Boogie—In appendices B and C, I make a frantic effort to bring newbies up to speed on fundamentals of numbers, math, logic, and electronics. *Bebop* addresses the same subjects at length and at leisure, with lots of supporting explanations. Lavishly illustrated. Hilariously written. Wonderful, valuable stuff that's very relevant to Stamp programmers. [Maxfield, Clive. 1995. *Bebop to the Boolean Boogie, an Unconventional Guide to Electronics Fundamentals, Components, and Processes.* High Text Publications]

Mobile Robots—Great starting point for those who want to build intelligent gadgets that skitter across the floor. Nice balance of cookbook and theory on subjects like electronic and mechanical assembly, motor-drive circuits, sensors, navigation, etc. [Jones, Joseph L. and Anita M. Flynn. 1993. *Mobile Robots, Inspiration to Implementation.* A. K. Peters]

CMOS Cookbook—Despite its age, this collection of digital-electronic recipes is very handy for Stamp users. For example, the 4060 oscillator divider circuit that forms the basis for the BS1 time/temperature project is adapted from a Cookbook circuit. In addition to fundamental circuits, there are many clever hints and tricks for digital electronics. [Lancaster, Don and Howard M. Berlin. 1988. *CMOS Cookbook,* 2d ed, Howard W. Sams and Co.]

Making Printed Circuit Boards—This is a great book for those who want to build reliable, professional projects based on printed-circuit boards. Covers the hows and whys of circuit-board layout and fabrication using inexpensive tools and materials. [Axelson, Jan. 1993. *Making Printed Circuit Boards (with projects and experiments),* TAB, McGraw-Hill]

GLOSSARY

Algorithm—A method for performing a computing task. For example, an efficient algorithm for extracting the ones-place digit of a decimal number is to take the remainder of that number divided by 10. An algorithm is different from a program or a subroutine in that it's considered to be the idea or logic that makes the program code work, but not necessarily the code itself. Stamp users often obtain useful algorithms for their programs by studying programs written for other computers, or from explanations of algorithms written in pseudocode.

Alias Variable—A variable name that is assigned to a previously defined variable (or portion thereof). Assigning an alias to a variable lets you recycle a variable that is used for one purpose in one part of a program, and a completely different purpose elsewhere. This is done primarily with temporary variables used for such things as FOR/NEXT counters. Such variables are not used for permanent storage and may be reused throughout the program. Instead of giving these variables generic names, it may be clearer to refer to them by different names in different contexts. Defining them is easy. In PBASIC, you simply give two different SYMBOLs to the same variable, e.g., SYMBOL temp = b2 and SYMBOL count = b2. In PBASIC2: temp VAR byte and count VAR temp. Use aliases with caution, as it's easy to forget that the same variable is accessible under more than one name.

Allocate—To set aside memory for some purpose. When you use the BS2's VAR directive to declare a variable, you are allocating RAM space to that variable as well as assigning a name to it. You are allocating EEPROM space when you use the BS2's DATA directive. In the strictest sense, using VAR (BS2) or SYMBOL (BS1) to rename the built-in variables like B3 or W1 isn't really allocating space, just renaming it.

Argument—A value provided to an instruction or subroutine for processing. For example, in "HIGH 3" the constant "3" is said to be the argument of HIGH.

Array—A variable with multiple cells of the same size that can be accessed by using an index number. Arrays are available only in the BS2. Arrays are useful when a single operation will be performed on several variables, or when you need to store a list of values.

Assembly Language—Human-readable version of machine language, which substitutes simple text tags (such as LD meaning "load") for the 1s and 0s of machine language.

Binary Operator—An operator that takes two arguments.

Binary—Number system that uses just two symbols, 0 and 1. Like other numbering systems, larger numbers can be constructed by adding more digits.

Bit—Shortened form of binary digit. The smallest unit of information in a digital system, a bit is capable of representing just two values, 0 or 1.

Brad—Binary radian. A unit of angular measure tailored to fit into a byte. Instead of dividing a circle into units of 0 to 359 degrees, the circle is divided into units of 0 to 255 brads. Each brad is approximately 1.406 degrees. The name is derived from another popular unit of angular measure, the radian, which divides the circle in terms of Pi (π) from 0 to 2π.

Branch—A detour within the normal top-to-bottom order of program execution. When a program executes instructions in the order in which they are written (from the top of the page or screen to the bottom), it is said to be running "straight-line." So when a program deviates from this straight line, as when it executes a GOTO or GOSUB, or the true condition of IF/THEN, BRANCH, or BUTTON, it is said to "branch" to the new destination in the program.

Bug—An error in a program.

Byte—A group of 8 bits, capable of representing a range of values from 0 to 255 decimal.

Carry—Shorthand for the carry-the-one procedure that must be performed when the result of an arithmetic operation exceeds the highest value that a given digit can hold.

Clear—To store 0 to a memory location. See also Set.

Cleared—A bit is cleared when a 0 is written to it. We also talk about larger pieces of memory—nibbles, bytes, words—as being cleared when all of their constituent bits are zeroed. To make this perfectly clear, many people write, "cleared to 0." Although this is redundant, it's forgivable, since it makes the meaning doubly clear.

Comment Out—To deactivate program instructions by placing a tick mark (') in front of them. This has the same effect as deleting the instructions, but lets you restore them by just deleting the tick. Some programmers also use commented-out instructions as a way of embedding optional code in a program listing.

Comment—Explanatory text embedded in a program. Comments are ignored by PBASIC and do not add to the size of the program downloaded to the Stamp. In both dialects of PBASIC, a comment begins with the tick-mark (') and ends at the end of the line of text.

Compile Time—The time during which the Stamp host software prepares a PBASIC program for downloading. At this time, the host software checks the program for syntax errors, carries out directives, evaluates constant expressions, and converts instructions into tokens. Note that certain PBASIC statements, like directives, are said to be "resolved during compile time." That means that the host computer, not the Stamp, does all the work of processing these statements. Examples would be the SYMBOL directive (BS1) or the VAR directive (BS2); although these affect the way that a program is compiled, they do not directly result in any program code (tokens) being downloaded to the Stamp.

Compiler—Software that analyzes a program written in a high-level language like BASIC and converts it into machine language. Compilers often run on a different computer than the programs they compile. For example, a BASIC compiler that creates machine language for PIC microcontrollers might run on a PC. The term compiler is sometimes loosely applied to the Stamp host programs, which convert BASIC into compressed tokens. That's technically not compilation, but tokenization.

Complement—To change bits to the opposite state: 1 to 0 and 0 to 1. For example, the complement of %11010000 is %00101111.

Condition—A requirement that must be true in order for an instruction to be carried out. For example, in the instruction IF A = 3 THEN doAction, A = 3 is the condition. If it is true that A = 3, then the program will continue at the label doAction.

Constant—A number that cannot change while a program runs. Constants may be represented by numbers embedded in a program, by expressions evaluated at compile time, or by symbols assigned to numbers or expressions using the Symbol directive (BS1) or the Con directive (BS2).

Corrupt—To change the contents of a variable in an unintended way. For example, in the BS2 if you use a mixture of fixed variables (B1, W3, etc.) and PBASIC-allocated variables (e.g., x VAR byte), some variables may overlap. Data written to one variable can end up in another, corrupting the contents of that variable. Electrical noise is also said to corrupt serial data by altering bits.

Debug—To track down and correct errors in a program.

Decimal—Familiar numbering system based on 10 symbols, 0–9.

Declare—To ask the compiler to assign a name and/or set aside space for a variable. With the BS1, you declare variables by assigning meaningful symbols to the default variable names, as in—"SYMBOL index = b2." With the BS2, you declare a variable by either assigning a new name to an existing variable, or requesting an entirely new variable. For example, "index VAR b2" assigns the name index to the predefined byte variable b2. Alternatively, "index VAR byte" asks the compiler to assign the name "index" to the next available byte. This latter approach is better, because it lets the Stamp host program optimize the arrangement of variables in memory to avoid wasteful gaps or bug-causing overlaps.

Default—The setting or action that applies when no instruction is given. For example, the values of all Stamp variables automatically default to 0 unless your program does something to change their values.

Directive—An instruction intended to be carried out by the Stamp host software at compile time rather than by the PBASIC at run time.

Don't Care State—In a truth table, a don't care entry—often distinguished by an X, shaded entry, or other special mark—is an element that does not influence the outcome of an operation. For instance, when a pin's direction is set to input, its output state is a "don't care" because it does not affect the state of the I/O pin.

EEPROM—Electrically erasable, programmable, read-only memory. Data stored in EEPROM is permanent; it is not lost when power is removed. However, EEPROM may be rewritten, replacing old data with new. EEPROM will only tolerate a finite number of writes, about 10 million is the current state of the art. Stamps use EEPROM to store BASIC programs and long-term data.

Evaluate—To solve the math/logic problem presented by an expression. The Stamp is said to evaluate the expression 4*3 when it computes the answer 12.

Expression—A combination of math/logic operators and constants/variables that make up an equation to be solved (evaluated) by a program. For example, b2/4 + offSet is a simple expression in which b2 is a variable, / and + are math operators, and offSet is a symbolic constant or named variable.

Fixed Point—Number format in which the decimal point does not move. For example, a single byte (range 0–255) might be used to represent voltages from 0 to 2.55, where each unit represented 0.01 volt. The decimal point would be fixed in position two places from the right.

Flag—A variable used to store an on/off, yes/no, or true/false condition. In some dialects of BASIC, you must use a byte or larger variable to store a flag; PBASIC allows you to use a bit variable.

Floating Input—An input that is not connected to anything is said to float between the positive supply and ground. A floating input cannot be relied upon to read as 0 or 1; it may read 0 now and 1 a moment later. Stamp inputs should not be allowed to float, as this can increase current draw. Set unused pins to output, or tie them to + or ground through a resistor.

Floating Point—Number format similar to scientific notation, consisting of a sign, mantissa, and exponent. For example, in the number -4.56×10^3, – is the sign, 4.56 is the mantissa, and 3 is the exponent. A wide range of values from very large to very small can be used in floating-point calculations. The name floating point comes from the fact that changing the exponent in effect moves the decimal point. While floating point is common in calculators and high-level programming languages, it is not available on microcontrollers like the Stamps.

Ground—The point in a circuit to which all currents flow.

Hexadecimal—Numbering system based on 16 symbols, consisting of the numbers 0 through 9 and letters A through F. Hex is often used as shorthand for binary numbers, since one hex digit represents exactly four bits.

Indirect Addressing—A way of accessing a variable based on an index value, indirect addressing is the idea behind arrays. The BS2 supports indirect addressing; the BS1 does not. See also array.

Initialization—Setting a variable to a known starting value. With some computers and programming languages, the value contained by a variable before anything is written to it is unknown and can be anything. As a result, it's considered to be proper programming practice to initialize variables before using them. However, the Stamps automatically initialize all RAM variables to 0 at startup (including reset, power-up, and reprogramming), so it's safe to skip initialization—provided that 0 is an acceptable starting value.

Integer—A whole number with no fractional part. For example, 1, 7, 192, and 6020 are all integers; .12, 3.14, 100.9, and 7981.001 are not. The Stamps (and most microcontrollers) handle only integers. A common way to work around this limitation is to convert values with fractional components into integer fractions. For instance, if a calculation calls for multiplying a number by 3.14, instead multiply it by 22, then divide by 7. (22/7 is very close to 3.14.) The BS2's star-slash instruction (*/) is an easy way to work with integer fractions, but a simple multiply-then-divide scheme works too.

Interpreter—Firmware that accepts high-level instructions one at a time and carries out appropriate sequences of machine-language instructions. The BASIC Stamps contain interpreter programs that work with BASIC instructions stored as compressed tokens in EEPROM.

G

Keyword—A word belonging to the built-in vocabulary of a programming language. For example LOW and TOGGLE are names of PBASIC instructions, so they are said to be keywords of the language. Keywords may not be used as labels, variable names, or constant names.

Label—In PBASIC, labels are used to mark points in a program that will be the destination for GOTO and GOSUB instructions. A label must begin with a letter (A-Z, upper- or lowercase) and may contain letters, numbers, and underscore characters. Labels end in a colon (:) to distinguish them from variables and other user-assigned names.

Least-significant Bit (lsb)—The rightmost bit of a particular grouping of bits (nibble, byte, or word). This bit is called the least significant because it makes the smallest contribution to the overall value of the number stored by the nibble, byte, or word. The value of a particular bit is its state (0 or 1) times a value associated by its position in the number. Consider our familiar decimal system—The number 549 is equal to (5 $\times$ 100) + (4 $\times$ 10) + (9 $\times$ 1). The 9 is the least-significant digit because its multiplier (also called weight) is 1, the smallest in the number. It's the same in binary—the number 1010 is equal to (1 $\times$ 8) + (0 $\times$ 4) + (1 $\times$ 2) + (0 $\times$ 1). The rightmost bit is the least-significant because its multiplier (weight) is only 1, the smallest in the number.

Literal Constant—Refers to a constant that is represented by its own normal symbol. For example, in the expression "b2 * 3" the "3" is a literal constant. The term is used to make a distinction between literal constants and "symbolic" constants, created with the Symbol directive (BS1) or Con directive (BS2). For example, you could assign the name "three" to the number "3" using these directives—SYMBOL three = 3 (BS1) or three CON 3 (BS2). Then the expressions "b2 * three" and "b2 * 3" would both result in variable b2 being multiplied by 3.

Logic Probe—A test instrument that indicates the logical state (0 or 1) of a point in a circuit. A decent probe can be set to match the logic thresholds of TTL or CMOS circuitry, can indicate the presence of very brief pulses, and has both visual (LED) and audible (two-tone buzzer) indicators. Most logic probes are designed to be connected to the power supply of the circuit being tested.

Logic Threshold—The dividing line between the voltage range that a system interprets as a 0, and the range seen as a 1. The Stamps' PIC microcontroller has a logic threshold of approximately 1.4 volts. Voltages below this threshold are read as 0; above it as 1.

Loop—A sequence of program instructions that is executed repeatedly. For example, the instructions between FOR... and NEXT in a BASIC program will be executed as many times as the FOR... instruction dictates.

Machine Language—The instructions that a microcontroller or microprocessor chip is designed to process and execute. Since machine language is made up of binary data (sequences of 1s and 0s), programmers generally work in assembly language, which substitutes human-readable tags called mnemonics for obscure codes.

Most-significant Bit (msb)—The leftmost bit of a particular grouping of bits (nibble, byte, or word). See least-significant bit.

Nibble (also Nybble)—A group of four bits, capable of representing a range of values from 0 to 15 decimal.

Nonvolatile—Refers to memory that retains its contents without power, such as the EEPROM used in the Stamp to hold programs and constants. The Stamps' EEPROM is unusual among nonvolatile storage devices in that it is also easily reprogrammable to replace old data with new.

Object Code—A file containing the machine-readable version of a program. See also Token.

Operand—A value (variable or constant) that a math or logic operator works on. For example, in the instruction y = kibbles & bits, the variables named kibbles and bits are the operands.

Operator—A symbol that causes the Stamp to carry out a math or logic operation. For example, in the expression "b2 * 3" the asterisk is the multiplication operator, causing the Stamp to multiply variable b2 by the constant 3.

Pointer—A variable containing the memory address of a value rather than the value itself. For example, the index value used to access a particular cell of an array, or the address supplied to an EEPROM Read or Write instruction may be considered a pointer. The term pointer is not often used in the context of BASIC, but it is an important concept in C, Pascal, assembly language, and pseudocodes used by people who program in these languages.

Polling—Reading an input within a loop. For example, if a program is supposed to respond to a switch closure by turning on a light, it may check the switch input a hundred times a second. Between checks of the switch, it can do something else. Many programs consist of a main loop that executes dozens or hundreds of times a second. Each trip through the loop, the controller will poll all of the inputs that it's expected to respond to.

Precedence—In many BASICs, certain math operations are performed before others, regardless of the order in which they are written. For example, multiplication is often given higher precedence than addition, so the expression 5+2*3 would be computed as 11. The 2*3 would be computed first because of multiplication's precedence. The Stamps do not support this kind of operator precedence; they solve math problems from left to right. So 5+2*3 would be 21. The BS2 lets you change this by using parentheses to raise the precedence of an operation. So to force the multiplication to go first, you would write 5+(2*3), and the BS2 would compute 11.

Prefix—A symbol that, when placed in front of an expression, variable name, etc., changes its meaning. For example, the symbol % is a PBASIC prefix meaning "this is a binary number."

Pseudocode—A description of a program or algorithm not written in a specific programming language, but in a combination of simplified English and math/logic notation. The idea of pseudocode is to explain a program without requiring the reader to understand all the details of a particular programming language. Many programmers sketch out the general structure of a program in pseudocode before writing an actual program.

Pulldown Resistor—A resistor connected between an I/O pin and ground. A pulldown resistor establishes a default low (0) state on the pin. Pulldowns are used to get input from a device that can only pull to +5V (e.g., a switch connected to +5V).

Pullup Resistor—A resistor connected between an I/O pin and the +5-volt power-supply output. A pullup resistor establishes a default high (1) state on the pin. Pullups are used to get input from a device that can only pull to ground (e.g., a switch connected to ground).

RAM—Random-access memory. Data may be easily and quickly written to and read from RAM an unlimited number of times. Stamps use RAM to store variables used by a program. Data stored in RAM is volatile, that is, it's lost when power is removed.

Run Time—The time during which the PBASIC program is being processed—run—by the PBASIC interpreter chip on the Stamp. This term is used to distinguish between processing that occurs on the host computer before downloading of a PBASIC program (compile time), and the processing that occurs once the program has been downloaded and is actually running (run time). For example, expressions that define constants are evaluated at compile time, while expressions involving variables occur at run time.

Set—A bit is set when a 1 is written to it. However, we also talk about the value of larger pieces of memory—nibbles, bytes, words—as being set to a particular value that may or may not be all ones. For example—"...this code sets byte b2 to 129." If the value written is 0, it's usually preferable to say that it has been cleared, rather than "set to 0," which will jar the delicate sensibilities of some computer-science types.

Signed Number—A number stored in such a way that it can be determined to be positive or negative. For example, the BS2 can treat a 16-bit word as either an unsigned value ranging from 0 to 65535, or as a signed number from –32767 to +32767.

Sink (as in "sinking current")—To provide a path for current to flow to ground. The Stamp is said to sink current when it is outputting a 0 (0V) to a load whose other end is connected to +5V (or a point in the circuit higher in potential than 0V). The Stamps are capable of sinking up to 25 mA per output pin as long as total sink current per port is no more than 50 mA. In the BS1, there's only one port. In the BS2, P0 through P7 make up one port; P8 through P15 are another.

Source (as in "sourcing current")—To provide a path for current to flow from the + power supply. The Stamp is said to source current when it is outputting a 1 (+5V) to a load whose other end is connected to ground (or a point in the circuit lower in potential than +5V). The Stamps are capable of sourcing up to 20 mA per output pin as long as total source current per port is no more than 40 mA. In the BS1, there's only one port. In the BS2, P0 through P7 make up one port; P8 through P15 are another.

Source Code—The human-readable version of a computer program. PBASIC source code is stored as a plain-text file that may be viewed and modified in the host software or other programs like word processors and text editors (including the text-editing portions of desktop publishing and e-mail programs). However, only the PBASIC host software can convert source code into tokens (also known as object code) for downloading to the Stamp.

String—A byte array or sequence of bytes used to store or represent text. The characteristic that makes an array into a string is the presence of a count of valid bytes, or an end-of-string marker (usually a byte value of 0, also known as an ASCII null). The BS2's DEBUG, SERIN, and SEROUT instructions all support both kinds of strings, counted and null-terminated.

Subscript—The value that selects a particular cell of an array for access. This book calls this value the index, but some other dialects of BASIC use the mathematical term subscript.

Symbolic Constant—A constant that is referred to by a name, rather than by a literal value. With the BS1, names are assigned using the Symbol directive. With the BS2, the Con directive assigns names.

Syntax—The rules of a programming language governing the correct way of arranging instructions, arguments, and other elements of the language into a program that the compiler (Stamp host program) can understand. When you break one of these rules, you make a syntax error. Of course, even if a program's syntax is correct, it may not run properly due to errors in the logic underlying the program.

Token—A symbol representing a BASIC instruction in machine-readable form. For example, the instruction FOR i = 1 to 10 might be represented as a series of bits like %01101000101010. When you run a PBASIC program, the host software converts your program into a series of tokens and loads these into the Stamp's EEPROM program memory. This process ignores your comments, and greatly reduces the storage requirement for a given program. A BASIC program consisting of 40kB of source code might reduce to less than 2kB of tokens. The process of converting program source code into tokens is called tokenization, but you may also see it referred to as compilation.

Truncation—Literally, "shortening by cutting off part." If you try to store a number into a variable that's too small to hold it, the number stored will be truncated. For example, if your program contains myNib = 29 (where myNib is a nibble variable with a range of 0 to 15), the value actually stored in myNib will be 13. To understand this, look at the bits. The value 29 can be expressed in 5 bits—%11101. A nibble holds only four bits, so when %11101 is stored to a nibble, the highest bit is lost (truncated), resulting in %1101, which is equivalent to 13.

Truth Table—A table that summarizes all the possible combinations of inputs and results of a particular logic operation.

Unary Operator—An operator that takes one argument.

Unsigned Number—A number stored without provisions for determining its sign (positive or negative).

User Interface—The hardware and software that an application uses to get input from a user and to convey status information to a user.

Variable—A temporary storage location in RAM for data used by a program. The data stored in a variable may be used interchangeably with constants in a program. It is customary (and smart!) to assign meaningful names to variables using the Symbol directive (BS1) or Var directive (BS2) to make a program easier to read and understand.

Volatile—Refers to memory that loses its contents when power is lost, e.g., RAM.

Word—In the Stamps, a group of 16 bits, capable of representing a range of values from 0 to 65535.

INDEX

About the Author

Scott Edwards is the owner of Scott Edwards Electronics in Sierra Vista, Arizona. A well-known member of the electronics hobbyist community, he is the author of the popular "Stamp Applications" column for *Nuts & Volts* magazine. He is also the author of the *PIC Microcontroller Sourcebook*.

CD-ROM WARRANTY

This software is protected by both United States copyright law and international copyright treaty provision. You must treat this software just like a book. By saying "just like a book," McGraw-Hill means, for example, that this software may be used by any number of people and may be freely moved from one computer location to another, so long as there is no possibility of its being used at one location or on one computer while it also is being used at another. Just as a book cannot be read by two different people in two different places at the same time, neither can the software be used by two different people in two different places at the same time (unless, of course, McGraw-Hill's copyright is being violated).

LIMITED WARRANTY

McGraw-Hill takes great care to provide you with top-quality software, thoroughly checked to prevent virus infections. McGraw-Hill warrants the physical CD-ROM contained herein to be free of defects in materials and workmanship for a period of sixty days from the purchase date. If McGraw-Hill receives written notification within the warranty period of defects in materials or workmanship, and such notification is determined by McGraw-Hill to be correct, McGraw-Hill will replace the defective CD-ROM. Send requests to:

McGraw-Hill
Customer Services
P.O. Box 545
Blacklick, OH 43004-0545

The entire and exclusive liability and remedy for breach of this Limited Warranty shall be limited to replacement of a defective CD-ROM and shall not include or extend to any claim for or right to cover any other damages, including but not limited to, loss of profit, data, or use of the software, or special, incidental, or consequential damages or other similar claims, even if McGraw-Hill has been specifically advised of the possibility of such damages. In no event will McGraw-Hill's liability for any damages to you or any other person ever exceed the lower of suggested list price or actual price paid for the license to use the software, regardless of any form of the claim.

McGRAW-HILL SPECIFICALLY DISCLAIMS ALL OTHER WARRANTIES, EXPRESS OR IMPLIED, INCLUDING, BUT NOT LIMITED TO, ANY IMPLIED WARRANTY OF MERCHANTABILITY OR FITNESS FOR A PARTICULAR PURPOSE.

Specifically, McGraw-Hill makes no representation or warranty that the software is fit for any particular purpose and any implied warranty of merchantability is limited to the sixty-day duration of the Limited Warranty covering the physical CD-ROM only (and not the software) and is otherwise expressly and specifically disclaimed.

This limited warranty gives you specific legal rights; you may have others which may vary from state to state. Some states do not allow the exclusion of incidental or consequential damages, or the limitation on how long an implied warranty lasts, so some of the above may not apply to you.